NEW GENERATION FORMULATIONS OF AGROCHEMICALS

Current Trends and Future Priorities

NEW GENERATION FORMULATIONS OF AGROCHEMICALS

Current Trends and Future Priorities

Tatiana G. Volova, DSc
Ekaterina I. Shishatskaya, DSc
Natalia O. Zhila, PhD
Svetlana V. Prudnikova, DSc
Anatoly N. Boyandin, PhD

Apple Academic Press Inc.
3333 Mistwell Crescent
Oakville, ON L6L 0A2
Canada USA

Apple Academic Press Inc.
1265 Goldenrod Circle NE
Palm Bay, Florida 32905
USA

First issued in paperback 2021

Exclusive worldwide distribution by CRC Press, a member of Taylor & Francis Group

ISBN 13: 978-1-77463-428-8 (pbk)
ISBN 13: 978-1-77188-749-6 (hbk)

CIP data on file with Canada Library and Archives

Library of Congress Cataloging-in-Publication Data

Names: Volova, T. G. (Tatyana Grigorievna), author. | Shishatskaya, Ekaterina I., author. | Zhila, Natalia O., author. | Prudnikova, Svetlana V., author. | Boyandin, Anatoly N., author.

Title: New generation formulations of agrochemicals : current trends and future priorities / authors: Tatiana G. Volova, Ekaterina I. Shishatskaya, Natalia O. Zhila, Svetlana V. Prudnikova, Anatoly N. Boyandin.

Description: Oakville, ON ; Palm Bay, Florida : Apple Academic Press, 2019. | Includes bibliographical references and index.

Identifiers: LCCN 2019018627 (print) | LCCN 2019020108 (ebook) | ISBN 9780429433610 (ebook) | ISBN 9781771887496 (hardcover : alk. paper)

Subjects: LCSH: Agricultural chemicals.

Classification: LCC S585 (ebook) | LCC S585 .V825 2019 (print) | DDC 632/.95--dc23

LC record available at https://lccn.loc.gov/2019018627

Apple Academic Press also publishes its books in a variety of electronic formats. Some content that appears in print may not be available in electronic format. For information about Apple Academic Press products, visit our website at **www.appleacademicpress.com** and the CRC Press website at **www.crcpress.com**

ABOUT THE AUTHORS

Tatiana G. Volova, DSc
Professor and Head, Department of Biotechnology, Siberian Federal University, Krasnoyarsk, Russia

Tatiana G. Volova, DSc, is a Professor and Head in the Department of the Biotechnology at Siberian Federal University, Krasnoyarsk, Russia. She is the creator and Head of the Laboratory of Chemoautotrophic Biosynthesis in the Institute of Biophysics, Siberian Branch of Russian Academy of Sciences. Professor Volova is conducting research in the field of physicochemical biology and biotechnology and is a well-known expert in the field of microbial physiology and biotechnology. Tatiana Volova has created and developed a new and original branch in chemoautotrophic biosynthesis, in which the two main directions of the 21st century technologies are conjugate, hydrogen energy, and biotechnology. The obtained fundamental results provided significant outputs and were developed by the unique biotechnical producing systems, based on hydrogen biosynthesis for single-cell protein, amino acids, and enzymes. Under the guidance of Professor Volova, the pilot production facility of single-cell protein, utilizing hydrogen, had been created and put into operation. The possibility of involvement of man-made sources of hydrogen into biotechnological processes as a substrate, including synthesis gas from brown coals and vegetable wastes, was demonstrated in the research of Professor Volova. She had initiated and deployed in Russia the comprehensive research on microbial degradable bioplastics; the results of this research cover various aspects of biosynthesis, metabolism, physiological role, structure, and properties of these biopolymers and polyhydroxyalkanoates (PHAs), and have made a scientific basis for their biomedical applications and allowed them to be used for biomedical research. Professor Tatiana Volova is the author of more than 300 scientific works, including 13 monographs, 16 inventions, and a series of textbooks for universities.

Ekaterina I. Shishatskaya, DSc

Professor and Head, Department of Medical Biology, Siberian Federal University, Krasnoyarsk, Russia

Ekaterina I. Shishatskaya, DSc, is a Professor and Head of the Department of Medical Biology at Siberian Federal University, Krasnoyarsk, Russia, and a leading researcher at the Institute of Biophysics, Siberian Branch of Russian Academy of Sciences, in Krasnoyarsk, where she supervises the direction of biomedical research of new materials. Dr. Shishatskaya's work includes an investigation of interaction mechanisms between biomaterials and biological objects and the development of high-tech biomedical devices. This actual science direction is oriented to the development of new reconstructive biomedical technologies, which includes cell biology and tissue engineering and biodegradable polymers of carbon acids – polyhydroxyalkanoates (PHAs), a new class of materials for medicine. Under her leadership, technologies for encapsulating of biologically active and medical substances in the polymer matrix have been developed. The release kinetics of substances in depending on the geometry of the polymer matrix and the loading of substances was investigated. Dr. Shishatskaya is a Russian leader in comprehensive medical and biological studies of this class of polymers. Her professional activity is focused on the implementation of the obtained results in practice, and she maintains communication with clinical centers and top specialists in regenerative medicine. Dr. Shishatskaya is the author of about 150 research works, including six monographs and eight patents. She is the winner of the President of Russian Federation for youth in the field of science and innovations and holds a State Prize of the Krasnoyarsk Region in the area of high education and science. She is also a laureate of L'Oréal-UNESCO for young women in science.

Natalia O. Zhila, PhD

Assistant Professor, Department of Biotechnology, Siberian Federal University, Krasnoyarsk, Russian Federation; Senior Researcher, Institute of Biophysics, Siberian Branch of Russian Academy of Sciences

Natalia O. Zhila, PhD, is an Assistant Professor in the Department of Biotechnology at Siberian Federal University, Krasnoyarsk, Russian Federation, and also a Senior Researcher at the Institute of Biophysics, Siberian Branch of Russian Academy of Sciences. She investigates biosynthesis of polyhydroxyalkanoates of different structure and their properties. She has conducted studies of biosynthesis of co-, ter and quarterpolymers containing various monomers: 3-hydroxyvalerate, 3-hydroxyhexanoate, 4-hydroxybutyrate, and diethylene glycol using chemolithoautotrophic bacteria. Moreover, her work also includes various aspects of physiology and biochemistry of microorganisms consisting in the study of the features of growth of bacteria, polymer synthesis and fatty acid composition of lipid of the cytoplasmic membrane and cell wall in different conditions of cultivation. She also studies the polyhydroxyalkanoates metabolism including a number and size of polymer granules in cells, the molecular weight of the polymer and the activity of enzymes of polymer biosynthesis and intracellular degradation. Dr. Zhila is an author of about 70 research works, including two monographs.

Svetlana V. Prudnikova, DSc

Professor, Biotechnology Department, Siberian Federal University, Krasnoyarsk, Russia

Svetlana V. Prudnikova, DSc, is a Professor in the Biotechnology Department at Siberian Federal University, Krasnoyarsk, Russia. She works in the area of microbiology: ecology and physiology of microorganisms, microbial biotechnology, biodegradable microbial polymers. She has conducted studies of microbiological degradation of polyhydroxyalkanoates in different climatic zones, described microbial communities of PHA-degrading microorganisms that dominate in tropical soil and seawater, in Siberian natural and agrogenically transformed soils. Besides PHA, Dr. Prudnikova is also studying the biosynthesis of bacterial cellulose, another microbial degradable and biocompatible polymer. The area of her research interests includes processes of microbial biodegradation of organic matter in soil ecosystems, soil microbiology, and bioremediation. Dr. Prudnikova is an author of about 50 scientific and methodical works, including 24 papers in peer-reviewed journals, two monographs, two patents and five textbooks for students of biological specialties of universities.

Anatoly N. Boyandin, PhD

Associate Professor, Department of Biotechnology, Siberian Federal University; Researcher, Institute of Biophysics, Siberian Branch of Russian Academy of Sciences, Krasnoyarsk, Russia

Anatoly N. Boyandin, PhD, is an Associate Professor in the Department of Biotechnology at Siberian Federal University, and also a researcher at the Institute of Biophysics, Siberian Branch of Russian Academy of Sciences, in Krasnoyarsk. His work includes different aspects of biosynthesis, modification, biodegradation, and application of bacterial polyhydroxyalkanoates (PHAs). Anatoly Boyandin has investigated mechanisms of polyhydroxyalkanoate biosynthesis by *Vibrio* and *Photobacterium* spp. He has studied biodegradability of PHAs in water and soil environments of different climate zones. New technologies of obtaining PHA composites have been developed by Dr. Boyandin, which have provided an improvement of mechanical properties of goods with preserving high biocompatibility. Also, he has worked out new approaches to biopolyester surface modification based on chemical and physical treatment and aimed at the increase of polymer hydrophilicity and biocompatibility. Dr. Boyandin has developed a new technology of casting solution processing biopolymeric materials into tube goods for further use of them as endoprostheses in urology. He has developed new agricultural formulations based on a structure "composite core—biopolymeric coating."

CONTENTS

ABBREVIATIONS

ADM	Archer Daniels Midland Company
AG	arabinogalactan
AMPA	aminomethylphosphonic acid
CD	cyclodextrin
CFU/g	colony-forming units per gram
CMC-KAO	carboxy methyl cellulose-kaolinite composite
DTA	differential thermal analysis
EE	encapsulation efficiency
FAO	Food and Agriculture Organization
FDA	Food and Drug Administration
FPA	fish-peptone agar
FTIR	Fourier transform infrared spectroscopy
HES	hydroxyethyl starch
ICI	imperial chemical industries
LCL	long-chain-length
MCL	medium-chain-length
MEA	malt extract agar
MET	metribuzin
NO3–	nitrate nitrogen
P(AA-co-AM)/kaolin	poly(acrylic acid-co-acrylamide)/kaolin
PCL	poly-ε-caprolactone
PCR	polymerase chain reaction
PEG	polyethylene glycol
PHAs	polyhydroxyalkanoates
PVA	polyvinyl alcohol
PVP	polyvinylpyrrolidone
SA	soil extract agar
SAA	starch and ammonia agar
SCL	short-chain-length
SDS	sodium dodecyl sulfate
SEA	soil extract agar
TBM	tribenuron-methyl

TEB	tebuconazole
WA	wort agar
WHO	World Health Organization

PREFACE

This monograph is devoted to the summary of the authors' results on the fundamental basis for the design and application of long-term and ecologically safe forms of formulations for protecting cultivated plants from pathogens and weeds, as well as supplying nitrogen fertilizers. The problems associated with the use and accumulation of xenobiotics in the biosphere and modern trends in the design of new-generation formulations are presented in the monograph. The potential of degradable polyesters polyhydroxyalkanoates (PHA) for the design of long-term formulations is described in the monograph. Original research results of authors on the properties of herbicides, fungicides, and nitrogen fertilizers deposited in a degradable polymer base and the effectiveness of the use of these formulations in laboratory ecosystems with higher plants infected with fusariosis and weeds are presented in the monograph.

This study was financially supported by the Project "Agropreparations of the New Generation: A Strategy of Construction and Realization" (Agreement No. 14.Y26.31.0023) in accordance with Resolution No. 220 of the Government of the Russian Federation of April 9, 2010; Russian Science Foundation (Grant No. 14-26-00039).

INTRODUCTION

The concept of sustainable development, which is the fundamental idea of the 21st century, involves introduction of new forms of economic management, which will reduce the consumption of non-renewable sources and preserve them for the future generations, enabling more effective use of energy resources, introduction of new functional and environmentally friendly recyclable materials, and development of fundamentally new tools and technologies for the protection of the environment and efficient environmental management.

Environmental protection is an integral component of sustainable development. Human activities endanger biotic and abiotic components of the environment. As the human population of the world expands, increasing amounts of chemical substances are produced and consumed. Therefore, the number of environmental problems grows steadily. Throughout human history, people have been upsetting the balance in nature by killing large animals, burning forests to create hunting grounds, pastures, and arable lands, and polluting soils and water bodies at their settlements. Environmental pollution has always been an issue of importance. Industrial, agricultural, and household activities cause various, sometimes adverse, changes in the state and properties of the environment. Rapid development of industry and agriculture in the 20th century showed that natural production of the biosphere is not limitless. Today people face exhaustion of natural resources and energy sources as well as food, clean water, and air deficiency. In many regions of the world, environmental pollution has reached critical levels.

Pollution of the planet by waste and the high rates of consumption of natural resources lead to the failure of the biosphere's self-regulating mechanism, entailing unpredictable consequences. Academician N.N. Moiseev wrote at the end of the 1970s: "The main problem of global ecology is sustainability of the biosphere; the loss of its sustainability actually means elimination of humankind on Earth."

Intensive farming involves the use of enormous amounts of various chemicals to control pests, weeds, and pathogens causing diseases of crops. However, no more than 10% of the pesticides applied reach their target; most of these substances accumulate in biological objects, contaminated soils and water environments, kill beneficial living organisms, and upset the balance of natural ecosystems. The newest trend in research is development and agricultural use of environmentally safe new-generation pesticides with targeted and controlled release of active ingredients embedded in biodegradable matrices or covered with biodegradable coatings, which are degraded in soil and other biological media by soil microflora to form products that are harmless to living and non-living nature and which are gradually released into the environment. The use of such formulations can reduce the amounts of the chemicals applied to soil and enable their sustained and controlled delivery over a plant growing season, preventing sharp releases into the environment that occur in the case of using free pesticides. The main condition for constructing such formulations is the availability of appropriate materials with the following properties: degradability, ecological compatibility with the environment and global biosphere cycles; safety for living and non-living nature; long-term presence in the natural environment (for weeks and months) and controlled degradation followed by formation of non-toxic products; chemical compatibility with pesticides and fertilizers; processability by available methods that are compatible with pesticide and fertilizer production technologies.

Development and use of new, environmentally friendly materials, which will be able to be degraded in the environment without producing toxic compounds, thus joining the global material cycles, is among the priorities for critical technologies of the 21st century. A major concern is that the annual production of synthetic plastics has exceeded 311 million tons; in developed countries, no more than 16–20% of them are recycled, and they largely accumulate in landfills. Up to 10,000 ha of the land, including arable fields, is annually occupied by new landfills. Plastic waste ruins municipal sewage and drainage systems and pollutes water environments. Greenpeace reports that up to 10% of the plastic produced every year gets into the Global Ocean.

Thus, the wide use of products of chemical synthesis, including chemical pesticides and synthetic plastics, based on nonrenewable resources has led to an excessive and progressive increase in the amounts of unrecycled wastes, causing environmental concerns and posing a global environmental problem. A way to alleviate the human-induced impact on ecosystems is to develop and use formulations, materials, and application processes that are harmless to the environment.

Scientific and technological advancements lead to wider use of products synthesized in biotechnological processes. Polymers of hydroxy-derived alkanoic acids, polyhydroxyalkanoates (PHAs), are valuable products of biotechnology, which have a number of useful properties including biocompatibility and biodegradability. These polymers are promising materials for fabricating biomedical devices, degradable packaging for food and drinks, personal hygiene products, and devices and formulations for municipal engineering and agriculture.

The authors of this book have considerable research experience and academic and practical expertise in synthesis, investigation, and use of degradable polyhydroxyalkanoates. Based on our knowledge of polyhydroxyalkanoate synthesis and kinetics of degradation in natural ecosystems and techniques developed for processing polymers into specialized products, we have formulated an innovative direction in using bioplastics. The goal of this line of research is to provide scientific background and develop scientific basis for constructing environmentally safe targeted formulations based on degradable polyhydroxyalkanoates for protecting crops against pests and agents of diseases with controlled release of the active ingredient. Since 2005, we have been engaged in research aimed at developing processes of constructing new-generation slow-release formulations. We have developed and investigated a series of fungicide and herbicide formulations and nitrogen fertilizers embedded in the degradable polymeric matrix.

This book summarizes the results of scientific research providing the basis for constructing and using environmentally safe slow-release formulations for protecting crops against pathogens causing diseases and weeds and for applying nitrogen fertilizers. The book describes results of our studies on the properties of experimental herbicide and fungicide formulations and their efficacy in laboratory higher plant ecosystems contaminated by Fusarium infection and weeds.

CHAPTER 1

AGROCHEMICALS: USAGE AND ASSOCIATED ENVIRONMENTAL ISSUES

The rapid development of chemistry and intensive farming have facilitated the production and use of a vast variety of chemicals intended for protecting crops from weeds, pests, and pathogens. Modern intensive farming is impossible without pesticides. The global losses of potential yields caused by pests reach 37%: 13% of them are due to insects, 12% due to weeds, and 12% due to diseases. The annual losses are assessed at US$2000 billion (Oerke, 2006; Pimentel, 2009). However, pesticide build-up in the biosphere, via accumulation and concentration in the food chains of biota in agroecosystems and natural ecosystems, poses a global environmental threat (Hansen, 2004; Kaplin, 2007). Only a small part of pesticides in the environment reach their goal; the remaining pesticides kill beneficial organisms, accumulate in biological objects, destroy the balance in natural ecosystems, and contaminate soil, water, and air. The scale of pesticide application is very large and keeps on growing.

The Food and Agriculture Organization (FAO) and the World Health Organization (WHO), in the Codex Alimentarius (1963), defined pesticide as any substance intended for preventing, destroying, or controlling any pest, including vectors of human or animal disease, unwanted species of plants or animals, causing harm during or otherwise interfering with the production, processing, storage, transport, or marketing of food, agricultural commodities, wood, and wood products or animal feedstuffs, or substances that may be administered to animals for the control of insects, arachnids, or other pests in or on their bodies. The term includes substances intended for use as a plant growth regulator, defoliant, desiccant, or agent for thinning fruit or preventing the premature fall of fruit. Also used as substances applied to crops either before or after harvest to protect the commodity from deterioration during storage and transport.

By 2016, the global pesticide market had reached US$ 60 billion and exceeded 3 million tons a year. This is about 0.27 kg of pesticides per hectare on the entire Earth's surface (Ippolito et al., 2015). Up to 40% of all pesticides produced are herbicides, 17% are insecticides, 10% are fungicides, and 33% are other biocides (Popp et al., 2013). Herbicides protecting cereals, soybeans, and corn are important market segments. Herbicides, which constitute up to 55–70% of all pesticides, are the 8th most harmful pollutants (Kaplin, 2007). Fungicides show the highest growth rates (+11%), outpacing insecticides.

Statistical data on pesticide use in different countries are often lacking. The total annual amount of pesticides used now is between 2 and 3 million tons (Atwood, Paisley-Jones, 2017). They are used mainly in Europe (45%) and U.S. (25%); India uses 4% and the rest of the world 26% (De et al., 2014). Almost 50% of the pesticides are sold in Latin America and Asia. The study of the global pesticide production market states that by 2019, global pesticide production is to reach 3.2 million tons, versus 2.3 million tons in 2013. Thus, the compound annual growth rate between 2014 and 2020 should reach 6.1%. Although North America is the most extensive pesticide producer, the Asia-Pacific region is expected to show the most significant annual growth rate, reaching 7.9% in 2014–2020. As reported by OECD and Eurostat, over the past decade, the amounts of pesticides used in the number of West European countries have been reduced. Between 2000 and 2010, the amounts of pesticides used by farmers decreased from 5 to 3.3 kg/ha in France, from 3.5 to 2.8 kg/ha in Great Britain, and from 20 to 13 kg/ha in Malta (FAOSTAT, 2014). In 2010, the most significant amounts of pesticides (16.0–17.8 kg/ha) were used in Israel, India, and China (FAOSTAT, 2014). In Russia, in 2013, the market of pesticides reached US$ 1.3 billion. According to other sources, about 71.36 billion RUB were used for plant protection measures in Russia in 2013. A total of 53.9 thousand tons of pesticides were used in Russia in 2013, including 31.1 thousand tons of herbicides, 8.4 thousand tons of fungicides, 5.5 thousand tons of insecticides, 4.9 thousand tons of seed protectants, and 3 thousand tons of desiccants. Over the past three years, the pesticide market has been increasing by 10–12% every year.

Environmental effects of pesticides are radically different from the effects of other types of chemicals, as pesticides are artificially synthesized compounds, which are intentionally introduced into the environment. Pesticides pose a severe threat to terrestrial and aquatic ecosystems, with their long-term effects remaining insufficiently understood (Boatman et al., 2007; Mineau, Whiteside, 2013; Grung et al., 2015). Moreover, the application of several pesticides may aggravate their adverse impacts or result in an antagonistic effect. For example, triazine herbicides can increase the toxic effect of organophosphorus insecticides or decrease the toxicity of the fungicide prochloraz, which is harmful to human health (Hernández et al., 2013). Moreover, pesticide degradation products can intensify the toxicity of the original pesticide. For instance, the herbicide glyphosate and the main product of its degradation, aminomethylphosphonic acid (AMPA), are often found in freshwater ecosystems (Székács, Darvas, 2012). Glyphosate interferes with the shikimate pathway of aromatic amino acid synthesis in plants (Steinrücken, Amrhein, 1980), which is not found in animal cells. Yet, very low concentrations of glyphosate cause apoptosis of animal cells, and the presence of AMPA substantially aggravates the toxic effect (Benachour, Séralini, 2009).

Pesticides are chemical compounds of various types, which are used to control harmful organisms in agriculture, medicine, industry, oil production, and other areas. Over 900 chemical compounds are used as ingredients of 1300 commercial pesticides, herbicides constituting 31%, insecticides 21%, fungicides 17%, acaricides 9%, and rodenticides 2%. The other 20% of pesticides include various biocides to control snails (molluscicides), algae (algicides), and nematodes (nematicides) as well as plant growth regulators (6%) and natural or artificial pheromones (5%). Further, 610 products, including organochlorine insecticides, which were used in the past, have been banned because of their high toxicity or low efficacy, due to development of resistance in target organisms (Sánchez-Bayo et al., 2011).

Pesticides are mostly used in agriculture for controlling arthropods (insecticides and acaricides), nematodes (nematicides), fungal (fungicides) and bacterial (bactericides) diseases of animals and plants, as well as for weed control (herbicides). By now, 227 weed species have been identified, which are responsible for damaging 90% of the crop yield; 18 of them are considered as the most lethal ones in the world. Pesticides are used as wettable aerosols, powders, emulsions, dusts, and granules. The "fate" of these compounds in the environment is determined by a combination of physical, chemical, and biological factors. The type of soil, its mineral and organic composition, moisture content, oxygen content, and temperature influence the rate of pesticide degradation, which is the result of oxidation, adsorption, hydrolysis, catalytic decomposition, and processing by soil microorganisms. Most pesticides are toxic to a wide range of plants and animals. Thus, the names insecticide, fungicide, herbicide are often misleading, as they disguise the actual effects of these compounds. Pesticides affect all living organisms and are very toxic to warm-blooded and poikilothermic vertebrates. The effect of pesticides is not dependent on the population density, but they are used only against large populations, i.e., their application is density dependent.

In pest control, the actual amount of pesticides applied is usually larger than the amount needed to kill the pest. Vast areas are treated with pesticides, which inevitably results in harmful effects. The harmfulness of pesticides is aggravated by their persistence in the soil for months or even years. They are also spread far beyond the regions where they are applied. Over 50% of the active ingredients of the pesticides applied become air-borne instantly. The pesticide may also be transported from foliage to soil by wind or rain. Soil application of several herbicides results in high residual herbicide contents. The residue that remains in the field after a crop is harvested may be another, though minor, the source of pesticides in soil. Pesticides resuspended in the air may be transported to the soil by wind or rain. No more than 25–50% of the pesticide sprayed from aircraft reaches its aim; the remaining part lands around the crop fields. Establishment of the unsprayed buffer areas can considerably decrease the drift to the adjacent regions. The effects of the sprays on non-target objects may decrease by 41% for herbicides, 21% for insecticides, and 14% for fungicides compared to the effects in the preceding years (Siebers et al., 2003; de Jong et al., 2008).

Herbicides are applied at rather high rates, which are toxic, although over time, in the soil, toxicity decreases to concentrations below the minimum efficacy level, and the treatment needs to be repeated. Not all pesticides will reach their target. On the soil surface, the pesticide may undergo chemical and/or biological degradation as well as photodegradation. Pesticide losses may also be caused by washout and leakage.

Residual amounts of pesticides are detected in water, soil, and air in all geographical regions, including areas very far away from the application site: oceans, deserts, and Polar Regions. The presence of pesticides has been detected at all levels of the food chain: from plankton to whales and Arctic wildlife. Most of the species accumulate these compounds, and their concentrations increase from link to link in the food chain. People are not protected from pesticides, which have been detected in various tissues and excretions of humans, even those living in areas far away from the pesticide application site. Studies conducted in the 1980s and 1990s showed that these substances had caused 1.0–1.5 million cases of severe intoxication in humans. The U.N. data show that every year, almost one million people are diagnosed with pesticide poisoning, and about 40,000 of them die.

The use of fungicides in plant protection is not safe either, as they change the structure of microbiota, which plays a significant role in processing organics in the soil. For instance, copper fungicides are toxic to earthworms and other soil animals. The ubiquitous use of herbicides causes loss of plant biomass and impoverishes the biodiversity of ecosystems. Insecticides are toxic to most soil invertebrates, birds, and small mammals, considerably decreasing their populations and disrupting the trophic structure of the communities. Persistent pesticides (DDT, etc.) accumulate in soil and groundwater and concentrate in the food chain, producing sublethal effects (Ribeiro et al., 2007; Sánchez-Bayo et al., 2011; Sopeña et al., 2009).

Thus, the environmental consequences of using pesticides are caused by their high toxicity. Chemical pesticides have the following drawbacks:

- phytotoxicity (inhibition of plant growth and fruiting);
- fruit drop risk;
- lowering plant resistance to pathogens;
- development of specific group resistance of pest populations to chemical pesticides;
- environmental pollution;
- unpredictable interactions between residues of various chemicals in the soil;
- eradication of beneficial insect fauna and, hence, development of outbreaks of the previously economically insignificant pests; and
- adverse effects on insect pollinators.

The effects of pesticides on undesirable biota can be divided into two categories: demo-ecological and ecosystemic effects. The effects of the first (demo-ecological) category are at the level of populations of the species that are sensitive to a pesticide. The effect may be immediate, killing the undesirable object, or the pesticide may gradually accumulate in food chains, or it may cause a decrease in the biotic potential, resulting in fecundity reduction or infecundity of individuals. The other (ecosystemic) category is more complicated. For example, the population density of a species that either is insensitive or shows low sensitivity to the effect of a given pesticide may decrease because of extermination of plants or animals this species feeds on. Another ecosystemic effect of pesticides may be an increase in the population density of a previously scarce species because of the death of the competing species or predators and parasites.

Pesticides are capable of migrating in the natural environment. They are transported from soil to the water of surface and subsoil flows, bottom sediments, and air. Humans receive them with plant- and animal-derived foods. In the areas where pesticides are used routinely, changes are observed in the population densities and species compositions of insects, birds, mammals, and soil inhabitants.

Most pesticides are cumulative poisons, whose toxic effects are determined not only by their concentrations but also by the exposure time. According to their toxicity to humans and warm-blooded animals, pesticides are classified as strong toxic agents (LD_{50} up to 50 mg/kg), highly toxic agents (LD_{50} up to 200 mg/kg), moderately toxic agents (LD_{50} up to 1000 mg/kg), and low-toxic agents (LD_{50} above 1000 mg/kg). A number of pesticides with mutagenic and carcinogenic properties pose a threat to the health when getting into the human body with food, as they can cause neuropathy and dysfunction of endocrine system and affect reproductive function (Giri et al., 2002; Olgun et al., 2004; Perez-Martinez et al., 2001; Damalas, Eleftherohorinos, 2011). Health effects of pesticides may be acute – nausea, headache, skin, and eye irritation – or chronic – cancer, neurological disorders, diabetes mellitus, reproductive dysfunction, congenital abnormalities, and cardiovascular diseases (Mostafalou, Abdollahi, 2013). Pesticides are especially dangerous for infants in case of chronic intoxication of the child's organism by micro-doses of certain pesticides contained in food and household chemical goods (Weiss et al., 2004). Pesticides have a number of properties that make them more harmful than other chemicals. Thus, more effective tools and approaches need to be found, which would not adversely affect humans and the environment (MacDougall et al., 1992).

Science and industry have put considerable effort into developing new types of pesticides and increasing their production, as the use of chemicals in agriculture is an effective way to increase crop yields. Hence, pesticide production and usage increase, too (MacDougall et al., 1992; Popp et al., 2013). New types of pesticides, which are being progressively used, control harmful organisms and weeds more effectively than previously used compounds. Recently there has been considerable research to create new pesticide formulations and study their behavior in the environment. The aim of this research is to produce less toxic and more selective pesticides and to decrease their application rates. Modern herbicides include sulfonylurea formulations for different regions and with different persistence and phenoxyphenoxy propionic and phenoxybenzoic acid derivatives, which are effective against a wide range of weeds, including monocotyledons. Important broad-spectrum herbicides are glyphosate and glyphosinate, which decompose in the soil to CO_2, H_2O, and phosphoric acid. Application of acryl-hydroxyphenoxy propionic acids manufactured as individual optic isomers reduces the rate of herbicide application considerably. A new pyridine insecticide—imidacloprid—is used to control pests resistant to other types of insecticides. New pesticides—benzyl urea derivatives—have a selective effect on insects, sparing beneficial insects. A wide range of synthetic pyrethroids and other compounds have been produced, including pheromones, anti-fungicides, and chemosterilizers – substances disrupting metamorphosis of insects or killing them. These groups of pesticides are particular: they can control definite insect species without damaging the ecosystem.

Unfortunately, the wide use of pesticides has not provided crops with effective protection (Datta, Kaviraj, 2003). A great number of insects and weeds remain uncontrolled and keep causing enormous damage to agriculture (Baligar, Kaliwal, 2003). Moreover, pests acquire resistance to pesticides. With the advent of synthetic chemicals whose complex molecules can gradually decompose, this resistance has become a frequent event: more than 200 species are known now to have this property. Frequent applications of pesticides lead to the selection of resistant pest species. Some of them can only be killed by ultrahigh doses of pesticides, which are higher than the originally used amounts of toxicants by a factor of several thousand. A few hundred arthropod species resistant to various pesticides (DDT, carbamates, pyrethroids, organophosphorus compounds) have been described in the literature (Soderlund, Knipple, 2003). Recently, many pathogenic fungi have developed resistance to a number of fungicides.

Excessive and uncontrolled use of pesticides has become a powerful factor in artificial selection, which leads to a change in the genetic composition of pests and an increase in the number of resistant pests and weeds. About 270 species of the weeds are resistant to herbicides, about 150 causative agents of plant diseases are resistant to fungicides, and more than 500 insect species have become resistant to insecticides (De et al., 2014). Limitations of chemical pesticides, their harmful effects on people and animals, and decreasing fertility of soil caused by the extermination of beneficial microflora have encouraged researchers to develop new formulations, with the lowest possible side effects.

Up-to-date pesticides must satisfy the following basic requirements (Agroekologiya, 2000):

- moderate persistence in the environment;
- low toxicity to humans, animals, and other useful organisms;
- rapid degradation in soil, water, air, and bodies of warm-blooded animals, with no toxic compounds produced;
- the ability not to accumulate in bodies of humans, animals, birds, and aquatic organisms; and
- absence of delayed adverse effects in case of prolonged use.

Mineral fertilizers, which are an indispensable part of modern intensive agriculture, can also pose a biological threat. Mineral fertilizers, as well as pesticides, have been considered pollutants since the start of the "green revolution." Only 30–50% of the fertilizers applied to soil are consumed by agricultural crops, while the remaining portion is accumulated in soil and is transported to water ecosystems by melted snow and rainwater. Violation of the rules of applying fertilizers leads to their accumulation in soil, plants, and, hence, food products. Increasing amounts of mineral fertilizers applied to soil can affect natural material cycles, cause eutrophication of water environments, and aggravate the nitrate problem.

Nitrogen fertilizers require special attention in terms of reducing their loss and decreasing the risk of their uncontrolled spread. First, the amounts of nitrogen fertilizers applied to soil are larger than those of other fertilizers because of high nitrogen requirements of plants; second, nitrogen is mobile in soil and is concentrated in crops. Thus, nitrogen fertilizers (nitrates and nitrites, in particular) are potentially hazardous to people. Moreover, calculations show that at least one-third of the fertilizers applied to soil are washed out by melted snow and rainwater, entering water bodies and water streams and causing their eutrophication, i.e., over-enrichment with nutrients, which causes blooms of blue-green algae and worsens water quality.

Thus, the traditional use of agrochemicals has come into conflict with global environmental concerns. Therefore, other, more effective, approaches to crop protection need to be found, which would not adversely affect humans and the environment. The solution to this problem could be a wide application of biotechnological methods that will, on the one hand, protect beneficial biota and increase productivity in agriculture and, on the other hand, reduce the toxic impact on individual ecosystems and the entire biosphere. There are biological (bacterial, fungal, and viral) pesticides, bio-herbicides, nonmedical antibiotics that kill pests and agents of diseases of farm animals and crops; targeted pesticide formulations, which can be applied at lower rates and which neither spread in soil and water nor accumulate in food chains.

Accumulation of synthetic plastic waste in natural environments also poses a global environmental problem. Synthetic polymer materials have become part and parcel of modern life, but they have also created a number of problems. First, resources used to produce synthetic polymers are nonrenewable and, second, the application of polymers that cannot decompose in the natural environment and their accumulation lead to environmental pollution, posing a global environmental problem. Up to 10,000 ha of the land, including agricultural fields, is annually occupied by new landfills, in which synthetic materials constitute up to 60–70%. Plastic wastes litter cities and ruin municipal sewage and drainage systems. Greenpeace data suggest that up to 10% of the annual plastic production leaks into the Global Ocean (Moore, 2002; Rios et al., 2007; Tanabe et al., 2004).

"Islands" consisting of polyethylene and plastic wastes have been formed in the Global Ocean, which accumulates contaminations from all air and water flows. There are five great garbage patches in the Global Ocean now: two in the Pacific Ocean, two in the Atlantic Ocean, and one in the Indian Ocean. These garbage vortices largely comprise plastic waste from the densely populated continental coastal zones (Bioticheskiy mekhanizm samoochishcheniya presnykh i morskikh vod, 2004; Markina, Aizdaicher, 2005; Zaitsev, 2006). One is as large as America. A giant "plastic island" stretches from coastal waters of California (about 500 miles offshore), through the North Pacific, by the Hawaiian Islands, to Japan's shores. About 100 million tons of floating garbage covers the area twice as large as the U.S. continental area; over 40 years, its mass has increased by a factor of 100. The "island" has been dubbed the Great Pacific Garbage Patch, the Eastern Garbage Patch, or the Pacific Trash Vortex. In 2001, the mass of plastic in it was six times as large as the mass of zooplankton in the ocean close to the "island." The size of the "island" is estimated at between 700,000 and 15 million km^2 or even more (0.41 to 8.1% of the total area of the Pacific Ocean) (Moore et al., 2001). As the Global Ocean is polluted more and more, the biodiversity of natural waters is decreasing, and critical contamination of the water is fraught with catastrophic consequences. Soon it may be unsafe to consume seafood. Each year, plastic garbage in the Pacific Ocean causes the death of over one million of seabirds and more than 100,000 marine mammals.

In order to reduce the risk of accumulation and uncontrolled spreading of chemicals in the biosphere, it is necessary to develop and use new-generation formulations that would be harmless to the environment and beneficial biota, including targeted formulations with slow and controlled release of the active ingredient.

1.1 CONSTRUCTION AND USE OF NEW-GENERATION PESTICIDES

Although traditional use of agrochemicals has come into conflict with the global environmental concerns and, thus, more effective approaches to crop protection need to be found, which would not adversely affect humans and the environment, research aimed at developing and using environmentally safe slow-release agricultural formulations is in the early stages of its development.

The effectiveness of agricultural technologies in food production is determined by various factors, including ecological, geographical, and economic ones; it also depends on renewable biological resources such as crops, domestic animals, and microorganisms. Research integrating various biological sciences has been devoted to enhancing biological productivity in agriculture. A promising approach to controlling pests and plant pathogens is to develop biological methods of plant protection (employing beneficial organisms to control harmful ones) and use them together with agrotechnical, quarantine, selection, and physiological measures (Agroekologiya, 2000).

The use of microorganisms as biopesticides is a relatively new line of biotechnology, but it can boast considerable achievements. Bacteria, fungi, and viruses have been increasingly used as commercial biopesticides. The processes involved in the production of these pesticides vary greatly, as microorganisms have diverse origins and physiological properties. However, all biopesticides must meet certain requirements such as selectivity and high efficacy, safety for people and beneficial plants and animals, long lifespans, ease of use, good wettability, and good adhesive properties. In addition to antibiotics, about 50 microbial formulations – bacterial, fungal, and viral ones – are used now to protect plants and animals from insects and rodents. Integrated biological crop protection can extend the plant growing period by 2 weeks, increase cucumber and tomato yields to 5 kg/m^2, improve the sanitary conditions in greenhouses, enable the production of health food, and lead to a 35% reduction in the cost of protective measures compared to pesticide use. Experts suggest that between 2014 and 2020, the world biopesticide market will be developing at the fastest rate.

Over 90 bacterial species infecting insects have been described by now. They mainly belong to the families Pseudomonadaceae, Enterobacteriaceae, Lactobacillaceae, Micrococcaceae, and Bacillaceae. Most of the commercial strains are of the genus Bacillus, and over 90% formulations are based on Bacillus thuringiensis (Bt), which has more than 22 serotypes. Bt-based formulations are intestinal toxins. Bacteria of the Bacillus thuringiensis group are effective against 400 insect species, including field, forest, orchard, and vineyard pests; these formulations are most effective against defoliators. More than 100 Bt strains are united in 30 groups by their serological and biochemical characteristics. Microbiological industries in various countries produce different formulations based on Bt, which are capable of forming spores, crystals, and toxic substances while growing.

Promising contact biopesticides are fungal preparations. Numerous species of entomopathogenic fungi commonly occur in nature. They affect a wide range of insects through various mechanisms, including contact, which makes them convenient to use. Fungi produce various bioactive substances, which enhance their pathogenic effect. Fungal preparations, however, are not used widely now. First, certain technological difficulties are associated with the cultivation of fungi, and, second, they can only be effective under limited environmental conditions (high and stable humidity). Hundreds of entomopathogenic fungi have been identified, but only two groups are regarded as very promising: muscardine fungi of the family Euascomycetes and entomophthoric fungi of the family Entomophthoraceae. Special consideration has been given to such pathogenic fungi as the causative agents of white muscardine disease (of the genus Beauveria) and green muscardine disease (of the genus Metarhizium), and Entomophthora, affecting sucking insects. Pathogenic fungi, unlike other microorganisms, are capable of infecting insects in different developmental stages (pupae or adults). Fungi grow quickly and have a high reproductive capacity. For fungal preparations to be effective, they should be applied in a certain season and at optimal concentrations, as their development is determined by weather. This condition is a limitation to the wide use of fungal preparations in the form of spores. One way to enhance and keep the physiological activity of fungal preparations in nature may be to immobilize the spores in polymeric carriers.

A new line of research, which is aimed at reducing the risk of uncontrolled spreading of pesticides and accumulation thereof in the biosphere, is development and agricultural use of environmentally safe new-generation pesticides with targeted and controlled release of active ingredients embedded in biodegradable matrices or covered by biodegradable coatings. The crucial component of constructing such preparations is the availability of appropriate materials with the following properties:

- ecological compatibility with the environment and global biospheric cycles, i.e., degradability;
- safety for a living and non-living nature;
- long-term (weeks and months) presence in the natural environment and controlled degradation followed by formation of non-toxic products;
- chemical compatibility with pesticides and fertilizers; and
- processability by available methods, compatible with pesticide and fertilizer production technologies.

Encapsulation of pesticides is a relatively new approach. First studies reporting production of microcarriers of polymeric materials loaded with pesticides were published only recently. The authors of those studies noted the following advantages of using pesticide controlled delivery systems:

- prolonged action due to the continuous release of a pesticide at a level sufficient for effective function over a long period;
- fewer treatments due to prolonged action after a single application; shorter time needed to apply such pesticides;
- lower contamination of the environment;
- longer activity of pesticides unstable in the aqueous medium;
- conversion of the liquid pesticide into the solid formulation, which simplifies shipping and decreases the flammability of the formulation; and
- lower toxicity to biota due to the reduction in pesticide mobility in soil and, hence, lower accumulation in the food chain (Kulkarni et al., 2000; Celis et al., 2002; Dubey et al., 2011; Roy et al., 2014).

Until recently, studies on this subject have been scant and incomplete. The first reviews devoted to the production of pesticide formulations embedded in polymer microspheres were published not long ago.

The key ingredient in constructing slow-release formulations is the availability of the appropriate biodegradable carrier. Thus, it is important to find and investigate materials with the necessary properties. The materials extensively studied as matrices for embedding agrochemicals are synthetic nondegradable polymers (polystyrene, polyacrylamide, polyethylene acrylate, polyamide, polyurethane, polycyanoacrylate) (Chang et al., 2005). Studies published in recent years report investigations of degradable materials, which can be decomposed by soil microflora; these materials do not pollute soil, and chemicals are released from them gradually. These are natural materials such as cellulose, agarose, dextran, carrageenan, starch, chitosan, alginate, gelatin, and albumins (Chang et al., 2006; Wang et al., 2007; Saravanan et al., 2008; Kumar et al., 2014). The shortcomings of these polymers are their low mechanical strength and rapid hydrolysis in liquid media, which is an obstacle to preparing slow-release pesticide or fertilizer formulations.

The search of the literature on pesticides embedded in matrices revealed rather few studies on this subject. The first studies on production of pesticide-loaded micro- and nano-spheres were published quite recently. Formulations of the pesticide norflurazon were prepared by encapsulating it in ethyl cellulose microspheres using the solvent evaporation technique; the authors studied pesticide encapsulation efficiency and release rate (Perez-Martinez et al., 2001). Shukla et al. (2002) described the preparation of microspheres from the aqueous solution of the pesticide monocrotophos by using polyurethane. Kulkarni et al. (2000) described encapsulation of pesticides by using sodium alginate crosslinked with glutaraldehyde. Pesticide release rates decreased after using the crosslinking agent. A technique of encapsulation of imidacloprid in calcium alginate has been developed (Kumar et al., 2014). Large-scale experiments were carried out to evaluate the rates of the polymeric matrix degradation and release of pesticides (atrazine, terbutryn, and fenamiphos) (Patterson et al., 2002).

Formulations of herbicides alachlor (Fernández-Urrusuno et al., 2000) and norflurazon (Sopeña et al., 2005) were prepared by encapsulating them in ethyl cellulose. The authors showed that the release rate of the active ingredient and control of weed growth might be regulated by the initial amount of encapsulated alachlor. Hashemi et al. (2001) described the preparation of polyurethane microcapsules loaded with Dursban and showed that capsule formation was influenced by the temperature, type, and amount of emulsifier, agitation speed, organic phase, etc. The sustained-release formulation of atrazine based on ethyl cellulose and prepared using solvent evaporation technique increased the range of use of this herbicide and minimized its adverse effects on the environment (Cea et al., 2010).

The data on the encapsulated slow-release fertilizers are even less representative. Jarosiewicz and Tomaszewska (2003) described a study evaluating the release of an NPK fertilizer encapsulated in a polymer matrix to water; the authors used polysulfone, ethyl cellulose, and polyacrylonitrile. They found that the structure of the polymer coating controlled diffusion of the elements from the inner part of fertilizer granules. Experiments showed that these polymer coatings reduced the solubility and release rate of components of this fertilizer. In another study, fertilizer granules were sprayed with a polysulfone solution (Liang, Liu, 2006). The authors reported that this technique of preparation of polymer coatings decreased the release rate of nutrients from the polymer-coated granules. Moreover, the increase in the number of sprayed polymer layers caused a decrease in the release of each major element by ca. 25%. The reduction in the release rate was associated with a more compact structure and lower porosity of the sprayed coating compared to the reference one. Abedi-Koupai et al. (2012) investigated green vitriol fertilizer coated by ethylene vinyl acetate and estimated fertilizer release to the soil. Iron ions were released at a slow rate, and after 168 h of exposure, about 90% of them were released. After 48 h, 58% of the active ingredient was released versus 80% release from uncoated microcapsules. Liang et al. (2007) described double-coated fertilizer constructions. The double-coated construction in the form of granules had a core/shell structure. Its core was urea formaldehyde and polyphosphate potassium fertilizers, and the shell was poly(acrylic acid-co-acrylamide)/kaolin (P(AA-co-AM)/kaolin) superabsorbent composite. Controlled release of the fertilizer from the formulations was achieved. Methods of the coating of lignin-urea with such polymers were described by Azzem et al. (2014).

Results of these few studies suggest the necessity of research in this field, aimed at developing new slow-release formulations that would enable targeted delivery of the active ingredient and its slow release over the plant growing period. The general goal of this research will be to reduce the amounts of agrochemicals and prevent their uncontrolled spread in the biosphere.

Among biodegradable polymers, the most actively studied ones are polymers synthesized by bacteria in biotechnological processes – polyhydroxyalkanoates (PHAs), which are thermoplastic, mechanically strong, and slowly degraded in biological media. As these polymers are disintegrated via true biological degradation and are not hydrolyzed in liquid media, products made of them can function in the soil for months. The rates of release and delivery of the preparation can be regulated by controlling the degradation rate of the PHA matrix by using products of different shapes that contain different amounts of preparations. The literature data on using PHAs to construct environmentally safe pesticide formulations are scant. Analysis of the available literature shows that research aimed at constructing such PHA-based formulation is only starting. The available studies only declared that PHAs are promising materials for constructing slow-release agrochemical formulations, but the authors noted that these polymers would be very useful as matrices for fertilizers and pesticides.

KEYWORDS

- **arthropod species**
- **fungicides**
- **polyhydroxyalkanoates**

REFERENCES

Abedi-Koupai, J., Varshosaz, J., Mesforoosh, M., & Khoshgoftarmanesh, A. H., (2012). Controlled release of fertilizer microcapsules using ethylene vinyl acetate polymer to enhance micronutrient and water use efficiency. *J. Plant. Nutr.*, *35*, 1130–1138.

Atwood, D., & Paisley-Jones, C., (2017). *Pesticides Industry Sales and Usage 2008–2012 Market Estimates.* US Environmental Protection Agency, DC Google Scholar: Washington.

Azeem, B., KuShaari, K., Man, Z. B., Bazit, A., & Thanh, T. H., (2014). Review on materials and methods to produce controlled release coated urea fertilizer. *J. Control. Release*, *181*, 11–21.

Baligar, P. N., & Kaliwal, B. B., (2003). Temporal effect of carbofuran, a carbamate insecticide in the interruption of estrous cycle and follicular toxicity in female Swiss albino mice. *Bullet Environ. Contam. Toxicol.*, *71*, 422–428.

Benachour, N., & Séralini, G. E., (2009). Glyphosate formulations induce apoptosis and necrosis in human umbilical, embryonic, and placental cells. *Chem. Res. Toxicol.*, *22*, 97–105.

Biotic Mechanism of Self-Purification of Fresh and Sea Water (Bioticheskiy Mechanism Samoochoscheniya Presnih i Morskih vod), (2004). MAKS-Press, Moscow, (In Russian).

Boatman, N., Parry, H., Bishop, J., & Cuthbertson, A., (2007). Impacts of Agricultural Change on Farmland Biodiversity in the UK. In: Hester, R., & Harrison, R., (eds.), *Biodiversity Under Threat* (pp. 1–32). RSC Publishing: Cambridge.

Cea, M., Cartes, P., Palma, G., & Mora, M. L., (2010). Atrazine efficiency in an andisol as affected by clays and nanoclays in ethylcellulose controlled release formulations. *R. C. Suelo Nutr. Veg.*, *10*, 62–77.

Celis, R., Hermosín, M. C., Carrizosa, M. J., & Cornejo, J., (2002). Inorganic and organic clays as carriers for controlled release of the herbicide hexazinone. *J. Agric. Food Chem.*, *50*, 2324–2330.

Chang, C. P., Chang, J. C., Ichikawa, K., & Dobashi, T., (2005). Permeability of dye through poly(urea-urethane) microcapsule membrane prepared from mixtures of di- and tri-isocyanate. *Colloids Surf. B: Biointerfaces*, *44*, 187–190.

Chang, C. P., Leung, T. K., Lin, S. M., & Hsu, C. C., (2006). Release properties on gelatin-gum Arabic microcapsules containing camphor oil with added polystyrene. *Colloids Surf. B: Biointerfaces*, *50*, 136–140.

Chernikov, V. A., & Chekeres, A. I., (2000). *Agroecology (Agroekologia)* (533 p.), Kolos: Moscow (in Russian). ISBN 5-10-003269-3.

Damalas, C. A., & Eleftherohorinos, I. G., (2011). Pesticide exposure, safety issues, and risk assessment indicators. *Int. J. Environ. Res. Public Health*, *8*, 1402–1419.

Datta, M., & Kaviraj, A., (2003). Acute toxicity of the synthetic pyrethroid deltamethrin to freshwater catfish Clarias gariepinus. *Bull. Environ. Contam. Toxicol.*, *70*, 296–299.

De Jong, F. M., De Snoo, G. R., & Van de Zande, J. C., (2008). Estimated nationwide effects of pesticide spray drift on terrestrial habitats in the Netherlands. *J. Environ. Manag.*, *86*, 721–730.

De, A., Bose, R., Kumar, A., & Mozumdar, S., (2014). Worldwide pesticide use. In: De, A., Bose, R., Kumar, A., & Mozumdar, S., (eds.), *Targeted Delivery of Pesticides Using Biodegradable Polymeric Nanoparticles* (pp. 5, 6). Springer, New Delhi.

Dubey, S., Jhelum, V., & Patanjali, P. K., (2011). Controlled release agrochemicals formulations: A review. *J. Sci. Ind. Res.*, *70*, 105–112.

FAO statistical yearbook 2014: Europe and Central Asia food and agriculture. http://www.fao.org/3/a-i3621e.pdf (accessed Feb 21, 2018).

Fernández-Urrusuno, R., Gines, J. M., & Morillo, E., (2000). Development of controlled release formulations of alachlor in ethylcellulose. *J. Microencapsulation, 17*, 331–342.

Food and Agriculture Organization of the United Nations, (2015). *Statistical Pocketbook.* World food and agriculture. http://www.fao.org/3/a-i4691e.pdf (accessed Feb 21, 2018).

Giri, S., Prasad, S. B., Giri, A., & Sharma, G. D., (2002). Genotoxic effects of malathion: An organophosphorus insecticide, using three mammalian bioassays *in vivo. Mutat. Res., 15*, 223–231.

Grung, M., Lin, Y., Zhang, H., Steen, A. O., Huang, J., Zhang, G., & Larssen, T., (2015). Pesticide levels and environmental risk in aquatic environments in China – A review. *Environ. Int., 81*, 87–97.

Hansen, L. J., Schwacke, L. H., Mitchum, G. B., Hohn, A. A., Wells, R. S., Zolman, E. S., & Fair, P. A., (2004). Geographic variation in polychlorinated biphenyl and organochlorine pesticide concentrations in the blubber of bottlenose dolphins from the US Atlantic coast. *Sci. Total Environ., 319*, 147–172.

Hashemi, S. A., & Zandi, M., (2001). Encapsulation process in synthesizing polyurea microcapsules containing pesticide. *Iran Polym. J., 10*, 265–270.

Hernández, A. F., Parrón, T., Tsatsakis, A. M., Requena, M., Alarcon, R., & López-Guarnido, O., (2013). Toxic effects of pesticide mixtures at a molecular level: Their relevance to human health. *Toxicology, 37*, 136–145.

Ippolito, A., Kattwinkel, M., Rasmussen, J. J., Schäfer, R. B., Fornaroli, R., & Liess, M., (2015). Modeling global distribution of agricultural insecticides in surface waters. *Environ. Pollut., 198*, 54–60.

Jarosiewicz, A., & Tomaszewska, M., (2003). Controlled-release NPK fertilizer encapsulated by polymeric membranes. *J. Agric. Food Chem., 51*, 413–417.

Kaplin, V. G., (2007). *Basics of Ecotoxicology (Osnovi Ekotoksikologii),* Kolos: Moscow, (In Russian).

Kulkarni, A. R., Soppimath, K. S., Aminabhavi, T. M., Dave, A. M., & Mehta, M. H., (2000). Glutaraldehyde cross-linked sodium alginate beads containing liquid pesticide for soil application. *J. Control. Release, 63*, 97–105.

Kumar, S., Bhanjana, G., Sharma, A., Sidhu, M. C., & Dilbaghi, N., (2014). Synthesis, characterization and on-field evaluation of pesticide loaded sodium alginate nanoparticles. *Carbohydr. Polym., 101*, 1061–1067.

Liang, R., & Liu, M., (2006). Preparation and properties of a double-coated slow-release and water-retention urea fertilizer. *J. Agric. Food Chem., 54*, 1392–1398.

Liang, R., Liu, M., & Wu, L., (2007). Controlled release NPK compound fertilizer with the function of water retention. *React. Funct. Polym., 67*, 769–779.

MacDougall, N., Hanemann, W. M., & Zilberman, D., (1992). *The Economics of Agricultural Drainage.* Report prepared for the Central Valley Regional Water Control Board (Standard Agreement No. 0–132–150–0) by the Department of Agricultural and Resource Economics: University of California, Berkeley.

Markina, Z. H. V., & Aizdaycher, N., (2005). A *Dunaliella salina (Chlorophyta)* as a test-object for assessing of the pollution of the marine environment by detergents (*Dunaliella salina (Chlorophyta)* kak test-obyect dlya ocenki zagryaznenia morskoy void detergentami). *Biology of the Sea, 31*, 274–279 (In Russian).

Mineau, P., & Whiteside, M., (2013). Pesticide acute toxicity is a better correlate of US grassland bird declines than agricultural intensification. *PLoS One.*, *8*, e57457.

Moore, C. A., Moore, S. L., Leecaster, M. K., & Weisberg, S. B., (2001). Comparison of plastic and plankton in the North Pacific Central Gyre. *Mar. Pollut. Bull.*, *42*, 1297–1300.

Moore, C., (2002). *Great Pacific Garbage Patch.* Santa Barbara News-Press.

Mostafalou, S., & Abdollahi, M., (2013). Pesticides and human chronic diseases: Evidences, mechanisms, and perspectives. *Toxicol. Appl. Pharmacol.*, *268*, 157–177.

Oerke, E. C., (2006). Crop losses to pests. *J. Agric. Sci.*, *144*, 31–43.

Olgun, S., Gogal, R. M., Adeshina, F., Choudhury, H., & Misra, H. P., (2004). Pesticide mixtures potentiate the toxicity in murine thymocytes. *Toxicology*, *196*, 181–195.

Patterson, B. M., Davis, G. B., & McKinley, A. J., (2002). Laboratory column experiments using polymer mats to remove selected VOCs, PAHs, and pesticides from groundwater. *Ground Water Monit. Rem.*, *22*, 99–106.

Pérez-Martínez, J. I., Morillo, E., Maqueda, C., & Gines, J. M., (2001). Ethyl cellulose polymer microspheres for controlled release of norfluazon. *Pest. Manag. Sci.*, *57*, 688–694.

Pimentel, D., (2009). Pesticides and pest controls. In: Peshin, R., & Dhawan, A. K., (eds.), *Integrated Pest Management: Innovation Development Process* (pp. 83–87). Springer Science + Business Media: Dordrecht.

Popp, J., Pető, K., & Nagy, J., (2013). Pesticide productivity and food security. A review. *Agron Sustain Dev.*, *33*, 243–255.

Ribeiro, M. L., Lourencetti, C., Yoshinaga, M. R., & Rodrigues De Marchi, M. R., (2007). Groundwater contamination by pesticides: Preliminary evaluation. *Quim. Nova.*, *30*, 688–694.

Rios, L. M., Moore, C., & Jones, P. R., (2007). Persistent organic pollutants carried by synthetic polymers in the ocean environment. *Mar. Pollut. Bull.*, *54*, 1230–1237.

Roy, A., Singh, S. K., Bajpai, J., & Bajpai, A. K., (2014). Controlled pesticide release from biodegradable polymers. *Central Europ. J. Chem.*, *12*, 453–469.

Sánchez-Bayo, F., Van den Brink, P. J., & Mann, R. M., (2011). Impacts of agricultural pesticides on terrestrial ecosystems. In: *Ecological Impacts of Toxic Chemicals* (pp. 63–87). Bentham Science Publishers Ltd.

Saravanan, M., & Rao, K. P., (2008). Pectin–gelatin and alginate–gelatin complex coacervation for controlled drug delivery: Influence of anionic polysaccharides and drugs being encapsulated on physicochemical properties of microcapsules. *Carbohydr. Polym.*, *80*, 808–816.

Shukla, P. G., Kalidhass, B., Shah, A., & Palaskar, D. V., (2002). Preparation and characterization of microcapsules of water-soluble pesticide monocrotophos using polyurethane as carrier material. *J. Microencapsul.*, *19*, 293–304.

Siebers, J., Binner, R., & Wittich, K. P., (2003). Investigation on downwind short-range transport of pesticides after application in agricultural crops. *Chemosphere.*, *51*, 397–407.

Soderlund, D. M., & Knipple, D. C., (2003). The molecular biology of knockdown resistance to pyrethroid insecticides. *Insect. Biochem. Mol. Biol.*, *33*, 563–577.

Sopeña, F., Cabrera, A., Maqueda, C., & Morillo, E., (2005). Controlled release of the herbicide norflurazon into water from ethylcellulose formulations. *J. Agric. Food Chem.*, *53*, 3540–3547.

Sopeña, F., Maqueda, C., & Morillo, E., (2009). Controlled release formulations of herbicides based on micro-encapsulation. *Cienc. Investig. Agrar.*, *36*, 27–42.

Sopeña, V. F., Maqueda, P. C., & Morillo, G. E., (2008). Microencapsulation of alachlor for reducing its pollution in soil-water system. In: *Abstracts of XVI International Conference on Bioencapsulation* (p. 58).

Steinrücken, H. C., & Amrhein, N., (1980). The herbicide glyphosate is a potent inhibitor of 5-enolpyruvylshikimic acid–3-phosphate synthase. *Biochem. Biophys. Res. Commun.*, *94*, 1207–1212.

Székács, A., & Darvas, B., (2012). Forty years with glyphosate. In: Hasaneen, M. N. A. E.-G., (ed.), *Herbicides-Properties, Synthesis, and Control of Weeds* (pp. 247–284). In Tech.: Rijeka, Croatia.

Tanabe, S., Watanabe, M., Minh, T. B., Kunisue, T., Nakanishi, S., Ono, H., & Tanaka, H., (2004). PCDDs, PCDFs, and coplanar PCBs in albatross from the North Pacific and Southern Oceans: Levels, patterns, and toxicological implications. *Environ. Sci. Technol.*, *38*, 403–413.

Wang, C., Ye, W., Zheng, Y., Liu, X., & Tong, Z., (2007). Fabrication of drug-loaded biodegradable microcapsules for controlled release by combination of solvent evaporation and layer-by-layer self-assembly. *Int. J. Pharm.*, *338*, 165–173.

Weiss, B., Amler, S., & Amler, R. W., (2004). Pesticides. *Pediatrics*, *113*, 1030–1036.

Zaitsev, Y. P., (2006). *Introduction to the Ecology of the Black Sea (Vvedenie v Ekologiyu Chernogo Morya)* (224 p.). Even: Odessa (in Russian). ISBN 966-8169-16-6.

CHAPTER 2

POLYHYDROXYALKANOATES: NATURAL DEGRADABLE BIOPOLYMERS

Polyhydroxyalkanoates (PHAs) are storage macromolecules of the cell (energy and carbon reserves) synthesized by prokaryotic organisms under specific conditions of unbalanced growth, when the synthesis of the major compounds (protein and nucleic acids) is reduced, but there is excess carbon in the medium. Very many microorganisms, including wild-type and genetically modified strains, are capable of storing PHAs (Madison and Huisman, 1999). The conditions under which constructive metabolism is directed to PHA synthesis and accumulation are determined by the redox state of the cytoplasm and intracellular concentrations of acetyl-CoA and free CoA (Senior and Dawes, 1973). Under unbalanced growth, in a medium lacking one of the constructive elements (nitrogen, phosphates, etc.) or under oxygen deficiency, acetyl-CoA does not enter the tricarboxylic acid cycle, and the level of free CoA is low. This is a favorable condition for activation of poly(3-hydroxybutyrate) (P(3HB)) synthesis enzymes. The conditions under which constructive anabolism is directed to PHA synthesis and accumulation are determined by the redox state of the cytoplasm and intracellular concentrations of pyruvate and free CoA (Senior and Dawes, 1973). Under unbalanced growth, in a medium lacking one of the constructive elements (nitrogen, phosphates, etc.) or under oxygen deficiency, acetyl-CoA does not enter the tricarboxylic acid cycle, and the level of free CoA is low. This is a favorable condition for activation of P(3HB) synthesis enzymes. During balanced growth, pyruvate, and reducing equivalents (NADH and NADPH) are generally expended in the tricarboxylic acid cycle to form amino acids and to generate energy in the cell. The level of free CoA remains high, which hinders P(3HB) synthesis. Under unbalanced growth, in a nitrogen-free or oxygen deficient medium, pyruvate does not enter the tricarboxylic acid cycle but is carboxylated, forming acetyl-CoA. The level of free CoA is low, which is a favorable condition for the activation of P(3HB) cycle enzymes.

Synthesis of PHAs by microorganisms has become an extensively studied subject, and this is not surprising, as these polymers have a high commercial value. PHAs are represented by polyesters consisting of homogenous monomers with different length of the carbon chain and by copolymers; there are high-crystallinity thermoplastic PHAs and thermolabile rubber-like elastomers (Philip et al., 2007). Synthesis of PHAs with tailored properties is a difficult technological task, and production of a PHA of a definite composition should be based on fundamental knowledge of how a certain PHA is synthesized and how the chemical structure of the polymer influences its physicochemical properties. Synthesis of PHAs consisting of monomers with different carbon chain lengths is effected using genetically modified PHA-producing strains, with genes controlling the key enzymes of PHA synthesis (β-ketothiolase, acetoacetyl-CoA reductase, PHA-synthase) taken from microorganisms of other taxa and introduced into them, and naturally occurring strains, which need to be thoroughly studied in order to understand their mechanisms of PHA synthesis and the effects of culture conditions on this process.

Various naturally occurring PHA producers such as *Ralstonia eutropha* (Peoples and Sinskey, 1989), *Alcaligenes latus* (Braunegg et al., 1985; Yamane et al., 1996; Chen et al., 1991), *Pseudomonas* (Timm et al., 1994), *Aeromonas* (Lee et al., 2000), *Rhodococcus ruber* (Pieper and Steinbüchel, 1992), and *Syntrophomonas wolfei* (McInerney et al., 1992) are able to synthesize a wide range of PHAs, depending on the carbon source and the cultivation conditions employed. There are more than 300 microorganisms capable of accumulating PHAs (Anderson and Dawes, 1990; Steinbüchel and Valentin, 1995; Braunegg et al., 1998; Madison and Huisman, 1999); these include naturally occurring and genetically modified strains. These microorganisms comprise aerobic and anaerobic bacteria; heterotrophs; chemoorganotrophs and chemotrophs; phototrophic prokaryotes, aerobic photobacteria, oligotrophic polyprosthecate bacteria, archaebacteria, anaerobic phototrophic bacteria, etc. Rather few of these species, however, are actually used as PHA producers. Among them are *Ralstonia* – a chemoorganotrophic organism capable of utilizing various carbon sources – and heterotrophic microorganisms such as *Methylotrophus, Methylobacterium,* and *Pseudomonas* (Anderson and Dawes, 1990; Lee, 1996a; Choi and Lee, 1997; 1999; Steinbüchel and Valentin, 1995; Braunegg et al., 1998; Madison and Huisman, 1999). Over the past few decades, extensive studies related to engineering and investigation of genetically modified organisms – PHA producers, including various microorganisms and higher plants, have been carried out. The choice of potential PHA producers is usually based on the following parameters: polymer chemical composition and yield, carbon substrate utilization, cell concentration in the culture, and process efficiency.

PHAs have very many attractive properties that make them promising materials for various applications, including biomedical ones. PHAs have significant advantages over other biomaterials (Dawes, 1990; Amass et al., 1998; Steinbüchel and Füchstenbusch, 1998; Sudesh et al., 2000; Steinbüchel, Hein, 2001; Volova, 2004; Chen, 2009; 2010a; Tan et al., 2014; Mozejko-Ciesielska and Kiewisz, 2016; Koller et al., 2017; Kourmentza et al., 2017; Sabbagh and Muhamada, 2017):

- The high biocompatibility of PHAs, polyhydroxybutyrate in particular, is accounted for by the fact that the monomers constituting this polymer – hydroxybutyric acid – are natural metabolites of body cells and tissues.
- PHAs are not hydrolyzed in liquid media as they undergo true biological degradation, which occurs via the cellular and the humoral pathways; the resulting monomers of hydroxybutyric acid do not cause abrupt acidification of tissues and, hence, do not give rise to any pronounced inflammatory reaction.
- PHA bioresorption rates are much lower than those of polylactides and polyglycolides; PHA-based implants can function *in vivo* for 2 to 3 years, depending on their form and implantation site; moreover, PHA degradation can be controlled.
- PHAs are produced by direct fermentation; no multistage technology is needed (monomer synthesis, polymerization, the addition of plasticizers and modifying components).
- PHAs can be synthesized on such feedstocks as sugars, organic acids, alcohols, mixtures of CO_2 and H_2, products of plant biomass hydrolysis, industrial wastes of sugar and palm oil production, hydrogen-containing products of processing of brown coals and hydrolysis lignin.
- PHAs constitute a family of polymers of various chemical structures, consisting of monomers containing 4 to 12 and more carbon units, including high-crystallinity thermoplastic materials and rubber-like elastomers.
- PHA properties (crystallinity, mechanical strength, temperature characteristics, and biodegradation rates) can be controlled by varying the composition of the culture medium and tailoring the chemical structure of the polymer.

- PHAs can be processed from various phase states (powder, solution, gel, melt) using conventional techniques.

PHAs are very promising polymers as, being thermoplastic, like polypropylene and polyethylene, they also have antioxidant and optical properties as well as piezoelectricity. PHAs are highly biocompatible and can be biodegraded in biological media. In addition to poly(3-hydroxybutyrate) (P(3HB)), there are various PHA copolymers, which, depending on their monomeric composition, have different basic properties (degree of crystallinity, melting point, plasticity, mechanical strength, biodegradation rate, etc.) (Madison and Huisman, 1999; Noda et al., 2010; Laycock et al., 2013; Bugnicourt et al., 2014).

The first of the discovered PHAs was poly(3-hydroxybutyrate) (P(3HB)). Now the number of identified, structurally different PHAs exceeds 150. PHAs are stored in bacterial cells in the form of granules, and the stained granules are distinctly visible under the microscope. The observation of the inclusions in bacterial cells goes back to Beijerinck, in 1888 (Cromwick, 1996), but the first determination of P(3HB) had to wait until 1927 and the work of Lemoigne (Lemoigne, 1927). During the following 30 years, interest in P(3HB) was scant. A convincing proposal for a functional role for P(3HB) first came from Macrae and Wilkinson in 1958, who observed that bacteria stored P(3HB) when the carbon-to-nitrogen source ratio in the medium was high and that subsequent degradation occurred in the absence of exogenous carbon and energy sources (Macrae and Wilkinson, 1958). They concluded that P(3HB) was an intracellular reserve material. This paper marks the time when the interest of microbiologists and biochemists in P(3HB) began to increase. The following 40 years saw increasing research attention to the process of poly(3-hydroxybutyrate) accumulation by various microorganisms. Until the end of 1973, interest in P(3HB) had been directed almost solely at its physiological significance in the functioning of microorganisms and at the influence of environmental factors on its synthesis and reutilization. The oil crisis of 1973 and the subsequent increase in the price of oil, as a non-renewable source of energy and feedstock, made OPEC members, which control the plastics market, realize the necessity of searching for alternative ways of producing plastic materials other than a petrochemical synthesis of polyolefins.

In 1976, Imperial Chemical Industries (ICI) of England started the first commercial investigation of microbiological production of poly(3-hydroxybutyrate) using sugar-containing substrates extracted from plant biomass (Senior, 1984). Not only could P(3HB) be synthesized from renewable sources, but some of its properties, such as thermoplasticity, resembled those of polypropylene (King, 1982). Other P(3HB) properties, like biodegradability and biocompatibility, piezoelectric ability, and the possibility of using it as a source of optically active molecules (Howells, 1982), were recognized early on as additional assets and kept ICI's interest in P(3HB) alive after the oil crisis had begun to pass.

In succeeding years, interest in the process of biological synthesis of polyhydroxybutyrate increased. It was found that P(3HB) could be synthesized by many prokaryotic microorganisms (more than 300 have been identified by now) with different efficiency, using various substrates. However, just a few species of microorganisms were chosen for commercial synthesis. They were the microorganisms that efficiently synthesized P(3HB) on a number of substrates: sugars, methanol, hydrocarbons, and mixed hydrogen and carbon dioxide (the hydrogen-oxidizing bacteria such as *Alcaligenes eutrophus,* which is now known as *Ralstonia eutropha,* and *Alcaligenes latus,* the nitrogen-fixing bacterium *Azotobacter vinelandii,* the Pseudomonas *Pseudomonas oleovorans,* the methylotrophs *Methylomonas* and *Methylobacterium organophilum* (Anderson and Dawes, 1990; Byron, 1987, 1994; Dawes, 1990; Braunegg et al., 1998).

Pure P(3HB) is brittle and has a low extension to break. The lack of flexibility and thermal stability limits its range of applications, and if P(3HB) were the only existing polyhydroxyalkanoate, it is doubtful that a large market niche could be found for PHAs. However, a polyester with properties not identical (but similar) to those of P(3HB) was isolated from activated sludge. Detailed chromatographic analysis revealed the presence of 3-hydroxybutyric acid as a major component and 3-hydroxyvaleric, 3-hydroxyhexanoic acid, and possibly 3-hydroxyheptanoic acid units as minor components of the new compound (Wallen and Rohwedder, 1974). This was the first report of a PHA heteropolymer.

The discovered ability of microorganisms to synthesize PHA heteropolymers gave an impetus to extensive investigations of these biopolymers. It was found that the presence of 3-hydroxyvalerate in PHA copolymers was a factor that significantly affected polymer properties, decreasing its melting point and crystallinity and making it more ductile, elastic, and processable than poly(3-hydroxybutyrate) (Luizier, 1992). Variation in monomer proportions of the PHA leads to considerable changes in its thermomechanical and fibrous properties. An extensive search for microorganisms capable of synthesizing PHA heteropolymers was conducted in many countries. Rather soon it was found that some microorganisms under certain growth conditions can synthesize not only homogenous polyhydroxybutyrate but also various PHAs containing copolymers of P(3HB) and other hydroxy-alkanoic acids as monomer units. By now, over 150 various PHAs have been described, but the PHAs that are practically produced and investigated are homogenous poly(3-hydroxybutyrate) and copolymers of 3-hydroxy-butyrate and 3-hydroxyvalerate, P(3HB/3HV), and hydroxybutyrate and hydroxyoctanoate, P(3HB/3HO).

It has been reported that PHAs of different chemical composition differ in their structure and basic physicochemical properties. There are data on the influence of fermentation conditions, mostly of the carbon source, on the total yield, composition, and some properties of polymers (molecular weight, degree of crystallinity, mechanical strength, degradation rates in natural and model environments) (Spyros et al., 1997; Kim et al., 1998; Pazur et al., 1998; Nagata et al., 1998; Avella et al., 2000a, b). This offers an opportunity for the synthesis of biopolymers with tailored properties. However, these very important investigations have been started rather recently, and the available data are fragmentary and mainly concern only a few PHAs.

Since the late 1980s – the early 1990s, the extensive body of information on PHA biosynthesis has been enriched by molecular genetic studies. The system of P(3HB) synthesis from *Alcaligenes, Pseudomonas,* and other organisms has been investigated and cloned; high-efficiency recombinant strains producing PHAs have been engineered (Madison and Huisman, 1999), including those based on *E. coli* (Lee and Chang, 1994; Lee, 1996b; Liu and Steinbüchel, 2000; Wong and Lee, 1998; Klinke et al., 1999); *Pseudomonas putida* (Boynton et al., 1999); yeast (Leaf et al., 1996); and transgenic higher plants with genes of PHA synthesis – *Arabidopsis thaliana* (Poirier et al., 1992, 1995), corn, cotton-plant, etc. (Rinehart et al. 1999).

2.1 PHA CHARACTERIZATION

Since 1974, when a polyester with properties not identical to those of poly(3-hydroxybutyrate) was isolated from activated sludge (Wallen, Rohwedder, 1974), extensive investigations have been conducted to find new PHAs and to study the conditions of their synthesis and their properties. By now, over 150 variously structured monomer units comprising PHA have been described. This line of investigations has attracted much attention because even a slight variation in the proportions of monomer units in a PHA can lead to a fundamental change in their properties, including thermomechanical ones, which is very important for practical purposes. The main PHA structures:

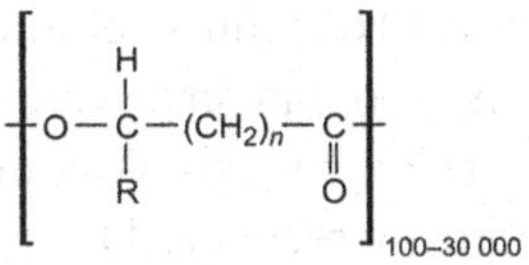

$$\left[-O-\overset{\overset{H}{|}}{\underset{\underset{R}{|}}{C}}-(CH_2)_n-\overset{}{\underset{\underset{O}{\|}}{C}}- \right]_{100\text{–}30\,000}$$

n = 1 R = hydrogen – poly(3-hydroxypropionate),
R = methyl – poly(3-hydroxybutyrate),
R = ethyl – poly(3-hydroxyvalerate),
R = propyl – poly(3-hydroxyhexanoate),
R = pentyl – poly(3-hydroxyoctanoate),
R = nonyl – poly(3-hydroxydodecanoate),
n = 2 R = hydrogen – poly(4-hydroxybutyrate),
n = 3 R = hydrogen – poly(5-hydroxyvalerate).

The diversity of PHAs is described in the review by Steinbüchel and Valentin (1995), which presents data that had been collected by the mid–1990s on the PHA types, mechanisms of their formation, and the nature of microorganisms producing PHAs. Most of the known hydroxyalkanoic acids have been detected as constituents of biosynthetic PHAs. Polymers with different compositions can include:

- hydroxyalkanoic acids with different lengths of hydroxycarbon chain, from 3-hydroxypropionic acid to 3-hydroxyhexadecanoic acid.
- unsaturated 3-hydroxyalkenoic acids with one or two double bonds in the R-pendant group.
- 3HAs with a methyl group at various positions of the R-pendant group in the polyester.
- non-3HAs such as 4-hydroxybutyric acid, 4-hydroxyvaleric acid, 4-hydroxyhexanoic acid, 4-hydroxyheptanoic acid, 4-hydroxyoctanoic acid, 4-hydroxydecanoic acid, 5-hydroxyvaleric acid, 5-hydroxyhexanoic acid, and 6-hydroxy-dodecanoic acid.
- 3-HAs with various different functional groups in the R-pendant group, including free carboxyl groups such as in malic acid or carboxyl groups esterified with alkyl groups or with benzoic acid.
- acetoxy groups bound to the R-pendant groups and phenoxy groups, para-cyanophenoxy groups or a para-nitrophenoxy group bound to the R-pendant group by ether linkages, and a phenyl group or a cyclohexyl group directly bonded by carbon-carbon linkages.

As monomer composition accepted by PHA synthase is determined by synthase substrate specificity and the ability of microorganisms to utilize various carbon compounds, new PHAs must certainly be detected by further studies. For instance, some microorganisms were found to be able to use malic acid as a substrate for PHA synthesis. Among PHAs described in the literature are polymers that were not recovered from axenic bacterial cultures in the laboratory but were rather detected in carbon-rich environments with high concentrations of natural microorganisms, which synthesized not only 3-hydroxybutyric acid but also many other hydroxyalkanoic acids. From activated sludge from a domestic sewage plant in Peoria, Wallen, and Rohwedder isolated PHAs consisting of 3HB, 3-hydroxyvalerate, 3-hexanoate and trace amounts of 3-hydroxyheptanoic acid (Wallen and Rohwedder, 1974).

Analysis of activated sludge from a sewage treatment plant in Veberöd (Sweden) revealed PHA consisting of 3HB, 3HHx, and 3-hydroxyoctanoate (Odham et al., 1986). A polymer containing 3HB, 3HV, 3-hydroxy–2-methylbutyric acid and 3-hydroxy–2-methylvaleric acid was recovered from activated sludge from a domestic sewage plant in Tokyo. A polymer containing 3HB and 5 other hydroxyalkanoic acids as its copolymers – 3HV, 3-HP, 3HO, 3-hydroxy–6-methylheptanoate, and 3-hydroxy–7-methyloctanoate – was recovered from estuarine sediment (Findlay and White, 1983). These data suggest that microbial communities developing on complex carbon substrate are capable of synthesizing complex, sometimes unusual, PHAs. Among the recently detected and identified HAs are the optically active ethyl 4-chloro–3-hydroxybutyrate, methyl (R)–4-chloro–3-hydroxybutyrate and (S)–3-hydroxy-γ-butyrolactone (Suzuki et al., 1996); sulfur-containing polymers with thioester linkages; copolymers of 3-hydroxybutyrate and 3-mercaptopropionate P(3HB-co-3MP) (Lütke-Eversloh et al., 2001).

Analysis of the diversity of PHAs showed that biosynthetic polymers contain HAs exhibiting the R-configuration but no HAs exhibiting the L-configuration. HAs with the hydroxy group in the position were only detected for malic acid incorporated in the polymer during the course of in vitro biosynthesis, and those with the hydroxy group in the ε position – for 6-hydroxydodecanoate.

PHAs can be divided into three groups depending on the number of carbon atoms in the monomer units:

1) short-chain-length (SCL) PHAs, which consist of 3 – 5 carbon atoms;
2) medium-chain-length (MCL) PHAs, which consist of 6 – 14 carbon atoms; and

3) long-chain-length (LCL) PHAs, which consist of 17 and 18 carbon atoms.

The best-known representative of SCL PHAs is poly-(R)–3-hydroxy-butyrate, P(3HB). Short-chain-length PHAs consist of poly-(R)-hydroxyal-kanoates containing 3 to 5 carbon units. These polyesters are synthesized by different bacteria (*A. latus, Bacillus cereus, Pseudomonas pseudoflava – Hydrogenophaga pseudoflava, Pseudomonas cepacia, Micrococcus halode-nitrificans, Azotobacter sp., Rhodospirillum rubrum, Ectothiorhodospira shaposhnikovii* and *Cupriavidus necator*). *C. necator* (previously known as *Wautersia eutropha, Ralstonia eutropha* or *Alcaligenes eutrophus*) is a universally recognized producer of short-chain-length PHAs (Nakamura et al., 1991; Bear et al., 1997; Ballistreri et al., 1999; Babel et al., 2001; Petti-nari et al. 2001; Zhang et al., 2004; Lenz and Marchessault, 2005; Stubbe et al., 2005; Hazer and Steinbüchel, 2007).

Medium-chain-length PHAs consist of aliphatic (R)-monomers of hydroxy acids, containing 6 to 14 carbon units. They are synthesized when the PHA producing strain is cultivated on the medium containing n-alkanoates or their precursors (i.e., alkanes) (Gross et al., 1989; Madison and Huisman, 1999; Witholt and Kessler, 1999).

Long-chain-length PHAs are usually synthesized in chemical synthesis reactions. This division of polymers into groups is based on the notion of substrate specificity of PHA synthases that can only accept certain hydroxy-alkanoic acids in the course of polymerization (Anderson, Dawes, 1990).

Earlier, it was assumed that the PHA synthase of *A. eutrophus* and of other natural strains could polymerize HAs consisting of 3–5 carbon atoms, but not those made of 6 or more. Thus, it was believed that wild-type *Alca-ligenes* strains were not capable of producing medium-chain-length PHAs. More recently, it has been reported that PHA synthases of some wild-type strains of *Ralstonia eutropha* (Volova et al., 1998; Green et al., 2002) and *Ectothiorhodospira shaposhnikovii* (Zhang et al., 2001) have broader substrate specificity and, thus, these microorganisms can accumulate both short- and medium-chain-length PHAs.

PHAs that contain a single type of repeat unit are known as homo-polymers, while PHAs containing a mixture of repeat units are known as copolymers (or heteropolymers) (Madison and Huisman, 1999; Olivera et al., 2001a; Luengo et al., 2003). Physical and mechanical properties of PHAs are determined by the molar percentages of monomers in the polymer. The properties of homopolymer PHAs are significantly different from those of copolymer PHAs (Matsusaki et al., 2000; Chen et al., 2006).

Short-chain-length PHAs are thermoplastics that have a high degree of crystallinity; they are rigid and brittle crystalline materials that have low elongation-to-break, while medium-chain-length PHAs are elastomers with low crystallinity and a low melting point (Nakamura et al., 1991; de Koning, 1995; Kim et al., 2001; Zinn et al., 2001; Lenz and Marchessault, 2005). Genetic construction of recombinant PHA producing strains, with genes from PHA produces introduced into these strains, have a great potential for synthesis of various polymers (Schmack et al., 1998; Steinbüchel and Hein, 2001; Lu et al., 2004; Chen et al., 2006; Wei et al., 2009).

As the diversity of PHAs is increasing, it has been proposed that they should be divided into two categories, based on the frequency of their occurrence: "usual" and "unusual" PHAs (Olivera et al., 2001a, b). Usual PHAs are PHAs intracellularly synthesized by microorganisms as storage macromolecules. This group includes polymer tailored from monomers (i.e., (R)–3-hydroxypropionate, (R)–3-hydroxybutyrate, (R)–3-hydroxyvalerate, (R)–3-hydroxyhexanoate, (R)–3-hydroxyoctanoate, (R)–3-hydroxydecanoate, and (R)–3-hydroxydodecanoate, or combinations thereof), which are obtained from different carbon sources (sugars, alkanes, aliphatic fatty acids, triacylglycerols, etc.) through general pathways (usually fatty acid synthesis and fatty acid β-oxidation) involving the synthesis of (R)–3-hydroxyacyl-CoAs (Steinbüchel and Füchstenbusch, 1998; Rehm et al., 1998; de Roo et al., 2000; Kessler and Witholt, 2001; Luengo et al., 2003; 2004; Velázquez et al., 2007). "Unusual" PHAs are polyesters rarely found in nature that are constituted by natural monomers with an unusual chemical structure (e.g., 4-hydroxyalkanoic acids, 5-hydroxyalkanoic acids, and 6-hydroxyalkanoic acids) that are synthesized by some microorganisms (Saito and Doi, 1994; Schmack et al., 1998; Choi et al., 1999; Olivera et al., 2010; Amirul et al., 2008); by unnatural monomers (generally obtained by chemical synthesis – xenobiotics) that can be taken up by the PHA producing microorganisms, activated to their CoA thioesteres, and used as substrates by the PHA polymerases.

"Unusual" PHAs with elongated backbones contain as monomers hydroxyalkanoic acids different from (R)–3-hydroxyalkanoates (Saito and Doi, 1994; Schmack et al., 1998; Choi et al., 1999; Amirul et al., 2008). Only a small group of wild-type bacteria able to produce P(3HB/4HB) have been reported to date: *Alcaligenes eutrophus* (Doi et al., 1989; Kunioka et al., 1989), *Cupriavidus spp. USMAA 1020* (Amirul et al., 2008), *Cupriavidus eutrophus* (Volova et al., 2011), *A. latus* (Hiramitsu et al., 1993; Kang et al., 1995), *Comamonas acidovorans* (Saito and Doi 1994; Lee et al., 2004), *Comamonas testosteroni* (Renner et al., 1996), *Rhodococcus sp.* (Haywood et al., 2001), and *H. pseudoflava* (Choi et al., 1999). *E. coli* strains have been engineered for the production of P(3HB/4HB): an *E. coli* strain coexpressing the 4-hydroxybutyric acid-CoA transferase from *C. kluyveri* and the PHA synthase from *C. necator* (Hein et al., 1997), a strain of *E. coli* expressing the succinate degradation pathway from *C. kluyveri* along with the PHA genes from *C. necator* (Valentin and Dennis, 1997); strains of *C. necator* and *P. putida* handicapped in PHA biosynthesis but harboring the PHA synthase genes from *Tuocapsa pfennigii* (Schmack et al., 1998; Gorenflo et al., 2001).

In view of the huge variety of PHAs, it is evident that these polyesters not only represent a hopeful alternative to replace traditional plastics, but also, owing to their special properties and physicochemical characteristics, they offer a variety of potential biotechnological applications (including ecological, medical, pharmaceutical, and industrial uses). Additionally, the description of new PHA producing strains isolated from unexplored habitats (Ayub et al., 2009; Kalyuzhnaya et al., 2008; Yan et al., 2008), the design of novel methods for the detection of PHAs in mixed microbial populations (Monteil-Rivera et al., 2007; Russell et al., 2007; Dias et al., 2008; Foster et al., 2008; Grubelnik et al., 2008; Serafim et al., 2008; Werker et al., 2008), and recent advances in genetic, metagenomic, and metabolic engineering (Olivera et al., 2001 a, b; Cowan et al., 2005; Solaiman and Ashby, 2005; Kung et al., 2007; Sandoval et al., 2007; Tsuge et al., 2007; Velázquez et al., 2007; Dias et al., 2008; Kalyuzhnaya et al., 2008; Ruth et al., 2008) together with new approaches based on chemical synthesis and blending (Hazer and Steinbüchel, 2007) will expand the number of PHAs and their potential applications.

The cost of the carbon source contributes significantly to the overall production cost of a PHA. Thus, one of the main objectives of PHA investigations is to find inexpensive substrates for PHA production. This objective can be attained by finding new substrates for the already known PHA-producing organisms, discovering new PHA-producing strains, and engineering recombinant strains that can utilize various substrates, including new ones. In principle, PHAs can be produced on various substrates. Among the best-known ones are individual compounds, such as carbon dioxide and hydrogen, sugars, alcohols, and organic acids; byproducts of alcohol, sugar, and hydrolysis industries and of olive and palm oil production; and unusual substrates, including toxic ones.

In sum, although PHAs were discovered in 1925 (Lemoigne, 1926) and have been actively studied since the mid-1980s – the early 1990s, these compounds are one of the most promising biomaterials of the 21st century.

2.2 PHA PROPERTIES

PHA properties vary depending on the structure of the side groups in the polymer chain and the distance between ester groups in a molecule. Only a few PHAs, however, has been thoroughly investigated by now. Characterization of several PHA types shows that their properties differ greatly, depending on the type and proportions of monomer units in the polymer chain (Table 2.1). Thus, PHAs can be used to produce various materials with different physical-mechanical properties for a variety of applications.

TABLE 2.1 Chemical Structures of PHAs, Their Producers and Substrates for Their Synthesis

PHA	Chemical structures
R–3-hydroxybutyrate	$[-O-CH(CH_3)-CH_2-C(=O)-]_x$
R–3-hydroxybutyrate 3 – hydroxypropionate	$[-O-CH(CH_3)-CH_2-C(=O)-]_x[-O-CH_2-CH_2-C(=O)-]_y$
R–3-hydroxybutyrate 4 – hydroxybutyrate	$[-O-CH(CH_3)-CH_2-C(=O)-]_x[-O-CH_2-CH_2-CH_2-C(=O)-]_y$
R–3-hydroxybutyrate 3 – hydroxyvalerate	$[-O-CH(CH_3)-CH_2-C(=O)-]_x[-O-CH(C_2H_5)-CH_2-C(=O)-]_y$
R–3-hydroxybutyrate 3 – hydroxyhexanoate	$[-O-CH(CH_3)-CH_2-C(=O)-]_x[-O-CH(C_3H_7)-CH_2-C(=O)-]_y$
R–3-hydroxybutyrate 3 – hydroxydecanoate	$[-O-CH(CH_3)-CH_2-C(=O)-]_x[-O-CH(C_7H_{11})-CH_2-C(=O)-]_y$

The first among the isolated PHAs and the most fully characterized one is poly(3-hydroxybutyrate) (P(3HB)). P(3HB) is a homopolymer of D(-)–3-β-hydroxybutyric acid, an isotactic polyester with regular ($C_4H_6O_2$) units. In contrast to complex synthetic polyesters, P(3HB) is a stereoregular optically active polymer, which forms helices in a solution and crystallizes in spherulites. Crystallization begins as soon as P(3HB) has been extracted from the cell mass with non-polar solvents (Anderson, Dawes, 1990; Doi, 1995). The crystallization of the polymer is thought to be under kinetic control and is inhibited by the submicron size of the particles, and their protein and phospholipid coat (Bonthrone et al., 1992).

P(3HB) is a colorless semi-crystalline hydrophobic substance. The crystalline structure of P(3HB) has been investigated by a number of researchers using the X-ray analysis of oriented polymer fibers. Analysis showed compact antiparallel chains packed in orthorhombic unit cells with a fiber repeat of 0.596 μm (Brückner et al., 1988). Typically, P(3HB) forms lath-shape crystals with dimensions of around 0.3–2.0 μm for the short and 5.10 μm for the long axes.

More than 50 published works about chemical structure and properties of PHAs were analyzed and database "Physico-chemical properties of PHAs of different structure. Biosynthesis's conditions and producers" (Certificate of the state registration in VNIIGPE No 2012620288–2012) (Volova et al., 2012). Results of the analysis are presented in Table 2.2.

The data obtained by X-ray structure analysis suggest a conclusion that the crystalline region is predominant in P(3HB). The degree of crystallinity of different P(3HB) samples ranges between 0.62 and 0.76, depending on the conditions of P(3HB) production (Doi, 1990; Madison and Huisman, 1999; Volova et al., 2000). P(3HB) is an isotactic polyester with regular identically oriented (head-to-tail) sequential units of D-(-)–3-β-hydroxybutyric acid: $(-O-CH(CH_3)-CH_2-CO-)_n$.

P(3HB), like many other polymers, has a heat distortion temperature somewhat lower than the thermal degradation temperature. Thus, polymers cannot exist in the gaseous state, and the main type of phase equilibrium in them is a condensed state – crystalline, glassy, viscoelastic, and liquid. The ability of P(3HB) to crystallize is determined by the inner properties of its chains and is characterized by crystallization temperature, T_c. In a number of polymers, crystallization develops only partly, and most of the polymers are semi-crystalline materials. P(3HB) is one of these. Thermal properties of P(3HB) and its ability to crystallize in the native state are its most important parameters because they determine the processability of the polymer.

TABLE 2.2 Physicochemical Properties of PHAs of Various Compositions (Certificate of the State Registration in VNIIGPE No 2012620288–2012)

Composition, mol.%	Molecular weight (Mn), kDa	Polydispersity	Crystallinity, %	Melting point, oC	Glass-transition temperature, oC	Strain, substrate	Reference
P(3HB)							
	768	1.9	59	177	4	Alcaligenes eutrophus H16 (ATCC 17699)	Nakamura et al., 1992
	166	2.6	–	175	4	Alcaligenes latus ATCC butyric acid	Kang et al., 1995
	218	–	78	182	–	R. eutropha 5786, (–)	Volova et al., 2001
	58	3.4	72	176	4	Comamonas acidovorans (IFO 13582), butyric acid	Mitomo et al.,2001
	510	2.2	60	175	1	Recombinant R. eutropha PHB–4 fructose and L–arabinose	Fukui et al., 2002
	254	1.18	–	175	3	Recombinant Pseudomonas putida KT2442, (–)	Liu and Chen, 2007
	1345	1.22	86	175	3	Recombinant R. eutropha PHB–4 gluconate	Luo et al.,2006
	350	1.75	–	162	–1.18	Recombinant Aeromonas hydrophila 4AK4, –	Zhao and Chen, 2007
	1580	2.9	–	159. 172	0	Recombinant Ralstonia eutropha PHB–4, fructose	Tanadchangsaeng et al., 2009

Composition, mol.%	Molecular weight (M_n), kDa	Polydispersity	Crystallinity, %	Melting point, °C	Glass-transition temperature, °C	Strain, substrate	Reference
P(3HB)							
	230	1.3	–	157. 171	0	Recombinant *Ralstonia eutropha* PHB–4, fructose	Tanadchangsaeng et al, 2009
	252	1.13	–	161.56	- 0.83	*Recombinant Pseudomonas putida* KT2442	Wang et al., 2009
	361	1.8	–	178.2	2.4	*Cupriavidus necator* A–04, 1,4–butanediol and fructose	Chanprateep et al., 2010
	530	2.6	60	159. 172	1.0	Recombinant *Ralstonia eutropha* PHB–4, fructose	Tanadchangsaeng et al., 2010
	847	1.93	–	163.3	–3.08	Recombinant *Pseudomonas putida* KT2442, –	Wang et al., 2011
P(3HB-co-4HB) (4HB, mol.%)							
3	58	6.9	–	165	1	*Alcaligenes latus* ATCC 29173, 4–hydroxybutyric acid	Kang et al., 1995
5	89.4	1.62	–	168.5	–2	*Cupriavidus necator* A–04, 1,4–butanediol and fructose	Chanprateep et al., 2010
6	494	2.1	56	162	–1	*Alcaligenes eutrophus* H16 (ATCC 17699), –	Nakamura et al., 1992

TABLE 2.2 *(Continued)*

Composition, mol.%	Molecular weight (Mn), kDa	Polydispersity	Crystallinity, %	Melting point, °C	Glass-transition temperature, °C	Strain, substrate	Reference
P(3HB-co-4HB) (4HB, mol.%)							
7	420	1.62	–	114	–6.69	Recombinant Aeromonas hydrophila 4AK4, 1,4–butanediol	Xie and Chen, 2008
10	395	3.0	46	159	–3	Alcaligenes eutrophus H16 (ATCC 17699),–	Nakamura et al., 1992
13	153	2.8	–	155	–1	Alcaligenes latus ATCC 29173, butyric, and 4–hydroxybutyric acids	Kang et al., 1995
16.7	260	3.4	–	140.6	2.5	R. eutropha H16 (ATCC 17699), 4–hydroxybutyric and butyric acids	Ishida et al., 2001
19	66	3.1	40	158	–4	Comamonas acidovorans (IFO 13582), 1,4–butanediol and butyric acid	Mitomo et al., 2001
23	590	1.8	–	152	–7	Cupriavidus sp, USMAA1020, γ–butyrolactone 48 h	Vigneswari et al., 2009
24	104	2.54	–	161.1	–5	Cupriavidus necator A–04, 1,4–butanediol and fructose	Chanprateep et al., 2010
29	110	5.0	–	144	–3	Alcaligenes latus ATCC 29173, butyric, and 4–hydroxybutyric acids	Kang et al., 1995
38	37	2,3	18	54	–6	Comamonas acidovorans (IFO 13582), 1,4–butanediol and butyric acid	Mitomo et al., 2001

Composition, mol.%	Molecular weight (Mn), kDa	Polydispersity	Crystallinity, %	Melting point, °C	Glass-transition temperature, °C	Strain, substrate	Reference
P(3HB-co-4HB) (4HB, mol.%)							
50.9	100	6.5	–	54.0	–43.4	R. eutropha H16 (ATCC 17699), 4–4-hydroxybutyric and butyric acids	Ishida et al., 2001
65.1	480	2.1	–	40.7	–34.1	R. eutropha H16 (ATCC 17699), 4–4-hydroxybutyric and butyric acids	Ishida et al., 2001
65	91	3.1	27	52	–40	Comamonas acidovorans (IFO 13582), 1,4–butanediol and butyric acid	Mitomo et., 2001
75	260	3.0	–	51	–45	Cupriavidus sp, USMAA1020, γ–butyrolactone	Vigneswari et al., 2009
76	126	4.1	–	46	–37	Alcaligenes latus ATCC 29173, butyric, and 4–hydroxybutyric acids	Kang et al., 1995
85	168	3.7	29	48	–41	Alcaligenes eutrophus H16 (ATCC 17699),–	Nakamura et al., 1992
94	42	2.8	54	54	–46	Comamonas acidovorans (IFO 13582), 1,4–butanediol	Mitomo al et., 2001
100	487	1.75	–	50.12	–45.67	Recombinant Pseudomonas putida KTHH06, γ–butyrolactone	Wang et al., 2011

TABLE 2.2 *(Continued)*

Composition, mol.%	Molecular weight (M_n), kDa	Polydispersity	Crystallinity, %	Melting point, °C	Glass-transition temperature, °C	Strain, substrate	Reference
P(3HB-co-3HV) (3HV, mol.%)							
4	107		65	172	–	*R. eutropha* B5786, CO_2, fructose +valerate	Volova et al., 2001
15	1353	1.74		161	–3	*Pseudomonas oleovorans NRRL B–778,* nonanoic acid	Ashby et al., 2002
14.3	216	–	64	162	–	*R. eutropha* B5786 CO_2+valerate	Volova and Kalacheva, 2005
25	400	–	55	162	–	*R. eutropha* B5786, CO_2 + valerate	Volova et al., 1992
29.9	280	–	51	160	–	*R. eutropha* B5786 CO_2+valerate	Volova and Kalacheva, 2005
43	370		54	159		*R. eutropha* B5786 CO_2+valerate	Volova and Kalacheva, 2005
100	62	3.5	–	103	–15.8	Recombinant *Aeromonas hydrophila* 4AK4, –	Shen et al., 2009
100	815	1.3	–	112.27	–15.09	Recombinant *Pseudomonas putida* KT2442, –	Wang et al., 2011

Composition, mol.%	Molecular weight (M_n), kDa	Polydispersity	Crystallinity, %	Melting point, °C	Glass-transition temperature, °C	Strain, substrate	Reference
P(3HB-co-3HHx) (3HHx, mol.%)							
1.5	540	2.8	48	161	–1	Recombinant *R. eutropha* PHB–4 fructose and L–arabinose	Fukui et al., 2002
5	460	3.3	–	143	–3	Recombinant *Wautersia eutropha,* palm oil	Loo et al., 2005
5	178	3.6	–	129.0	–0.6	Recombinant *C. necator* PHB–4, unrefined palm oil	Chia et al., 2010
12	210	2.03	–	97	–1.75	Recombinant *Aeromonas hydrophila* 4AK4, dodecanoic acid	Zhao and Chen, 2007
12	200	1.13	–	69	–1.58	Recombinant *Pseudomonas putida* KT2442	Wang et al., 2009
35	229.3	1.64	–	–	–	Recombinant *Pseudomonas putida* GPp104, oleate, and capronate	Asrar et al., 2002
96	51	5.9	–	–	–20.8	Recombinant *Aeromonas hydrophila* 4AK4	Jian et al., 2010
100	206	1.32	–	–	–28.19	Recombinant *Pseudomonas putida* KT2442	"

Note: "–" = unclear.

The macroscopic properties of polymers are determined by their chemical structure and phase state, which in P(3HB) and other PHAs vary widely. The dynamics of the temperature of sequential phase transitions has been investigated in P(3HB) by the method of differential thermal analysis (DTA) (Yuan, 1997; Ashraf et al., 1999). Thermograms show a peak corresponding to the glass transition of the amorphous phase in the temperature range $-10°C< T <+10°C$. The next peak is in the range of +50°C – a crystallization peak, and then there is a melting point peak. For homogenous P(3HB) the melting temperature is in the range 176 to 180°C; the temperature of the onset of crystallization is about 47°C (Volova et al., 2007).

The second, after P(3HB), most extensively studied PHA is P(3HB-co-3HV) (Table 2.1). P(3HB-co-3HV) copolymers have been shown to exhibit isodimorphism (Scandola et al., 1992) with the 3HB and 3HV units co-crystallizing. The amorphous and the crystalline densities in this copolymer are 1.16 and 1.2 g/cm^3, respectively (Waddington, 1994; Poirer et al., 1995). Mechanical properties of P(3HB-co-3HV) can vary significantly, depending on the monomer ratio. With an increase in the HV fraction, the material becomes less crystalline and more elastic. The crystallization rate and the spherulite size can be varied by changing the percentage of HV (Akhtar et al., 1992).

Another PHA type, a copolymer of 3-hydroxybutyrate and 4-hydroxybutyrate, in contrast to P(3HB), is a highly elastic and flexible polymer, of tensile strength up to 1000% (2 orders of magnitude more than that of P(3HB)). P(3HB-co-4HB)s are materials varying in their physical and mechanical properties. Copolymers containing 20–40 mol.% 4HB are similar to elastic rubber; with an increase in the 4HB fraction from 0 to 49 mol.% the crystallinity of the polymer decreases from 60 to 14% (Saito and Doi, 1994; Saito et al., 1996; Volova et al., 2011). This type of the copolymer (with 0 to 29 mol.% 4HB) has only one, 3HB-type, crystal lattice. In contrast, the copolymers containing a large fraction of 4HB (78–100 mol.%) have a 4HB-type crystal lattice. In the latter case, the growth rate of crystals decreases, probably because 4HB units are not incorporated into the 3HB lattice (Saito and Doi, 1994; Saito et al., 1996). The lattice parameters of P(4HB) reported by Mitomo et al. (2001) are as follows: a=7.75 Å, b=4.79 Å, and c=11.94 Å. The correlation between the 3HB/4HB ratio and the molecular weight of the material was not significant, but the fraction of 4HB strongly affected the thermal properties of the polymer (Mitomo et al., 2001). The melting points of these copolymers can decrease from 178 to 130°C and lower with an increase in the 4HB fraction. With an increase in the 4HB fraction from

0 to 100 mol.%, the glass transition temperature decreases sharply, from 4 to 46–48°C. With an increase in the 4HB fraction from 0 to 16 mol.%, the tensile strength of the films drops from 43 to 26 MPa, while the elongation at break increases from 5 to 144%. However, when the 4HB fraction increases from 64 to 100 mol.%, the tensile strength grows from 17 to 104 MPa (Saito et al., 1996).

Another PHA, a copolymer of 3-hydroxyhexanoate and 3-hydroxyoctanoate (P(3HHx-co-3HO)) is also a medium-chain-length PHA. It has low melting points and is similar to elastic rubber. The properties of a copolymer of P(3HB) and P(3HHx) are different. For instance, the crystallinities of these copolymers decrease from 40 to 18% as the 3HHx fraction increases from 0 to 25 mol.%. The crystallographic parameters of the P(3HB-co-3HHx) copolyesters are hardly influenced by the presence of the 3HHx units, suggesting that the 3HHx units cannot be incorporated into the P(3HB) crystal lattice. The rates of crystal growth for P(3HB-co-3HHx) are markedly reduced with an increase in the 3HHx fraction, indicating that the randomly distributed 3HHx units in P(3HB-co-3HHx) lead to a remarkable decrease in the rate of deposition of the 3HHx segments at the growing front of P(3HB) crystalline lamellae (Doi et al., 1995).

Thus, the description of just a few PHA types shows that their properties differ considerably. The employment of parametric control and molecular-genetic methods in PHA biosynthesis makes it possible to produce various PHAs, including 1-, 2-, 3-, and multi-component ones, that would exhibit various physicochemical and mechanical properties.

Properties of PHAs with different chemical compositions still remain insufficiently studied. Although PHA studies are numerous, very few of them investigate the properties of various PHA copolymers using state-of-the-art physicochemical methods, and, moreover, the data reported in them are rather contradictory. Analysis of the literature on the physicochemical properties of PHAs shows that some important data are missing. Moreover, authors of PHA studies report different data on the effect of PHA composition on polymer molecular weight, degree of crystallinity, and temperature characteristics, even when they analyze PHAs of similar chemical composition (Table 2.2). This can be due to a number of reasons. For instance, the molecular weight is determined by such factors as the microorganism used to produce it, cultivation conditions, and polymer recovery technique. The data on P(3HB) molecular weight differ greatly, sometimes by two orders of magnitude. The lowest P(3HB) M_n (58 kDa) is given for the polymer produced by *Comamonas acidovorans* in a study by Mitomo et al. (2001), and the highest (1580 kDa) – for the P(3HB) synthesized by recombinant *Ralstonia eutropha* PHB–4 harboring the PHA synthase gene from *Pseudomonas sp.* 61–3 when grown on fructose. Differences in the M_n of the P(3HB-co-3HV) copolymers containing equal fractions of 3HV vary from several tens kDa to 1000 kDa. For instance, the M_n of the copolymer synthesized by *Cupriavidus sp.* and containing 36 mol.% 4HB amounted to 540 kDa, while the M_n of the copolymer synthesized by *Comamonas acidovorans* and containing the same fraction of 4HB was much lower (37 kDa) (Table 2.2).

Another significant parameter determining polymer properties is the degree of crystallinity. The available literature data on the degrees of crystallinity of PHAs with different chemical compositions are, however, rather contradictory. Different authors report P(3HB) degrees of crystallinity ranging from 59 to 86% (Table 2.2). Data presented by Noda et al. (2005) suggest that the 3HV fraction below 20–22 mol.% does not change the crystallinity of the copolymer. On the other hand, Dai et al. (2007) reported a significantly lower degree of crystallinity of the copolymer containing 29 and 32 mol.% 3HV (5 and 9%, respectively). The addition of acetate to the culture medium did not change polymer composition but increased its C_x (to 28–34%) (Poirier, 1999). The available data on degrees of crystallinity of poly(3-hydroxybutyrate-co-3-hydroxyhexanoate) are fragmentary. Noda et al. (2005) showed that the degree of crystallinity of the copolymer containing 12–18 mol.% 3HHx is 38–40%, while Fukui et al. (2002) reported a similar degree of crystallinity for the copolymer containing only 1.5 mol.% 3HHx. Melting point (T_{melt}) and thermal degradation temperature (T_{degr}) are very important characteristics of PHA, which determine the conditions of PHA processing from melts. The T_{melt} of P(3HB) reported by different authors ranges from 162 to 197°C. The data on the T_{melt} of P(3HB-co-3HV) vary widely. For the copolymer containing 6 mol.% 3HV, the T_{melt} was 186°C (Akhtaret et al., 1992), but the T_{melt} of the copolymer with the same composition examined by Zhao and Chen (2007) was 170°C; in another study (Zhang et al., 2009) it was determined as 56°C. Even more contradictory is the data on P(3HB-co-4HB) T_{melt}. The T_{melt} of the copolymers containing 2 to 7 mol.% 4HB ranges from 114 to 172°C, as reported by different authors. A decrease in T_{melt} was recorded when the molar fraction of 4HB was increased t 75–100% (Vignesvari et al., 2009; Chanprateep, Kulpreecha, 2006). Doi et al. (1990) showed, however, that the T_{melt} of the copolymer with 84 mol.% 4HB was 130°C.

The Institute of Biophysics SB RAS and the Siberian Federal University (Russia) have devoted considerable research effort to revealing the relationship between the monomer unit composition and proportions of PHAs and their physicochemical properties. Researchers have developed and implemented processes of biosynthesis of PHA bipolymers (Volova and Kalacheva, 2005; Volova et al., 2011, 2013a, 2016a;), terpolymers (Volova et al., 2013b; 2014), quaterpolymers (Volova et al., 2016b), and block copolymers, which contain, in addition to 3-hydroxybutyrate and 3-hydroxyvalerate monomer units, ethylene glycol (Volova et al., 2016c) and methyl valerate monomers (Vinogradova and Volova, 2016). This research suggests that the composition of PHA influences its molecular weight and temperature characteristics: the polymer may contain several fractions that considerably differ in weight average molecular weight and melting temperature. Monomer units of 4-hydroxybutyrate produce the strongest effect on copolymers containing them; some of the PHA samples have degrees of crystallinity as low as 10–20%. A number of studies report variations in the physical/mechanical properties of PHA products as dependent upon the PHA chemical composition (Volova, 2004; Volova et al., 2008, 2013a, c; 2017a, b).

Thus, PHAs are thermoplastic polymers with various chemical compositions, which do not undergo abiotic hydrolysis in an aqueous medium and whose properties (molecular weight, crystallinity, mechanical strength, and biodegradability) vary considerably depending on their monomer composition and proportions of monomer units. The diversity of PHAs that have widely varying properties, the possibility of constructing hybrids with different substances, and the availability of numerous techniques for processing them provide the basis for producing an extensive range of novel materials with new valuable properties.

2.3 PHA PRODUCTION AND APPLICATIONS

Many companies have been engaged in the commercialization of PHA production technologies. Since the 1980s, they have been producing PHAs on a pilot scale or industrially. The best-known companies and corporations that have been engaged in PHA activities are: Monsanto Co., Metabolix Inc., Procter & Gamble, Berlin Packaging Corp., Bioscience Ltd., BioVentures Alberta Inc., and Merck, which produce polymers with the trademarks Biopol®, Biopol™, TephaFLEX™, DegraPol/btc®, Nodax™, etc.

ICI of England was the first industrial corporation to start commercial production of PHAs. Since 1992, Zeneka Seeds and Zeneka BioProducts (U.K.) began commercialization of a family of poly-3-hydroxybutyrate and P(3HB/3HV) copolymers with the tradename of Biopol®. The cost of Biopol®, which was produced at 10–15,000 tons per year, reached US$ 16,000/t. That was an order of magnitude higher than the global market value of polypropylene.

A world leader in PHA commercialization is Metabolix Inc. (U.S.), which was founded in Cambridge (MA) in 1992. The company has more than 500 owned and licensed patents and applications worldwide. The company is producing polymers using a recombinant strain – *E. coli* K12 – and sugars as substrate. The commercial names of Metabolix polymers are Biopol® and Biopol™. Their outputs reached 90 t in 2005 and 907 t in 2006. The company has many branches in different countries of the world. In 2004, Metabolix formed a strategic alliance with "Archer Daniels Midland Company" (ADM) to commercialize PHAs using the large fermentation capacity of ADM. In 2009, Metabolix marketed a new family of biodegradable plastics – Mirel. ADM has begun construction of the world's first Mirel biorefinery located in Clinton, Iowa. This new facility will produce 55,000 t of three Mirel varieties per year. Metabolix has signed a collaborative agreement with Australia's Cooperative Research Centre for Sugar Industry Innovation Through Biotechnology. Metabolix works in cooperation with British Petroleum to further develop direct production of bioplastics in switchgrass. Metabolix has also received government support for its technology from the U.S. Department of Agriculture and the U.S. Department of Commerce's Advanced Technology Program.

Tepha Inc. (U.S.), founded in 1998 as a sister company of Metabolix, is engaged in the medical applications of PHAs. The company has over 30 licenses for the production of PHA-based articles. Polymer production is performed using a patented process of fermentation of transgenic microorganisms. The product has been trademarked as TephaFLEX™.

Procter & Gamble, Chemicals (U.S.) has been developing and producing PHA heteropolymers consisting of monomer units that contain 4 to 12 carbon atoms. Unlike Metabolix, this company is engaged in the development of PHAs produced by fermentation of sugars and fatty acids; the trademark of the product is Nodax™. Structurally, Nodax polymers are similar to LDPE; they have branched chains, and, thus, the melt and glass-transition temperatures and crystallinity of Nodax plastics are lower than those of Biopol, which makes them easier to process.

Almost all industrially developed countries are to a greater or lesser extent engaged in PHA production, but large-scale production and application of PHAs are impossible without reducing their cost (Chen, 2009, 2010b).

Austrian companies produce a homopolymer of 3-hydroxybutyric acid using *Alcaligenes latus,* which can accumulate P(3HB) at a concentration of 90%. P(3HB) is produced from different substrates, including waste materials. The company produced P3HB in a quantity of 1,000 kg per week in a 15-m^3 fermentor. The P(3HB) production and the processing technology are now owned by Biomer, Germany. In 1995, the Brazilian sugar mill Copersucar assembled a pilot-scale P3HB production plant. The goal of this pilot plant was to produce enough P(3HB) to supply the market for tests and trials. Also, this pilot plant was intended as a training facility for future operators, and it is currently providing data for scale-up and economic evaluation of the process. Copersucar managed to produce 120–150 g/L CDW containing 60–65% P(3HB) with a productivity of 1.44 kg P3HB m^3 an hour and a P3HB yield of 3.1 kg sucrose per kilogram of P(3HB). In 2001, Copersucar began to produce PHA from sugar cane processing waste (Natano et al., 2001). In China, commercial production of PHAs is based on the use of efficient PHA producers such as naturally occurring and engineered strains of *Ralstonia eutropha* and other species. Chinese researchers have developed a process that can produce poly-3-hydroxybutyrate/3-hydroxyvalerate in high efficiency. Without a supply of pure oxygen, *R. eutropha* grew to a density of 160 g/L CDW within 48 h in a 1,000-L fermentor (Chen, 2010b).

Rubber-like P(3HB/4HB) copolymers have recently attracted considerable attention. The wild-type *Ralstonia eutropha* and recombinant *Ralstonia* and *E. coli* strains are used by Chinese and U.S. companies to produce P(3HB/4HB). With the addition of 1,4-butanediol in different amounts, 4-hydroxybutyrate can be accumulated to 5–40 mol.% in the copolymer, thus generating copolymers with various thermal and mechanical properties for various applications. Facilities with capacities of 10,000 and 50,000 tons of P(3HB/4HB) have been built in China and the U.S., respectively (Chen, 2009). P3HB/4HB may be the PHA available in the greatest quantity on the market. At the same time, companies in both countries are working to develop various bulk applications.

The least commercialized PHAs today are medium-chain-length copolymers consisting of monomer units containing 4 to 12 carbon atoms. The early development of this class of PHA copolymers was initiated by Procter & Gamble (USA) around the late 1980s. These PHAs consisted of various medium-chain-length monomer units, including copolymers without 3-hydroxybutyrate. A family of PHAs consisting of medium-chain-length monomer units with different chain lengths were synthesized and trademarked Nodax™ at Meredian (Bainbridge, GA, U.S.) (Poliakoff and Noda,

2004; Noda et al., 2005). In 2007, Meredian took over the Nodax™ technology from Procter & Gamble for the full commercialization of this class of bioplastics. A pilot facility was used to validate production and process design in 2009, prior to construction of its first full-scale PHA production facility, which was established in 2010. The planned annual output is about 300 million tons of PHA (Chen, 2009). Chinese and Korean researchers, in collaboration with Procter & Gamble, U.S., have been engaged in commercial production of this copolymer. The range of potential applications for this copolymer is getting wider, but its current production cost is still too high for real commercial applications.

In Russia, little research has been done on degradable polymers (Fomin and Guzeev, 2001; Shtilman, 2006). Biodegradable polymers have not been produced commercially yet. There are plans to establish polylactide production. Only a few research teams are engaged in PHA-related studies. These are the Institute of Microbiology RAS, the A.N. Bakh Institute of Biochemistry RAS, the Institute of Physiology and Biochemistry of Microorganisms RAS, and the Institute of Petrochemical Synthesis RAS. The Institute of Biophysics SB RAS was the first in Russia to establish pilot production of poly-3-hydroxybutyrate and 3-hydroxybutyrate/3-hydroxyvalerate copolymers in cooperation with the Biokhimmash company (within the framework of the ISTC project) in 2005 (Volova et al., 2006a). Since that time, PHA synthesis processes have been considerably improved. During implementation of the mega projects supported by the Russian Government (2010–2014), the team of researchers at the Siberian Federal University widened the range of 2- and 3-component PHAs with different chemical structures that contained major fractions of short- and medium-chain-length monomer units (Volova et al., 2013b, 2014, 2016a) and established a new high-productivity pilot production facility equipped with an automatic fermentation system (Bioengineering, Switzerland) (Kiselev et al., 2014) (Figure 2.1). The trademark "BIOPLASTOTAN" was registered for PHAs of different chemical compositions and PHA based products (Trademark "BIOPLASTOTAN").

As mentioned previously, PHAs have physicochemical properties similar to those of some synthetic polymers such as polypropylene, which are produced in large quantities and cannot be degraded in the environment. At the present time, the use of PHAs is limited by their rather high cost, but the range of their applications is wide. The growing environmental concern on the one hand, and the possibility of reducing the cost of biopolymers by increasing the production efficiency, on the other, make PHAs promising materials of the 21st century.

FIGURE 2.1 Pilot production of PHA – Siberian Federal University, Krasnoyarsk, Russia (photo of T. Volova).

There are two possible ways to increase production and broaden applications of PHAs. One way is to develop large-scale PHA production, i.e., to increase PHA outputs and reduce their cost by manufacturing inexpensive items such as packaging materials, everyday articles, films, and pots for agriculture, etc. Researchers in the U.S., Japan, EU countries, India, Malaysia, etc. are conducting extensive studies of PHAs for a variety of applications. They are mostly used to manufacture packaging items and garbage containers, food, and cosmetic containers, and agricultural items (Plastics from Bacteria. Natural Functions and Applications, 2010). These are extruded vials, jars, bottles, containers, and boxes for shampoos, lotions, etc. PHAs were initially used to make everyday articles such as shampoo bottles and packaging materials by Wella AG in Germany (Weiner, 1997). Short-chain-length PHAs are used to fabricate packaging films, shopping bags, containers, and paper coatings, everyday use items such as housings for TV sets and computers, toys, sports equipment, disposable dishes, hygiene products, etc. by Biomers and Metabolix, and several other companies (Clarinval and Halleux, 2005; Noda, 2005). Some PHAs can be used to form gels and latexes, as a basis for producing glues and filling agents, including ones intended for stabilization of dyes. PHA laminates with paper and other polymers are successfully used as materials for producing garbage bags. PHAs can also be used to produce non-woven materials, various personal hygiene articles, etc. PHAs are used to produce dairy cream substitutes and flavor delivery agents in foods. There is a market for PHA depolymerization and hydrolysis products. These polymers can be converted to optically pure multifunctional hydroxy acids (Chen, 2009). Potential nutritional qualities of PHAs have been discussed. Several research teams have evaluated monomers of (R)–3-hydroxybutanoic acid, as an alternative to the sodium salt of the monomer, for potential nutritional and therapeutic uses, such as treatment of metabolic acidosis. Use of these polymeric forms might provide controlled release systems for the monomer and, importantly, overcome the problems associated with administering large amounts of sodium ions *in vivo.* Tasaki et al. (1998) reported the results of infusing dimers and trimers of (R)–3-hydroxybutyrate into rats. PHAs have important applications for agriculture, such as those related to the production of packaging films for food and fertilizers, pots, nets, ropes, etc. A new and environmentally important PHA application may be the delivery of agricultural chemicals, which are used to protect cultivated plants from pathogens and pests. Researchers of the Siberian Federal University were the first to prove that PHAs can be used as a degradable matrix enabling controlled release of pesticides and herbicides during the growing season of plants; pre-emergence formulations were developed, i.e., ones that can be buried in soil together with seeds (Prudnikova et al., 2013).

Another way is to establish small-scale facilities for the production of high-cost specialized items. PHAs show the greatest potential in medicine and pharmaceutics. The mild immune response to PHA implants and the sufficient duration of PHA degradation in biological media make these polymers attractive candidates for use as drug carriers in controlled-release drug delivery systems, implants, and grafts for tissue and organ regeneration, materials for tissue engineering and designing of bioartificial organs (Pouton, 2001; Sudesh, 2004; Volova, 2004; Williams and Martin, 2002; Volova and Shishatskaya, 2011; Amass et al., 1998; Sudesh et al., 2000; Volova et al., 2003, 2006b, 2013b, 2017c; Loo and Sudesh, 2007; PHAs: Biosynthesis, Industrial Production, and Applications in Medicine, 2014; Luef et al., 2015).

Press releases of such well-known companies as Tepha Inc., Metabolix, Procter & Gamble, and Monsanto suggest growing interest in PHAs and intensive research aimed at production, modification, and investigation of PHAs for cardiovascular surgery, dentistry, orthopedics, and pharmacology. Although PHAs are attracting more and more attention, there are rather few results of biomedical studies reported in the literature, as can be inferred from the analysis of biomedical studies of novel biomaterials, including PHAs, that are available in the major databases. Many aspects of PHA biotechnology and materials science remain unclear. Among them are the production of high-purity PHAs and processes used to fabricate various PHA-based special biomedical items. Kinetics and mechanisms of PHA *in vivo* biodegradation need to be better understood. More studies should be performed to gain insight into mechanisms of *in vivo* interaction of PHA devices with cells and tissues and their medical and technical parameters in living organisms. The Food and Drug Administration (FDA) in the U.S. has approved the use of several products manufactured by Tepha Inc. in clinical trials. These are suture material, mesh grafts, and films for uroplastic surgery.

Analysis of available literature suggests that PHAs have been extensively studied as materials for cardiovascular surgery. These polymers are proposed as candidates for fabricating vascular grafts and coatings for grafts fabricated from synthetic materials (Marois et al., 1999a, b, 2000). Perhaps the most remarkable results with PHA polymers have been obtained in the development of cell-seeded tissue-engineered heart valves. Stock et al. (2000) successfully replaced the polyglycolide-polylactide valves in animals with poly-3-hydroxyoctanoate (P3HO) pulmonary conduits. Porous scaffolds prepared from elastic poly-3-hydroxyoctanoate were used to fabricate heart valves (Sodian et al., 1999, 2000). For 8 days, the cells proliferated and filled the pores; they also synthesized collagen and generated connective tissue between the outer and inner surfaces of the scaffold. Constructs of porous P(3HO/3HHx) were seeded with cells of vascular tissue and evaluated under pulsatile flow *in vitro* (Sodian et al., 1999).

One of the most advanced applications of PHA polymers in cardiovascular products has been the development of a regenerative PHA patch that can be used to close the pericardium after heart surgery, without formation of adhesions between the heart and sternum (Malm et al., 1992, 1994; Martin and Williams, 2003). Researchers of the Department of Thoracic and Cardiovascular Surgery at the University Hospital in Uppsala, Sweden, examined P(3HB) patches used as pericardial substitutes (Duvernoy et al., 1995). The patch was gradually degraded and replaced by native tissues strong enough to prevent the development of postoperative adhesions between the patch and the cardiac surface.

There are very few data on using PHAs to enhance the biocompatibility of vascular stents. Unverdorben et al. (2002) tried to use poly-3-hydroxybutyrate stents in experiments on rabbits, but the experiment was not quite successful. Tepha (U.S.) researchers together with their German colleagues (Institute of Biomedical Engineering, Rostock) have tested polymer stents prepared from poly(tetrafluoroethane) and a blend of polylactide/poly-3-hydroxybutyrate

(Grabow et al., 2007), comparing them with metallic stents. For the stents to withstand the pressure in vessels, they were rather bulky. This caused an adverse response of the vessel wall exhibited as neointima growth and stronger inflammatory reaction than that caused by metallic stents.

PHA solutions can be used to prepare fibers. In one of the first studies of PHA fibers, Miller, and Williams proved that P(3HB/3HV) monofilaments are not biodegraded *in vitro* and *in vivo* (Miller and Williams, 1987). Tepha (U.S.) has developed suture material based on P(3HB/4HB). The filaments are melt-spun in a single-screw extruder by passing the material through 4 zones of the extruder, with temperatures 140, 190, 200, and 205°C, and multi-stage orientation (Martin et al., 2000). The extrusion process and subsequent orientation yield filaments with a tensile strength over 126 MPa, which retain their properties for long periods of time. Having conducted all necessary tests, Tepha received the permission of FDA and marketed mono- and poly-filament fibers, meshes, and films under the trademark of TephaFLEX®.

PHAs are attractive materials for use in the surgical reconstruction of bone tissues. There are data on the preparation of mechanically strong PHA/HA composites, proving that incorporation of HA into a PHA enhances polymer strength. In recent years, much research effort has been devoted to the use of PHAs as matrices for drug delivery and as scaffolds for cell cultures intended to fabricate grafts for tissue engineering applications.

In Russia, biomedical studies of PHAs and experimental prototypes of PHA products were initiated by the researchers of the Institute of Biophysics SB RAS and the Siberian Federal University united in Research-and-Education Center "Yenisei." In cooperation with the Institute of Transplant Surgery and Artificial Organs (currently the V.I. Shumakov Federal Research of Transplant Surgery and Artificial Organs Center), the research team studied films of poly-3-hydroxybutyrate and P(3HB/3HV) copolymers and proved that they did not produce any cytotoxic effect when contacting directly with the cultured cells and when implanted as sutures *in vivo*; high-purity specimens were suitable for contact with blood. Results of these studies were summarized in the two editions of the first Russian book on PHAs: "PHAs– biodegradable polymers for medicine" (Volova et al., 2003; 2006b). Since then, the research team has considerably widened the scope of its PHA studies by synthesizing PHAs with different chemical compositions and by designing and studying experimental films, barrier membranes, granules, filling materials, ultrafine fibers produced by electrospinning, solid, and porous 3D implants for bone tissue defect repair, tubular biliary stents, mesh implants modified by PHA coating, microparticles for drug delivery, etc. In cooperation with the V.F. Voino-Yasenetsky Krasnoyarsk Federal Medical University, the research team has conducted pioneering clinical trials. Results have been covered by several Russian patents, reported in papers published in peer-reviewed Russian and international journals, summarized in books and reviews (Volova, 2004; Shishatsky and Shishatskaya, 2010; Volova and Shishatskaya, 2011; Shishatsky et al., 2010; Volova et al., 2013c, 2017b, c).

Another direction is to develop large-scale production of PHAs, i.e., to increase PHA outputs and reduce their cost by manufacturing affordable, inexpensive items such as packaging materials, everyday articles, films, and pots for greenhouses, etc. Researchers in the U.S., Japan, EU countries, India, Malaysia, etc. are conducting extensive studies of PHAs for manufacturing packaging items and garbage containers, food, and cosmetic containers, and agricultural items (Plastics from Bacteria. Natural functions and applications, 2010). There is a market for PHA products in cosmetology. These are extruded vials, jars, bottles, containers, and boxes for shampoos, lotions, etc. PHAs were initially used to make everyday articles such as shampoo bottles and packaging materials by Wella AG in Germany (Weiner, 1997). Short-chain-length PHAs are used to fabricate packaging films, shopping bags, containers, and paper coatings, everyday use items such as housings for TV sets and computers, pens, toys, sports equipment, disposable dishes, hygiene products, etc. by Biomers and Metabolix (Clarinval and Halleux, 2005; Noda, 2005). Some PHAs can be used to form durable gels and latexes, as a basis for producing glues and filling agents, including ones intended for stabilization of dyes. PHA laminates with paper and other polymers are successfully used as materials for producing garbage bags. PHAs can also be used to fabricate non-woven materials, various personal hygiene articles, etc. PHAs are used to produce dairy cream substitutes and flavor delivery agents in foods.

One more way to use PHAs is to manufacture biofuel. Methyl esters of 3-hydroxybutyrate and medium-chain-length PHAs produced by etherification can be used as biofuel. Combustion temperature of these compounds is about 20–30 kJ/g, which is comparable with the ethanol combustion temperature (27 kJ/g). Supplementation of ethanol with 10% 3HB methyl esters increases the ethanol combustion temperature to 30 kJ/g but decreases the combustion temperatures of propanol and butanol as well as gasoline and diesel. The cost of one ton of PHA-based biofuel is estimated at about US $ 1200. As biofuel produced today, including ethanol and biodiesel, is considered as "food vs. fuel" and "fuel vs. arable land," PHA-based biofuel, which can be manufactured from various industrial wastes including wastewater and mud, may offer a new opportunity for PHAs in power industry.

PHAs are used to manufacture agricultural devices. These are films for greenhouses, packages for fertilizers and vegetables, pots, nets, ropes, etc. A new and environmentally important PHA application may be delivery of pesticides, which are intended to protect crops from pathogens and pests, and fertilizers. Researchers of the Institute of Biophysics SB RAS and the Siberian Federal University (Volova et al., 2016d) were the first to prove that PHAs can be used as a degradable matrix enabling controlled release of pesticides and herbicides during the growing season of plants; pre-emergence formulations were developed, i.e., ones that can be buried in soil together with seeds. That provided the basis for the new important use of PHAs – construction of new-generation slow-release targeted formulations, in which chemicals for crop protection would be embedded in the matrix of these degradable polymers.

KEYWORDS

- **monomers**
- **PHA characterization**
- **PHA production and applications**
- **PHA properties**
- **poly(3-hydroxybutyrate)**
- **polyhydroxyalkanoates**

REFERENCES

Akhtar, S., Pouton, C. W., & Notarianni, L. J., (1992). Crystallization behavior and drug release from bacterial polyhydroxyalkanoates. *Polymer.*, *33*, 117–126.

Amass, W., Amass, A., & Tighe, B., (1998). A review of biodegradale polymers: Uses, current developments in the synthesis and characterization of biodegradable polyesters, blends of biodegradable polymers and recent advances in biodegradation studies. *Polymer Int.*, *47*, 89–144.

Amirul, A. A., Yahya, A. R. M., Sudesh, K., Azizan, M. N. M., & Majid, M. I. A., (2008). Biosynthesis of poly(3-hydroxybutyrate-co-4-hydroxybutyrate) copolymer by a

Cupriavidus spp. USMAA 1020 isolated from Lake Kulim, Malaysia. *Bioresour. Technol., 99*, 4903–4909.

Anderson, A. J., & Dawes, E. A., (1990). Occurrence, metabolism, metabolic role, and industrial uses of bacterial polyhydroxyalkanoates. *Microbiol. Rev., 54*, 450–472.

Ashby, R. D., Solaiman, D. K. Y., & Fogila, T. A., (2002). The synthesis of short- and medium-chain-length poly(hydroxyalkanoate) mixtures from glucose- or alkanoic acid-grown *Pseudomonas oleovorans. J. Ind. Microbiol. Biotechnol., 28*, 147–153.

Ashraf, A. M., Gamal, S. S., Amany, H. H. II., (1999). Dielectric investigation of cold crystallization of poly(3-hydroxybutyrate) and poly(3-hydroxybutyrate-co-3-hydroxyvalerate). *Polymer., 40*, 5377–5391.

Asrar, J., Valentin, H. E., Berger, P. A., Tran, M., Padgette, S. R., & Garbow, J. R., (2002). Biosynthesis and properties of poly(3-hydroxybutyrate-co-3-hydroxyhexanoate) polymers. *Biomacromolecules, 3*, 1006–1012.

Avella, M., La Rota, G., Martuscelli, E., Raimo, M., Sadocco, P., Elegir, G., & Riva, R., (2000b). Poly(3-hydroxybutyrate-co-3-hydroxyvalerate) and wheat straw fiber composites: Thermal, mechanical properties and biodegradation behavior. *J. Mater. Sci., 35*, 829–836.

Avella, M., Martuscelli, E., & Raimo, M., (2000a). Review properties of blends and composites based on poly(3-hydroxy)butyrate (PHB) and poly(3-hydroxybutyrate-hydroxyvalerate) (PHBV) copolymers. *J. Mater. Sci., 35*, 523–545.

Ayub, N. D., Tribelli, P. M., & López, N. I., (2009). Polyhydroxyalkanoates are essential for maintenance of redox state in the antartic bacterium *Pseudomonas sp.* 14–3 during low temperature adaptation. *Extremophiles., 13*, 59–66.

Babel, W., Ackerman, J. U., & Breuer, U., (2001). Physiology, regulation, and limits of the synthesis of poly(3HB). *Adv. Biochem. Eng. Biotechnol., 71*, 125–157.

Ballistreri, A., Giuffrida, M., Impallomeni, G., Lenz, R. W., & Fuller, R. C., (1999). Characterization by mass spectrometry of poly(3-hydroxyalkanoates) produced by *Rhodospirillum rubrum* from 3-hydroxyacids. *Int. J. Biol. Macromol., 26*, 201–211.

Bear, M. M., Leboucher-Durand, M. A., Langlois, V., Lenzb, R. W., Goodwinc, S., & Guérin, P., (1997). Bacterial poly-3-hydroxyalkenoates with epoxy groups in the side chains. *React. Funct. Polym., 34*, 65–77.

Bonthrone, K. M., Clauss, J., Horowitz, D. M., Hunter, B. K., & Sanders, J. K. M., (1992). The biological and physical chemistry of polyhydroxyalkanoates as seen by NMR spectroscopy. *FEMS Microbiol., 103*, 269–277.

Boynton, Z. L., Koon, J. J., Brennan, E. M., Clouart, J. D., Horowitz, D. M., Gerngross, T. U., & Huisman, G. W., (1999). Reduction of cell lysate viscosity during processing of poly(3-hydroxyalkanoates) by chromosomal integration of the staphylococcal nuclease gene in *Pseudomonas putida. Appl. Microbiol. Biotechnol., 65*, 1524–1529.

Braunegg, G., & Bogensberger, B. (1985). On the kinetics of growth and storage of poly-D(-)-3-hydroxybutyric acid in *Alcaligenes latus* (Zur kinetik des wachstums und der speicherung von. Poly-D(-)-3-hydroxybutersaure bei *Alcaligenes latus). Acta biotechnol., 5(4),* 339-345 (in German).

Braunegg, G., Lefebvre, G., & Genzer, K. F., (1998). Polyhydroxyalkanoates, biopolyesters from renewable resources: Physiological and engineering aspects (Review article). *J. Biotechnol., 65*, 127–161.

Brückner, S., Meille, S. V., Malpezzi, L., Cesàro, A., Navarini, L., & Tombolini, R., (1988). The structure of Poly(D-(-)-β-polyxydroxybutyrate). A refinement based on the Rietrveld method. *Macromolecules, 21*, 967–972.

Bugnicourt, E., Cinelli, P., Lazzeri, A., & Alvarez, V., (2014). Polyhydroxyalkanoate (PHA): Review of synthesis, characteristics, processing and potential applications in packaging. *Express Polym. Lett.*, *8*, 791–808.

Byron, D., (1987). Polymer synthesis by microorganisms: Technology and economics. *Trends Biotechnol.*, *5*, 246–250.

Byron, D., (1994). Polyhydroxyalkanoates. In: Mobley, D. P., (ed.), *Plastics From Microbes: Microbial Synthesis of Polymers and Polymer Precursors* (pp. 5–33). Hanser Munich.

Chanprateep, S., & Kulpreecha, S., (2006). Production and characterization of biodegradable terpolymer poly(3-hydroxybutyrate-co-3-hydroxyvalerate-co-4-hydroxybutyrate) by *Alcaligenes* sp. A–04. *J. Biosci. Bioeng.*, *101*, 51–56.

Chanprateep, S., Buasri, K., Muangwong, A., & Utiswannakul, P., (2010). Biosynthesis and biocompatibility of biodegradable poly(3-hydroxybutyrate-co-4-hydroxybutyrate*). Polym. Degrad. Stab.*, *95*, 2003–2012.

Chen, G. Q., & Steinbüchel, A., (2010). *Plastic From Bacteria, Natural Functions and Applications.* Springer-Verlag, Berlin Heidelberg.

Chen, G. Q., (2009). A microbial polyhydroxyalkanoates (PHA) based bio- and materials industry. *Chem. Soc.*, *38*, 2434–2446.

Chen, G. Q., (2010a). Plastics completely synthesized by bacteria: polyhydroxyalkanoates. In: Chen, G. Q., & Steinbüchel, A., (eds.), *Microbiol. Monogr. Plastics From bacteria. Natural Functions and Applications* (Vol. 14, pp. 17–38). Springer.

Chen, G. Q., (2010b). Industrial production of PHA. In: Chen, G. Q., & Steinbüchel, A., (eds.), *Microbiol. Monogr. Plastics From Bacteria. Natural Functions and Applications* (Vol. 14, pp. 121–132). Springer.

Chen, G. Q., Konig, K. H., & Lafferty, R. M., (1991). Production of poly-D-(-)-3-hydroxybutyrate and poly-D-(-)-3-hydroxyvalerate by strains of *Alcaligenes latus*. *Antonie van Leeuwehoek.*, *60*, 61–66.

Chen, J. Y., Song, G., & Chen, G. Q., (2006). A lower specificity of PhaC2 synthase from *Pseudomonas stutzeri* catalyzes the production of copolyesters consisting of short-chain-length and medium-chain-length 3-hydroxyalkanoates. *Antonie Van Leeuwenhoek.*, *89*, 157–167.

Chia, K. H., Ooi, T. F., Saika, A., Tsuge, T., & Sudesh, K., (2010). Biosynthesis and characterization of novel polyhydroxyalkanoate polymers with high elastic property by *Cupriavidus* necator PHB–4 transformant. *Polym. Degrad. Stab.*, *95*, 2226–2232.

Choi, J., & Lee, S. Y., (1997). Process analysis and economic evaluation for poly(3-hydroxybutyrate) production by fermentation. *Bioprocess Eng.*, *17*, 335–342.

Choi, J., & Lee, S. Y., (1999). Efficient and economical recovery of poly-(3-hydroxybutyrate) from recombinant *Escherichia coli* by simple digestion with chemicals. *Biotechnol. Bioeng.*, *62*, 546–553.

Choi, M. H., Yoon, S. C., & Lenz, R. W., (1999). Production of poly(3-hydroxybutyric acid-co-4-hydroxybutyric acid) and poly(4-hydroxybutyric acid) without subsequent degradation by *Hydrogenophaga pseudoflava*. *Appl. Environ. Microbiol.*, *65*, 1570–1577.

Clarinval, A. M., & Halleux, J., (2005). Classification of biodegradable polymers. In: Smith, R., (ed.), *Biodegradable Polymers for Industrial Applications* (pp. 3–31). Woodhead Publishing: Cambridge.

Cowan, D., Meyer, Q., Stafford, W., Muyanga, S., Cameron, R., & Wittwer, P., (2005). Metagenomic gene discovery: Past, present and future. *Trends Biotechnol.*, *23*, 321–329.

Cromwick, A. M., Foglia, T., & Lenz, R. V., (1996). The microbial production of poly(hydroxyalkanoates) from tallow. *Appl. Microbiol. Biotechnol.*, *46*, 464–469.

Dai, Y., Yuan, Z., Jack, K., & Keller, J., (2007). Production of targeted poly(3-hydroxyalkanoates) copolymers by glycogen accumulating organisms using acetate as sole carbon source. *J. Biotechnol.*, *129*, 489–497.

Dawes, E. A., (1990). *Novel Biodegradable Microbial Polymers* (p. 287). Kluwer Academic: Dordrecht.

De Koning, G. J. M., (1995). Physical properties of bacterial poly((R)–3-hydroxyalkanoates). *Can. J. Microbiol.*, *41*, 303–309.

De Roo, G., Ren, Q., Witholt, B., & Kessler, B., (2000). Development of an improved *in vitro* activity assay for medium chain length PHA polymerase based on coenzyme A release measurements. *J. Microbiol. Methods.*, *41*, 1–8.

Dias, J. M., Oehmen, A., Serafim, L. S., Lemos, P. C., Reis, M. A., & Oliveira, R., (2008). Metabolic modeling of polyhydroxyalkanoate copolymers production by mixed microbial cultures. *BMC Syst. Biol.*, *8*, 2–59.

Doi, Y., (1990). *Microbial Polyesters* (p. 156). VCH: New York.

Doi, Y., (1995). Microbial synthesis, physical properties, and biodegradability of polyhydroxyalkanoates. *Macromol. Symp.*, *98*, 585–599.

Doi, Y., Kitamure, S., & Abe, H., (1995). Microbial synthesis and characterization of poly(3-hydroxybutyrate-co-3-hydroxyhexanoate). *Macromolecules*, *28*, 4822–4828.

Doi, Y., Segawa, A., & Kunioka, M., (1989). Biodegradable poly(3-hydroxybutyrate-co-4-hydroxybutyrate) produced from gammabutyrolactone and butyric acid by *Alcaligenes eutrophus*. *Polym. Commun.*, *30*, 169–171.

Doi, Y., Segawa, A., & Kunioka, M., (1990). Biosynthesis and characterization of poly(3-hydroxybutyrate-co-4-hydroxybutyrate) in *Alcaligenes eutrophus*. *Int. J. Biol. Macromol.*, *12*, 106–111.

Duvernoy, O., Malm, T., Ramström, J., & Bowald, S., (1995). A biodegradable patch used as a pericardial substitute after cardiac surgery: 6- and 24-month evaluation with CT. *Thorac. Cardiovasc. Surg.*, *43*, 271–274.

Findlay, R. H., & White, D. C., (1983). Polymeric beta-hydroxyalkanoates from environmental samples and *Bacillus megaterium*. *Appl. Environ. Microbiol.*, *45*, 71–78.

Fomin, V. A., & Guzeev, V. V., (2001). Biodegradable polymers their current and potential uses. *Plasticheskiye Massy (Plastics)*, *2*, 42–46 (in Russian).

Foster, L. J., Schwahn, R., Pipich, D., Holden, P. J., & Richter, D., (2008). Small-angle neutron scattering characterization of polyhydroxyalkanoates and their bioPEGylated hybrids in solution. *Biomacromolecules*, *9*, 314–320.

Fukui, T., Abe, H., & Doi, Y., (2002). Engineering of *Ralstonia eutropha* for production of poly(3-hydroxybutyrate-co-3-hydroxyhexanoate) from fructose and solid-state properties of the copolymer. *Biomacromolecules*, *3*, 618–624.

Gorenflo, D., Schmack, G., Vogel, R., & Steinbüchel, A., (2001). Development of a process for the biotechnological large-scale production of 4-hydroxyvalerate-containing polyesters and characterization of their physical and mechanical properties. *Biomacromolecules*, *2*, 45–57.

Grabow, N., Bünger, C. M., Schultze, C., Schmohl, K., Martin, D. P., Williams, S. F., Sternberg, K., & Schmitz, K. P., (2007). A biodegradable slotted tube stent based on poly(L-lactide) and poly(4-hydroxybutyrate) for rapid balloon-expansion. *Ann. Biomed. Eng.*, *35*, 2031–2038.

Green, P. R., Kemper, J., Schechtman, L., Guo, L., Satkowski, M., Fiedler, S., Steinbüchel, A., & Rehm, B. H., (2002). Formation of short chain/medium chain length polyhydroxyalkanoate copolymers by fatty acid *β*–oxidation inhibited *Ralstonia eutropha. Biomacromolecules, 3*, 208–213.

Gross, R. A., DeMello, C., Lenz, R. W., Brandl, H., & Fuller, R. C., (1989). Biosynthesis and characterization of poly(β-hydroxyalkanoates) produced by *Pseudomonas oleovorans. Macromolecules, 22*, 1106–1115.

Grubelnik, A., Wiesli, L., Furrer, P., Rentsch, D., Hany, R., & Meyer, V. R., (2008). A simple HPLC-MS method for the quantitative determination of the composition of bacterial medium chain-length polyhydroxyalkanoates. *J. Separ. Sci., 31*, 1739–1744.

Haywood, G. W., Anderson, A. J., Williams, G. A., Dawes, E. A., & Ewing, D. F., (1991). Accumulation of a poly(hydroxyalkanoate) copolymer containing primarily 3-hydroxyvalerate from simple carbohydrate substrates by *Rhodococcus. Int. J. Biol. Macromol., 13*, 83–87.

Hazer, B., & Steinbüchel, A., (2007). Increased diversification of polyhydroxyalkanoates by modification reactions for industrial and medical applications. *Appl. Microbiol. Biotechnol., 74*, 1–12.

Hein, S., Söhling, B., Gottschalk, G., & Steinbüchel, A., (1997). Biosynthesis of poly(4-hydroxybutyric acid) by recombinant strains of *Escherichia coli. FEMS Microbiol. Lett., 153*, 411–418.

Hiramitsu, M., Koyama, N., & Doi, Y., (1993). Production of poly (3-hydroxybutyrate-co-4-hydroxybutyrate) by *Alcaligenes latus. Biotechnol. Lett., 15*, 461–464.

Howells, E. R., (1982). Opportunities for biotechnology for the chemical industry. *Chem. Ind., 8*, 508–511.

Ienczak, J. L., Schmidell, W., & Aragão, G. M. F., (2013). High-cell-density culture strategies for polyhydroxyalkanoate production: A review. *J. Ind. Microbiol. Biotechnol., 40*, 275–286.

Ishida, K., Wang, Y., & Inoue, Y., (2001). Comonomer unit composition and thermal properties of poly(3-hydroxybutyrate-co-4-hydroxybutyrate)s biosynthesized by *Ralstonia eutropha. Biomacromolecules, 2*, 1285–1293.

Jian, J., Li, Z. J., Ye, H. M., Yuan, M. Q., & Chen, G. Q., (2010). Metabolic engineering for microbial production of polyhydroxyalkanoates consisting of high 3-hydroxyhexanoate content by recombinant *Aeromonas hydrophila. Bioresour. Technol., 101*, 6096–6102.

Kalyuzhnaya, M. G., Lapidus, A., & Ivanova, N., (2008). High–resolution metagenomics targets specific functional types in complex microbial communities. *Nat. Biotechnol., 26*, 1029–1034.

Kang, C. K., Kusaka, S., & Doi, Y., (1995). Structure and properties of poly(3-hydroxybutyrate-co-4-hydroxybutyrate) produced by *Alcaligenes latus. Biotechnol. Lett., 17*, 583–588.

Kaur, G., & Roy, I., (2015). Strategies for large-scale production of polyhydroxyalkanoates. *Chem. Biochem. Eng., 29*, 157–172.

Kessler, B., & Witholt, B., (2001). Factors involved in the regulatory network of polyhydroxyalkanoate metabolism. *J. Biotechnol., 86*, 97–104.

Kim, D. Y., Baek, Y., & Rhee, Y. H., (1998). Bacterial poly(3-hydroxyalkanoates) bearing carbon-carbon triple bonds. *Macromolecules, 32*, 4760–4763.

Kim, D. Y., Jung, S. B., & Choi, G. G., (2001). Biosynthesis of polyhydroxyalkanoate copolyester containing cyclohexyl groups by *Pseudomonas oleovorans*. *Int. J. Biol. Macromol.*, *29*, 145–150.

King, P. P., (1982). Biotechnology, an industrial view. *J. Chem. Technol. Biotechnol.*, *32*, 2–8.

Kiselev, E. G., Demidenko, A. D., Baranovsky, S. V., & Volova, T. G., (2014). Scaling-up the process of the synthesis of biodegradable polyhydroxyalkanoates in a pilot facility. *Journal of Siberian Federal University, Biology Series*, *2*, 134–147 (in Russian).

Klinke, S., Ren, Q., Witholt, B., & Kessler, B., (1999). Production of medium–chain–length poly(3–hydroxyalkanoayes) from gluconate by recombinant *Escherichia coli*. *Appl. Environ. Microbiol.*, *65*, 540–548.

Koller, M., Maršálek, L., De Sousa Dias, M. M., & Braunegg, G., (2017). Producing microbial polyhydroxyalkanoate (PHA) biopolyesters in a sustainable manner. *New Biotechnol.*, *37*, 24–38.

Kourmentza, C., Plácido, J., Venetsaneas, N., Burniol-Figols, A., Varrone, C., Gavala, H. N., & Reis, M. A. M., (2017). Recent advances and challenges towards sustainable polyhydroxyalkanoate (PHA) production. *Bioengineering*, *4*, 55–97.

Kung, S. S., Chuang, Y. C., Chen, C. H., & Chien, C. C., (2007). Isolation of polyhydroxyalkanoates-producing bacteria using a combination of phenotype and genotype approach. *Lett. Appl. Microbiol.*, *44*, 364–371.

Kunioka, M., Kawaguchi, Y., & Doi, Y., (1989). Production of biodegradable copolyesters of 3-hydroxybutyrate and 4-hydroxybutyrate by *Alcaligenes eutrophus*. *Appl. Microbiol. Biotechnol.*, *30*, 569–573.

Laycock, B., Halley, P., Pratt, S., Werker, A., & Lant, P., (2013). The chemomechanical properties of microbial polyhydroxyalkanoate. *Prog. Polym. Sci.*, *38*, 536–583.

Leaf, T. A., Peterson, M. S., Stoup, S. K., Somers, D., & Srienc, F., (1996). *Saccharomyces cerevisiae* expressing bacterial polyhydroxybutyrate synthase produces poly-3-hydroxybutyrate. *Microbiology*, *42*, 1169–1180.

Lee, S. H., Oh, D. H., Ahn, W. S., Lee, Y., Choi, J., & Lee, S. Y., (2000). Production of poly(3-hydroxybutyrate-co-3-hydroxyhexanoate) by high–cell–density cultivation of *Aeromonas hydrophila*. *Biotechnol. Bioeng.*, *20*, 240–244.

Lee, S. Y., & Chang, H. N., (1994). Effect of complex nitrogen source on the synthesis and accumulation of poly(3–hydroxybutyric acid) by recombinant *Escherichia coli* in flask and fed–batch cultures. *J. Environ. Polym. Degrad.*, *2*, 169–176.

Lee, S. Y., (1996a). Bacterial polyhydroxyalkanoates. *Biotechnol. Bioeng.*, *49*, 1–14.

Lee, S. Y., (1996b). Plastic bacteria? Progress and prospects for polyhydroxyalkanoate production in bacteria. *Trends Biotechnol.*, *14*, 431–438

Lee, W. H., Azizan, M. N. M., & Sudesh, K., (2004). Effects of culture conditions on the composition of poly(3-hydroxybutyrate-co-4-hydroxybutyrate) synthesized *by Comamonas acidovorans*. *Polym. Degrad. Stab.*, *84*, 129–134.

Lemoigne, M., (1926). Products of dehydration and polymerization of β-oxobutyric acid (Produits de déshydration and polymerisation de l´acide *β*-oxobutyrique). *Bull. Soc. Chim. Biol.*, *8*, 770–782 (in French).

Lemoingne, M., (1927). Studies on microbial autolysis: origin of β-oxybutyric acid formed by autolysis (Etudes sur l'autolyse microbienne: origine de l'acide β-oxybutyrique forme par autolyse). *Ann. Inst. Pasteur.*, *41*, 148–165 (in French).

Lenz, R. W., & Marchessault, R. H., (2005). Bacterial polyesters: Biosynthesis, biodegradable plastics and biotechnology. *Biomacromolecules*, *6*, 1–7.

Liu, S. J., & Steinbüchel, A., (2000). A novel genetically engineered pathway for synthesis of po-ly(hydroxyalkanoic acids) in *Escherichia coli*. *Appl. Environ. Microbiol.*, *66*, 739–743.

Liu, W., & Chen, G. Q., (2007). Production and characterization of medium-chain-length polyhydroxyalkanoates with high 3-hydroxytetradecanoate monomer content by fadB and fadA knockout mutant of *Pseudomonas putida* KT2442. *Appl. Microbiol. Biotechnol.*, *76*, 1153–1159.

Loo, C. Y., & Sudesh, K., (2007). Polyhydroxyalkanoates: Bio-based microbial plastics and their properties. *Malays. Polym. J.*, *2*, 31–57.

Loo, C. Y., Lee, W. H., Tsuge, T., Doi, Y., & Sudesh, K., (2005). Biosynthesis and characterization of poly(3-hydroxybutyrate-co-3-hydroxyhexanoate) from palm oil products in a *Wautersia eutropha* mutant. *Biotechnol. Lett.*, *27*, 1405–1410.

Lu, X. Y., Wu, Q., & Chen, G. Q., (2004). Production of poly(3-hydroxybutyrate-co-3-hydroxyhexanoate) with flexible 3-hydroxyhexanoate content in *Aeromonas hydrophila* CGMCC 0911. *Appl. Microbiol. Biotechnol.*, *64*, 41–45.

Luef, K. P., Stelzer, F., & Wiesbrock, F., (2015). Poly(hydroxyalkanoate)s in medical application. *Chem. Biochem. Eng.*, *29*, 287–2912.

Luengo, J. M., Arias, S., Sandoval, A., Arias-Barrau, E., Arcos, M., Naharro, G., & Olivera, E. R., (2004). From aromatic to bioplastic: The phenylacetyl-CoA catabolon as a model of catabolic convergence. In: Pandalai, S. G., (ed.), *Recent Research Developments in Biophysics and Biochemistry* (Vol. 4, pp. 257–292). Research Signpost: Kerala.

Luengo, J. M., García, B., & Sandoval, A., (2003). Bioplastics from microorganisms. *Curr. Opin. Biotechnol.*, *6*, 251–260.

Luizier, W. D., (1992). Materials derived from biomass/biodegradable materials. *Proc. Natl. Acad. Sci. USA.*, *89*, 839–842.

Luo, R., Chen, J., Zhang, L., & Chen, G. Q., (2006). Polyhydroxyalkanoate copolyesters produced by *Ralstonia eutropha* PHB–4 harboring a low-substrate-specify PHA synthase PhaC2Ps from *Pseudomonas stutzeri* 1317. *Biochem. Eng. J.*, *32*, 218–225.

Lütke–Eversloh, T., Bergander, K., Luftmann, H., & Steinbüchel, A., (2001). Identification of a new class of biopolymer: Bacterial synthesis of a sulfur–containing polymer with thioester linkages. *Microbiology*, *147*, 11–19.

Macrae, R. M., & Wilkinson, J. F., (1958). The influence of cultural conditions on poly-*β*-hydroxybutyrate synthesis in *Bacillus megaterium*. *Proc. R. Soc. Edinb. A-MA.*, *27*, 73–78.

Madison, L. L., & Huisman, G. W., (1999). Metabolic engineering of poly(3-hydroxyalkanoates): From DNA to plastic. *Microbiol. Mol. Biol. Rev.*, *63*, 21–53.

Malm, T., Bowald, S., Bylock, A., Busch, C., & Saldeen, T., (1994). Enlargement of the right ventricular outflow tract and the pulmonary artery with a new biodegradable patch in transannular position. *Eur. Surg. Res.*, *26*, 298–308.

Malm, T., Bowald, S., Karacagil, S., Bylock, A., & Busch, C., (1992). A new biodegradable patch for closure of atrial septal defect. An experimental study. *Scand. J. Thorac. Cardiovasc. Surg.*, *26*, 9–14.

Marois, Y., Zhang, Z., Vert, M., Beaulieu, L., Lenz, R. W., & Guidoin, R., (1999b). *In vivo* biocompatibility and degradation studies of polyhydroxyoctanoate in the rat: A new sealant for the polyester arterial prosthesis. *Tissue Eng.*, *5*, 369–386.

Marois, Y., Zhang, Z., Vert, M., Deng, X., Lenz, R. W., & Guidoin, R., (1999a). Effect of sterilization on the physical and structural characteristics of polyhydroxyoctanoate (PHO). *J. Biomater. Sci. Polym. Edn.*, *10*, 469–482.

Martin, D. P., Peoples, O. P., & Williams, S. F., (2000). *Nutritional and Therapeutic Uses of 3-Hydroxyalkanoate Oligomers*. PCT Patent Application No. WO 00/04895.

Martin, D., & Williams, S., (2003). Medical application of polyhydroxybutyrate: A strong flexible absorbable biomaterial. *Biochem. Eng. J.*, *16*, 97–105.

Matsusaki, H., Abe, H., & Doi, Y., (2000). Biosynthesis and properties of poly(3-hydroxybutyrate-co-3-hydroxyalkanoates) by recombinant strains of *Pseudomonas sp.* 6 1–3. *Biomacromolecules*, *1*, 17–22.

McInerney, M. J., Amos, D. A., Kealy, K. S., & Palmer, J. A., (1992). Synthesis and function of polyhydroxyalkanoates in anaerobic syntrophic bacteria. *FEMS Microbiol. Rev.*, *103*, 195–206.

Miller, N. D., & Williams, D. F., (1987). On the biodegradation of poly-β-hydroxybutyrate (PHB) homopolymer and poly-*β*-hydroxybutyrate-hydroxyvalerate copolymers. *Biomaterials*, *8*, 129–137.

Mitomo, H., Hsieh, W. C., Nishiwaki, K., Kasuya, K., & Doi, Y., (2001). Poly(3-hydroxybutyrate-co-4-hydroxybutyrate) produced by *Comamonas acidovorans*. *Polymer.*, *42*, 3455–3461.

Monteil-Rivera, F., Betancourt, A., Van Tra, H., Yezza, A., & Hawari, J., (2007). Use of headspace solid-phase microextraction for the quantification of poly(3-hydroxybutyrate) in microbial cells. *J. Chromatogr.*, *1154*, 34–41.

Mozejko-Ciesielska, J., & Kiewisz, R., (2016). Bacterial polyhydroxyalkanoates: Still fabulous? *Microbiol. Res.*, *192*, 271–282.

Nagata, M., Machida, T., Sakai, W., & Tsutsumi, N., (1998). Synthesis, characterization, and enzymatic degradation studies on novel network aliphatic polyester. *Macromolecules*, *32*, 6450–6454.

Nakamura, S., Doi, Y., & Scandola, M., (1992). Microbial synthesis and characterization of poly(3-hydroxybutyrate-co-4-hydroxybutyrate. *Macromolecules*, *25*, 4237–4241.

Nakamura, S., Kunioka, M., & Doi, Y., (1991). Biosynthesis and characterization of bacterial poly(3-hydroxybutyrate-co-3-hydroxypropionate). *J. Macromol. Sci.*, *28*, 15–24.

Natano, R. V., Mantelatto, P. E., & Rossell, C. E., (2001). Integrated production of biodegradable plastic, sugar and ethanol. *Appl. Microbiol. Biotechnol.*, *57*, 1–5.

Noda, I., (2005). *Plastic Articles Digestible by Hot Alkaline Treatment*. Patent No. 6,872,802. B2. US.

Noda, I., Green, P., Satkowski, M., & Schechtman, L. A., (2005). Preparation and properties of novel class of polyhydroxyalkanoate copolymers. *Biomacromolecules*, *6*, 580–586.

Noda, I., Lindsey, S. B., & Caraway, D., (2010). Nodax™ class PHA copolymers: Their properties and applications. In: Chen, G. Q., & Steinbüchel, A., (eds.), *Plastics From Bacteria. Natural Functions and Applications* (pp. 237–256). Springer-Verlag, Berlin Heidelberg.

Odham, G., Tunlid, A., Westerdahl, G., & Mårdén, P., (1986). Combined determination of poly-*β*-hydroxyalkanoic and cellular fatty acids in starved marine bacteria and sewage sludge by gas chromatography with flame ionization or mass spectrometry detection. *Appl. Environ. Microbiol.*, *52*, 905–910.

Olivera, E. R., Arcos, M., Naharro, G., & Luengo, J. M., (2010). Unusual PHA biosynthesis. In: Chen, G. Q., & Steinbüchel, A., (eds.), *Plastics From Bacteria. Natural Functions and Applications* (pp. 18–34). Springer-Verlag, Berlin Heidelberg.

Olivera, E. R., Carnicero, D., García, B., Miñambres, B., Moreno, M. A., Cañedo, L., et al., (2001b). Two different pathways are involved in the β-oxidation of n-alkanoic and n-phenylalkanoic acids in *Pseudomonas putida* U: Genetic studies and biotechnological applications. *Mol. Microbiol., 39*, 863–874.

Olivera, E. R., Carnicero, D., Jodrá, R., Miñambres, B., García, B., Abraham, G. A., et al., (2001a). Genetically engineered *Pseudomonas*: A factory of new bioplastics with broad applications. *Environ. Microbiol., 3*, 612–618.

Pazur, R. J., Hocking, P. J., Raymond, S., & Marchessault, R. H., (1998). Crystal structure of syndiotactic poly(β-hydroxybutyrate) from X-ray fiber and powder diffraction analyses and molecular modeling. *Macromolecules, 32*, 6485–6492.

Peoples, O. P., & Sinskey, A. J., (1989). Poly-β-hydroxybutyrate biosynthesis in *Alcaligenes eutrophus* H16. Identification and characterization of the P(3HB) polymerase gene (phbC). *J. Biol. Chem., 264*, 15298–15303.

Pettinari, M. J., Vázquez, G. J., Silberschmidt, D., Rehm, B., Steinbüchel, A., & Méndez, B. S., (2001). Poly(3-hydroxybutyrate) genes in *Azotobacter sp.* strain FA8. *Appl. Environ. Microbiol., 67*, 5331–5334.

Philip, S., Keshavarz, T., & Roy, I., (2007). Polyhydroxyalkanoates: Biodegradable polymers with a range of applications. *J. Chem. Technol. Biotechnol., 82*, 233–247.

Pieper, U., & Steinbüchel, A., (1992). Identification, cloning and sequence analysis of the poly(3-hydroxyalkanoic acid) synthase gene of the gram-positive bacterium *Rhodococcus ruber*. *FEMS Microbiol. Lett., 96*, 73–79.

Poirier, Y., (1999). Production of new polymeric compounds in plants. *Curr. Opin. Biotechnol., 10*, 181–185.

Poirier, Y., Dennis, D. E., Klomparents, K., & Somerville, C., (1992). Polyhydroxybutyrate, a biodegradable thermoplastic, produced in transgenic plants. *Science, 256*, 520–523.

Poirier, Y., Nawrath, C., & Somerville, C., (1995). Production of polyhydroxyalkanoates, a family of biodegradable plastics and elastomers, in bacteria and plants. *Biotechnology, 13*, 142–150.

Poliakoff, M., & Noda, I., (2004). Plastic bags, sugar cane and advanced vibrational spectroscopy: Taking green chemistry to the third world. *Green Chem., 6*, G37–G38.

Pouton, C. W., (2001). Polymeric materials for advanced drug delivery. *Adv. Drug. Deliv. Rev., 53*, 1–3.

Prudnikova, S. V., Boyandin, A. N., Kalacheva, G. S., & Sinskey, A. J., (2013). Degradable polyhydroxyalkanoates as herbicide carriers. *J. Polym. Environ., 21*, 675–682.

Rehm, B. H. A., Krüger, N., & Steinbüchel, A., (1998). A new metabolic link between fatty acid de novo synthesis and polyhydroxyalkanoic acid synthesis. *J. Biol. Chem., 273*, 24044–24051.

Renner, G., Pongratz, K., & Braunegg, G., (1996). Production of poly(3-hydroxybutyrate-*co*-4-hydroxybutyrate) by *Comamonas testosteronii* A3. *Food Technol. Biotechnol., 34*, 91–95.

Rinehart, J. A., Petersen, M., & Jonh, M., (1999). Tissue-specific and developmental regulation of cotton gene Fb12A-demonstration of promoter activity in transgenic plants. *Plant Physiol., 112*, 1331–1341.

Russell, R. A., Holden, P. J., Wilde, K. L., Hammerton, K. M., & Foster, L. J., (2007). Production and use of deuterated polyhydroxyoctanoate in structural studies of PHO inclusions. *J. Biotechnol., 132*, 303–305.

Ruth, K., De Roo, G., Egli, T., & Ren, Q., (2008). Identification of two acyl-CoA synthetases from *Pseudomonas putida* GPo1: One is located at the surface of polyhydroxyalkanoate granules. *Biomacromolecules, 9*, 1652–1659.

Sabbagh, F., & Muhamada, I. I., (2017). Production of poly-hydroxyalkanoate as secondary metabolite with main focus on sustainable energy. *Renewable Sustain. Energy Rev.*, *72*, 95–104.

Saito, Y., & Doi, Y., (1994). Microbial synthesis and properties of poly(3-hydroxybutyrate-co-4-hydroxybutyrate) in *Comamonas acidovorans. Int. J. Biol. Macromol.*, *16*, 99–104.

Saito, Y., Nakamura, S., Hiramitsu, M., & Doi, Y., (1996). Microbial synthesis and properties of poly(3-hydroxybutyrate-co-4-hydroxybutyrate). *Polym. Int.*, *39*, 169–174.

Sandoval, A., Arias-Barrau, E., Arcos, M., Naharro, G., Olivera, E. R., & Luengo, J. M., (2007). Genetic and ultrastructural analysis of different mutants of *Pseudomonas putida* affected in the poly-3-hydroxy-n-alkanoate gene cluster. *Environ. Microbiol.*, *9*, 737–751.

Scandola, M., Ceccorulli, G., Pizzoli, M., & Gazzano, M., (1992). Study of the crystal phase and crystallization rate of bacterial poly(β-hydroxybutyrate-co-β-hydroxyvalerate). *Macromolecules*, *25*, 1405–1410.

Schmack, G., Gorenflo, V., & Steinbüchel, A., (1998). Biotechnological production and characterization of polyesters containing 4-hydroxyvaleric acid and medium-chain-length hydroxyalkanoic acids. *Macromolecules*, *31*, 644–649.

Senior, P. J., & Dawes, E. A., (1973). The regulation of poly-β-hydroxybutyrate metabolism in *Azotobacter beijerinckii. Biochem. J.*, *134*, 225–238.

Senior, P. J., (1984). Polyhydroxybutyrate, a specialty polymer of microbial origin. In: Dean, A., Ellwood, D., & Evans, C., (eds.), *Continuous Culture* (Vol. 8, pp. 266–271). Ellis Horwood: Chichester.

Serafim, L. S., Lemos, P. C., Torres, C., Reis, M. A., & Ramos, A. M., (2008). The influence of process parameters on the characteristics of polyhydroxyalkanoates produced by mixed cultures. *Macromol. Biosci.*, *8*, 355–366.

Shen, X. W., Yang, Y., Jian, J., Wu, Q., & Chen, G. Q., (2009). Production and characterization of homopolymer poly(3-hydroxyvalerate) (PHV) accumulated by wild type and recombinant *Aeromonas hydrophila* strain 4AK4. *Bioresour. Technol.*, *100*, 4296–4299.

Shishatsky, O. N., & Shishatskaya, E. I., (2010). *Analysis of the Market for Materials and Devices for Medical Applications* (106 p.). Krasnoyarskii Pisatel: Krasnoyarsk (in Russian). ISBN 978-5-905203-01-5.

Shishatsky, O. N., Shishatskaya, E. I., & Volova, T. G., (2010). *Degradable Biopolymers: Need, Production, Applications* (156 p.). Novyye informatsionnyye tekhnologii (New information technologies): Krasnoyarsk (in Russian). ISBN 978-5-905203-01-5.

Shtilman, M. I., (2006). *Polymers Intended for Biomedical Applications* (399 p.). Akademkniga Publishers: Moscow (in Russian). ISBN 978-5-94628-239-0.

Sodian, R., Hoerstrup, S. P., Sperling, J. S., Daebritz, S. H., Martin, D. P., Schoen, F. J., et al., (2000). Tissue engineering of heart valves: *In vitro* experiences. *Annal. Thorac. Surg.*, *70*, 140–144.

Sodian, R., Sperling, J. S., Martin, D. P., Stock, U., Mayer, J. E. Jr., & Vacanti, J. P., (1999). Tissue engineering of a trileaflet heart valve-early *in vitro* experiences with a combined polymer. *Tissue Eng.*, *5*, 489–493.

Solaiman, D. K., & Ashby, R. D., (2005). Rapid genetic characterization of poly(hydroxyalkanoate) synthase and its applications. *Biomacromolecules*, *6*, 532–537.

Spyros, A., Kimmich, R., & Briese, B., (1997). 1H NMR imaging study of enzymatic degradation in poly(3-hydroxybutyrate) and poly(3-hydroxybutyrate-co-3-hydroxyvalerate). Evidence for preferential degradation of amorphous phase by PHB depolymerase B from *Pseudomonas lemoignei. Macromolecules*, *30*, 8218–8225.

Steinbüchel, A., & Füchstenbusch, B., (1998). Bacterial and other biological systems for polyesters production. *Trends Biotechnol.*, *16*, 419–427.

Steinbüchel, A., & Hein, S., (2001). Biochemical and molecular basis of microbial synthesis of polyhydroxyalkanoates in microorganisms. *Adv. Biochem. Eng. Biotechnol.*, *71*, 81–123.

Steinbüchel, A., & Valentin, H. E., (1995). Diversity of bacterial polyhydroxyalkanoic acids. *FEMS Microbiol. Lett.*, *128*, 219–228.

Stock, U., Nagashima, M., Khalil, P. N., Nollert, G. D., Herden, T., Sperling, J. S., et al., (2000). Tissue-engineered valve conduits in the pulmonary circulation. *J. Thorac. Cardiovasc. Surg.*, *119*, 732–740.

Stubbe, J., Tian, J., Sinskey, A. J., Lawrence, A. G., & Liu, P., (2005). Nontemplate-dependent polymerization processes: Polyhydroxyalkanoate synthases as a paradigm. *Ann. Rev. Biochem.*, *74*, 433–480.

Sudesh, K., (2004). Microbial polyhydroxyalkanoates (PHAs): An emerging biomaterial for tissue engineering and therapeutic applications. *Med. J. Malaysia.*, *59*, 55–66.

Sudesh, K., Abe, H., & Doi, Y., (2000). Synthesis, structure and properties of polyhydroxyalkanoates: Biological polyesters. *Prog. Polym. Sci.*, *25*, 1503–1555.

Suzuki, T., Idogaki, H., & Kasai, N., (1996). A novel generation of optically active ethyl 4-chloro–3-hydroxybutyrate as C4 chiral building unit using microbial dechlorinattion. *Tetrahedron: Asymmetry*, *11*, 3109–3112.

Tan, G. Y. A., Chen, C. L., Li, L., Ge, L., Wang, L., Razaad, I. M. N., Li, Y., Zhao, L., Mo, Y., & Wang, J. Y., (2014). Start a research on biopolymer polyhydroxyalkanoate (PHA): A review. *Polymers.*, *6*, 706–754.

Tanadchangsaeng, N., Kitagawa, A., Yamamoto, T., Abe, H., & Tsuge, T., (2009). Identification, biosynthesis, and characterization of polyhydroxyalkanoate copolymer consisting of 3-hydroxybutyrate and 3-hydroxy–4-methylvalerate. *Biomacromolecules*, *10*, 2866–2874.

Tanadchangsaeng, N., Tsuge, T., & Abe, H., (2010). Co-monomer compositional distribution, physical properties, and enzymatic degradability of bacterial poly(3-hydroxybutyrate-co-3-hydroxy–4-methylvalerate) copolyesters. *Biomacromolecules*, *11*, 1615–1622.

Tasaki, O., Hiraide, A., Shiozaki, T., Yamamura, H., Ninomiya, N., & Sugimoto, H., (1998). The dimer and trimer precursor of 3-hydroxybutyrate oligomers as of ketone bodies for nutritional care. *Parent. Enteral. Nutr.*, *23*, 321–325.

Timm, A., Wiese, S., & Steinbüchel, A., (1994). A general method for identification of polyhydroxyalkanoic acid synthase genes from *Pseudomonads* belonging to the rRNA homology group I. *Appl. Microbiol. Biotechnol.*, *40*, 669–675.

Trademark "BIOPLASTOTAN" Registration Certificate No. 315652 of the Federal Institute for Patent Examination for Application No. 2006703271/50, Priority of 15.02.2006.

Tsuge, T., Watanabe, S., Sato, S., Hiraishi, T., Abe, H., Doi, Y., & Taguchi, S., (2007). Variation in copolymer composition and molecular weight of polyhydroxyalkanoate generated by satutation mutagenesis of *Aeromonas caviae* PHA synthase. *Macromol. Biosci.*, *7*, 846–854.

Unverdorben, M., Spielberger, A., Schywalsky, M., Labahn, D., Hartwig, S., Schneider, M., et al., (2002). A polyhydroxybutyrate biodegradable stent: Preliminary experience in the rabbit. *Cardiovasc. Intervent. Radiol.*, *25*, 127–132.

Valentin, H. E., & Dennis, D., (1997). Production of poly(3-hydroxybutyrate-co-4-hydroxybutyrate) in recombinant *Escherichia coli* grown on glucose. *J. Biotechnol.*, *58*, 33–38.

Velázquez, F., Pflüger, K., Cases, I., De Eugenio, L. I., & De Lorenzo, V., (2007). The phosphotransferase system formed by PtsP, PtsO, and PtsN proteins controls production of polyhydroxyalkanoates in *Pseudomonas putida. J. Bacteriol.*, *189*, 4529–4533.

Vigneswari, S., Vijaya, S., Majid, M. I. A., Sudesh, K., Sipaut, C. S., Azizan, M. N., & Amirul, A. A., (2009). Enhanced production of poly(3-hydroxybutyrate-co-4-hydroxybutyrate) copolymer with manipulated variables and its properties. *J. Ind. Microbiol. Biotechnol.*, *36*, 547–556.

Vinogradova, O. N., & Volova, T. G., (2016). Biosynthesis and properties of PHA containing monomers 3-hydroxy-4-methylvalerate. *Journal of Siberian Federal University, Biology*, *9*, 145–152 (in Russian).

Volova, T. G., & Kalacheva, G. S., (2005). The synthesis of hydroxybutyrate and hydroxyvalerate copolymers by the bacterium *Ralstonia eutropha. Microbiology*, *74*, 54–59.

Volova, T. G., & Shishatskaya, E. I., (2011). *Biodegradable Polymers: Synthesis, Properties, Applications* (389 p.). Krasnoyarskii Pisatel: Krasnoyarsk (in Russian). ISBN 978-5-98997-059-9.

Volova, T. G., (2004). *Polyhydroxyalkanoates – Plastic Materials of the 21st Century: Production, Properties, Application* (282 p.). Nova Science Pub. Inc.: New York. ISBN-10: 1590339924 ISBN-13: 978-1590339923.

Volova, T. G., Kalacheva, G. S., & Plotnikov, V. F., (1998). Biosynthesis of heteropolymeric polyhydroxyalkanoates by chemolithoautotrophic bacteria. *Microbiology*, 512–517.

Volova, T. G., Kalacheva, G. S., & Steinbüchel, A., (2008). Biosinthesis multi-component polyhydroxyalkanoates by the bacterium *Wautersia eutropha. Macromol. Symposia*, *269*, 1–7.

Volova, T. G., Kiselev, E. G., Vinogradova, O. N., Nikolaeva, E. D., Chistyakov, A. A., Sukovatyi, A. G., & Shishatskaya, E. I., (2014). A glucose-utilizing strain, *Cupriavidus eutrophus* B–10646: Growth kinetics, characterization and synthesis of multicomponent PHAs. *PloS One.*, *9*, 87551–87566.

Volova, T. G., Lukovenko, S. G., & Vasiliev, A. D., (1992). Production and investigation of physicochemical properties of microbial polyhydroxyalkanoates. *Biotekhnologiya (Biotechnology)*, *1*, 19–22 (in Russian).

Volova, T. G., Mironov, P. V., & Vasiliev, A. D., (2007). Physicochemical properties of multicomponent polyhydroxyalkanoates. *Biofizika (Biophysics)*, *52*, 460–465 (in Russian).

Volova, T. G., Sevastianov, V. I., & Shishatskaya, E. I., (2006b). *Polyhydroxyalkanoates–Biodegradable Polymers for Medicine* (286 p.). Platina Publishers: Krasnoyarsk (in Russian). ISBN 5-7638-0645-X.

Volova, T. G., Sevastyanov, V. I., & Shishatskaya, E. I., (2003). In: Shumakov, V. I., (ed.), *Polyhydroxyalkanoates – Biodegradable Polymers for Medicine)* (330 p.). SB RAS Publishers: Novosibirsk (in Russian). ISBN 5-7692-0608-X.

Volova, T. G., Shishatskaya, E. I., & Sinskey, A. J., (2013c). *Degradable Polymers: Production, Properties and Applications* (380 p.). Nova Science Pub. Inc.: New York. ISBN 978-1-62257-832-0.

Volova, T. G., Shishatskaya, E. I., Gordeev, S. A., & Zeer, E. P., (2001). A study of the structure and properties of poly(3-hydroxybutyrate) – a thermoplastic biodegradable polymer. *Perspektivnyye Materialy (Advanced Materials)*, *2*, 40–48 (in Russian).

Volova, T. G., Syrvacheva, D. A., Zhila, N. O., & Sukovatiy, A. G., (2016a). Synthesis of P(3HB-co-3HHx) copolymers containing high molar fraction of 3-hydroxyhexanoate monomer by *Cupriavidus eutrophus* B10646. *J. Chem. Technol. Biotechnol.*, *91*, 416–425.

Volova, T. G., Vasiliev, A. D., & Zeer, E. P., (2000). Investigation of molecular structure of poly(3-hydroxybutyrate), a thermoplastic and degradable polymer. *Biofizika (Biophysics), 45*, 33–439 (in Russian).

Volova, T. G., Vinnik, Yu. S., Shishatskaya, E. I., Markelova, N. M., & Zaikov, G. E., (2017c). *Natural-Based Polymers for Biomedical Applications,* Apple Academic Press: Toronto.

Volova, T. G., Vinogradova, O. N., Zhila, N. O., Kiselev, E. G., Peterson, I. V., Vasil'ev, A. D., et al., (2017a). Physicochemical properties of multicomponent polyhydroxyalkanoates: Novel aspects. *Polymer Science, Ser. A., 59*, 98–106.

Volova, T. G., Vinogradova, O. N., Zhila, N. O., Peterson, I. V., Kiselev, E. G., Vasiliev, A. D., et al., (2016b). Properties of a novel quaterpolymer P(3HB/4HB/3HV/3HHx). *Polymer, 101*, 67–74.

Volova, T. G., Voinov, N. A., Muratov, V. S., Bubnov, N. V., Gurulev, K. V., Kalacheva, G. S., et al., (2006a). Pilot production of degradable biopolymers. *Biotekhnologiya (Biotechnology), 6*, 28–34 (in Russian).

Volova, T. G., Zhila, N. O., Kalacheva, G. S., Sokolenko, V. A., & Sinskey, A. J., (2011). Synthesis of 3-hydroxybutyrate-co-4-hydroxybutyrate copolymers by hydrogen oxidizing bacteria. *Appl. Biochem. Microbiol., 47*, 497–499.

Volova, T. G., Zhila, N. O., Kiselev, E. G., & Shishatskaya, E. I., (2017b). Polyhydroxyalkanoates – natural degradable polymers. In: *Chimiya Biomassi: Biotopliva I Novie Materiali (Chemistry of Biomass: Biofuels and Novel Materials)* (pp. 652–725). Nauchnii mir: Moscow.

Volova, T. G., Zhila, N. O., Prudnikova, S. V., Boyandin, A. N., & Shishatskaya, E. I., (2016d). *Fundamental Principles of Design and Application of New Generation Agricultural Products* (220 p.). Krasnoyarsk (in Russian). ISBN: 978-5-906740-01-4.

Volova, T. G., Zhila, N. O., Shishatskay, E. I., Mironov, P. V., Vasilev, A. D., Sukovatyi, A. G., & Sinskey, A. J., (2013a). The physicochemical properties of polyhydroxyalkanoates with different chemical structures, *Polymer Science, Ser. A., 55*, 427–437.

Volova, T. G., Zhila, N. O., Shishatskaya, E. I., & Sukovatyi, A. G., (2012). *Database Physicochemical Properties of Polyhydroxyalkanoates with Different Structure*. Biosynthesis conditions and producers. Certificate of VNIIGPE No. 2012620288 of the state record of the database of (in Russian).

Volova, T., Kiselev, E., Shishatskaya, E., Zhila, N., Boyandin, A., Syrvacheva, D., et al., (2013b). Cell growth and PHA accumulation from CO2 and H2 of a hydrogen-oxidizing bacterium, *Cupriavidus eutrophus* B–10646. *Bioresour. Technol., 146*, 215–222.

Volova, T., Zhila, N., Kiselev, E., & Shishatskaya, E., (2016c). A study of synthesis and properties of poly-3-hydroxybutyrate/diethylene glycol copolymers. *Biotechnol. Prog., 32*, 1017–1028.

Waddington, D. S., (1994). *Polyhydroxyalkanoates and Film Formation There From.* US Patent No. WO 94/16000.

Wallen, L. L., & Rohwedder, W. K., (1974). Poly-3-hydroxyalkanoate from activated sludge. *Environ. Sci. Technol., 8*, 576–579.

Wang, H. H., Li, X. T., & Chen, G. Q., (2009). Production and characterization of homopolymer polyhydroxyheptanoate (P3HHp) by a fad BA knockout mutant *Pseudomonas putida KTOY06* derived from–*P. putida* KT2442. *Process Biochem., 44*, 106–111.

Wang, H. H., Zhou, X. R., Li, X. T., & Chen, G. Q., (2011). Biosynthesis of polyhydroxyalkanoates homopolymers by *Pseudomonas putida. Appl. Microbiol. Biotechnol., 89*, 1497–1507.

Wei, X., Hu, Y. J., Xie, W. P., Lin, R. L., & Chen, G. Q., (2009). Influence of poly(3-hydroxybutyrate-co-4-hydroxybutyrate-co-3-hydroxyhexanoate) on growth and osteogenic differentiation of human bone marrow-derived mesenchymal stem cells. *J. Biomed. Mater. Res. A.*, *90*, 894–905.

Weiner, R. M., (1997). Biopolymers from marine prokaryotes. *Trends Biotechnol.*, *15*, 390–394.

Werker, A., Lind, P., Bengtsson, S., & Nordström, F., (2008). Chlorinated-solvent-free gas chromatographic analysis of biomass containing polyhydroxyalkanoates. *Water Res.*, *42*, 2517–2526.

Williams, S. F., & Martin, D. P., (2002). Applications of PHAs in medicine and pharmacy. In: Steinbüchel, A., (ed.), *Series of Biopolymers* (Vol. 4, pp. 91–121). Willey-VCH Verlag GmbH.

Witholt, B., & Kessler, B., (1999). Perspectives of medium chain length poly (hydroxyalkanoates), a versatile set of bacterial bioplastics. *Curr. Opin. Biotechnol.*, *10*, 279–285.

Wong, H. H., & Lee, S. L., (1998). Poly(3-hydroxybutyrate) production from whey by high density cultivation of recombinant *Escherichia coli*. *Appl. Microbiol. Biotechnol.*, *50*, 30–33.

Wu, L. P., (2014). *Polyhydroxyalkanoates (PHAs): Biosynthesis, Industrial Production and Applications in Medicine* (349 p.). Nova Sciences Publ. Inc.: New York ISBN-13: 978-1633216228 ISBN-10: 1633216225.

Xie, W. P., & Chen, G. Q., (2008). Production and characterization of terpolyester poly (3-hydroxybutyrate-co-4-hydroxybutyrate-co-3-hydroxyhexanoate) by recombinant *Aeromonas hydrophila* 4AK4 harboring genes *phaPCJ*. *Biochem. Eng. J.*, *38*, 384–389.

Yamane, T., Fukunage, M., & Lee, Y. W., (1996). Increased PHB productivity by high cell density fed batch culture of *Alcaligenes latus*, a growth-associated PHB producer. *Biotechnol. Bioeng.*, *50*, 197–202.

Yan, S., Subramanian, S. B., Tyagi, R. D., & Surampalli, R. Y., (2008). Polymer production by bacterial strains isolated from activated sludge treating municipal wastewater. *Water Sci. Technol.*, *57*, 533–539.

Yuan, Y., & Ruckenstein, E., (1997). Miscibility and transesterification of phenoxy with biodegradable poly(3-hydroxybutyrate). *Polymer.*, *39*, 1893–1897.

Zhang, H. F., Ma, L., Wang, Z. H., & Chen, G. Q., (2009). Biosynthesis and characterization of 3-hydroxyalkanoate terpolyesters with adjustable properties by *Aeromonas hydrophila*. *Biotechnol. Bioeng.*, *104*, 582–589.

Zhang, S., Kamachi, M., Takagi, Y., Lenz, R. W., & Goodwin, S., (2001). Comparative study of the relationship between monomer structure and reactivity for two polyhydroxyalkanoate synthases. *Appl. Microbiol. Biotechnol.*, *56*, 131–136.

Zhang, S., Kolvek, S., Goodwin, S., & Lenz, R. W., (2004). Poly(hydroxyalkanoic acid) biosynthesis in *Ectothiorhodospira shaposhnikovii*: Characterization and reactivity of a type III PHA synthase. *Biomacromolecules*, *5*, 40–48.

Zhao, W., & Chen, G. Q., (2007). Production and characterization of terpolyester poly (3-hydroxybutyrate-co-3-hydroxyvalerate-co-3-hydroxyhexanoate) by recombinant *Aeromonas hydrophila* 4AK4 harboring genes phaAB. *Process Biochem.*, *42*, 1342–1347.

Zinn, M., Witholt, B., & Egli, T., (2001). Occurrence, synthesis and medical applications of bacterial polyhydroxyalkanoate. *Adv. Drug. Rev.*, *53*, 5–21.

CHAPTER 3

BIODEGRADATION BEHAVIOR OF POLYHYDROXYALKANOATES

As the human population of the world grows and economic activity intensifies, increasing amounts of chemical substances are produced and consumed, creating an avalanche of environmental problems. Synthetic polymer materials have become part and parcel of modern life, but their increasing outputs pose an environmental threat. This problem can be solved by developing new, environmentally friendly materials, which can be involved in biospheric cycling (Kijchavengkul and Auras, 2008). Synthetic plastics accumulate in landfills and are disposed of underground, polluting the land and natural waters, including vast areas of the Global Ocean. This is detrimental to aquatic ecosystems and can cause increased mortality of freshwater and marine biota.

Polyhydroxyalkanoates (PHAs), representing the developing industry of degradable bioplastics, are good candidates to replace synthetic polymers gradually. As the outputs of PHAs increase, studies examining the degradation of these polymers in natural environments acquire increasing significance. Results obtained in laboratory experiments cannot be used to construct prognostic models and predict PHA behavior and degradation in diverse and changeable natural ecosystems. This can only be achieved in integrated studies, which should answer the following key questions:

- How does the microbial community composition in a given environment influence the process of PHA degradation and what microorganisms are the most effective PHA degraders under given conditions?
- How do the chemical composition of a PHA, the process used to prepare PHA-based devices, and the shape and size of the devices influence the PHA degradation rate?
- How do the macro- and microstructure of PHAs and their properties (crystallinity, molecular weight, polydispersity) change during degradation?

- Do the physicochemical conditions of the environment (temperature, pH, oxygen availability, salinity, etc.) considerably affect this process?
- How will the process of PHA degradation be affected by weather and climate of different regions?

Analysis of the available literature shows that rather few authors reported integrated studies of various aspects of PHA degradation, which is a very complex process.

Most of the studies were performed in the laboratory, and they mainly addressed the mechanism of interaction between the PHA supramolecular structure and PHA-depolymerizing enzymes, the structure and molecular organization of various depolymerases (Kim et al., 2007) and microorganisms secreting extracellular PHA depolymerases.

PHAs are degraded in biological media to form products innocuous to the environment: carbon dioxide and water under aerobic conditions or methane and water under anaerobic conditions. PHA biodegradation is performed by microorganisms that secrete intra- or extracellular PHA depolymerases, which differ in their molecular organization and substrate specificity (Jendrossek, Handrick, 2002). Six hundred PHA-depolymerases from various microorganisms have been identified by now; comparison of their amino-acid sequences provided a basis for uniting them in 8 superfamilies including 38 families (Knoll et al., 2009). Most PHA-degrading microorganisms contain only one depolymerase, but there are species that have several depolymerases. For example, *Pseudomonas lemoignei,* one of the best-studied PHA degraders, has at least six extracellular PHA depolymerases, encoded by phaZ1 – phaZ6 genes (Schöber et al., 2000). Three of them are specific for P(3HB) and P(3HB/3HV) with low 3HV fractions (P(3HB) depolymerases A, B, and D encoded by the phaZ1, phaZ2, and phaZ3 genes).

The activity of these enzymes with the homopolyester P(3HV) is below 5% of the activity obtained with P(3HB) as a substrate. None of the three P(3HB) depolymerases is able to produce clearing zones on opaque P(3HV)-granule-containing agar. The two remaining PHA depolymerases (P(3HB) depolymerase C and P(3HV) depolymerase, encoded by phaZ5 and phaZ4, respectively) also degrade P(3HB), but are additionally able to hydrolyze P(3HV), with higher activity compared with depolymerases A, B, and C. The sixth, P(3HV) depolymerase, encoded by phaZ6, is an active degrader of 3HV or carbon sources with an odd number of carbon atoms.

PHAs are degraded intracellularly by intracellular depolymerases. It is assumed that intracellular depolymerases do not hydrolyze partially crystalline

PHAs isolated from the biomass and that extracellular depolymerases are not substrate-specific towards the polymer present in cells as granules.

PHA chains can be degraded by cleaving. The attack starts at chain-folding surfaces and then degradation goes on perpendicularly to the lamella, until it reaches the solid lamellar center and stops. Some researchers assume that single crystals are preferentially attacked at the crystal ends rather than at chain-folding surfaces (Urmeneta et al., 1995; Hocking et al., 1996). Another degradation mechanism was proposed more recently (Nobes et al., 1998): the predominant effect observed with all crystals was a significant narrowing of the lamellae, suggesting an edge attack mechanism. Biodegradation of polyhydroxyalkanoates is performed by microorganisms, which inhabit different natural environments. Therefore, ecological, and taxonomic studies need to be conducted to investigate the diversity of microorganisms degrading polymers in different biological media.

3.1 MICROBIAL DEGRADATION OF POLYHYDROXYALKANOATES

PHA biodegradation is a complex multistage process, which is influenced by a number of factors, including PHA chemical composition and properties, the type of the polymer product and the technique employed to fabricate it, climate, and weather, and the structure of the microbial community. PHA biodegradation is to a great extent determined by the composition and metabolic activity of the microbial community.

Microorganisms degrading poly(3-hydroxybutyrate) were first isolated more than 40 years ago (Chowdhury, 1963). The microorganisms identified belonged to several taxa: *Bacillus, Pseudomonas, Streptomyces.* Two years later, 16 other microorganisms degrading P(3HB) extracellularly were described (Delafield et al., 1965). Then, as the range of the investigated PHAs broadened, microorganisms degrading not only homogenous hydroxybutyrate but also short- and medium-chain-length heteropolymers were isolated and described (Brandl, Puchner, 1991; Briese et al., 1994; Mergaert et al., 1993, 1994, 1995).

Aerobic and anaerobic PHA degrading bacteria have been isolated from various ecosystems such as soil, compost, aerobic, and anaerobic sewage sludge, fresh, and seawater, estuarine sediments, and air (Kumagai et al., 1992; Imam et al., 1999; Kusaka et al., 1999; Quinteros et al., 1999; Volova et al., 2006; Shah et al., 2007; Volova et al., 2010). Thus, PHA degrading microorganisms, including P(3HB) degrading bacteria, are present in all terrestrial and aquatic ecosystems.

PHAs can be degraded by fungi. PHA degrading fungi have been isolated from soil and compost (Matavulj and Molitoris, 1992) and from freshwater, seawater, and sludge (Mergaert et al., 1996). Lee et al. (2005) isolated P(3HB) degrading micromycetes from 15 natural habitats. Most of these were Deuteromycetes (fungi imperfecti). Ninety-five fungal species have been investigated, among them 16 *Ascomycetes* species, 46 *Basidiomycetes* species, 26 *Deuteromycetes* species, 1 *Mastigiomycetes* species, 2 *Myxomycetes* species, and 2 *Zygomycetes* (Jendrossek, Handrick, 2002). The contribution of fungi to microbial degradation of PHAs in soil has been investigated in a number of recent studies. The screening of P(3HB) degrading strains showed that more than 70% of the strains of the genera *Penicillium, Absidia, Gilbertella, Mucor, Rhizopus* are able to utilize P(3HB) (Kozlovsky et al., 1999). Sang et al. (2002) studied biodegradation of P(3HB) films in soil and revealed a considerable growth of fungal population on the surface of the films and domination of *Fusarium oxysporium, Paecilomyces lilacinus,* and *Paecilomyces farinosus.* The dominant role of *Aspergillus sp.* in PHA biodegradation was reported by Sanyal et al. (2006); the authors showed that *Aspergillus sp.* exhibited higher degrading activity towards P(3HB) compared with P(3HB/3HV).

By now, aerobic and anaerobic P(3HB)-degrading microorganisms have been isolated from various natural ecosystems (soil, compost, sludge, fresh, and salt water, estuary sediments) (Boyandin et al., 2012a; Boyandin et al., 2012b; Boyandin et al., 2013; Doi et al., 1992; Imam et al., 1999; Kusaka et al., 1999; Quinteros et al., 1999; Shah et al., 2007; Volova et al., 2007; Volova et al., 2010). However, there has been little research reported on degradation of PHAs with different chemical compositions, including degradation of such promising low-crystallinity elastomeric PHAs as P(3HB/4HB) and P(3HB/3HHx); only a few authors have addressed this subject recently (Morse et al., 2011; Rodriguez-Contreras et al., 2012; Salim et al., 2012; Wang et al., 2004; Weng et al., 2011; Weng et al., 2013).

In most of the published studies, PHA degradation was investigated in laboratory experiments. Rather less attention has been paid to PHA degradation in nature. Moreover, the data on the effect of PHA composition on its degradation rate is contradictory. Several studies reported more rapid degradation of P(3HB/3HV) compared with P(3HB) (Madden et al., 1998; Sridewi et al., 2006). On the other hand, results obtained in experiments with PHA-depolymerases from *P. lemoignei* and soil studies performed *in situ* showed an opposite relationship (Mergaert et al., 1994). The possible reasons for the contradictory data that were obtained by different researchers at different times may be dissimilarities in the properties of the PHA specimens used in the studies. PHAs were synthesized by different producers on different media; the specimens could differ in the amounts of residual impurities present in them (such as lipids); the techniques employed to process PHAs were not the same; moreover, the polymers were degraded under different conditions.

Specificity of microorganisms degrading PHAs with different chemical compositions remains a poorly studied issue. Isolation of PHA degraders is usually performed by analyzing the media (soil, compost, water) in which polymer specimens have been maintained and/or microorganisms isolated from biofilms on the surface of polymers, by inoculating them onto standard microbiological media. However, among the microorganisms isolated, there may be not only true primary PHA degraders but also commensal organisms, which utilize degradation products of high-molecular-weight PHAs. True PHA degraders should be isolated by the clear zone technique (Mergaert et al., 1993), which involves inoculation of the isolates onto mineral agar that contains polymer powder as a sole carbon source. Clear zones are formed around colonies of microorganisms with PHA-depolymerase activity on the surface of the agar medium, as a result of polymer degradation. This method was used to isolate primary P(3HB) degraders in different natural environments – soils in Siberia and tropics and seawater (Boyandin et al., 2012a; Boyandin et al., 2013; Volova et al., 2007; Volova et al., 2010).

Thus, analysis of the available literature shows that there is no unanimity of opinion regarding biodegradation behavior of a given polymer in a given biological environment. The question of how the composition and properties of the PHA and environmental conditions (temperature, pH, salinity, and composition of the microbial community) influence the PHA biodegradation rate still remains unanswered.

3.2 MICROBIAL DEGRADATION OF PHAS WITH DIFFERENT CHEMICAL COMPOSITION

The study addresses degradation of PHAs with different chemical compositions – the polymer of 3-hydroxybutyric acid [P(3HB)] and copolymers of P(3HB) with 3-hydroxyvalerate [P(3HB/3HV)], 4-hydroxybutyrate [P(3HB/4HB)], and 3-hydroxyhexanoate [P(3HB/3HHx)] (10–12 mol.%) – in the agro-transformed field soil of the temperate zone (Volova et al., 2017).

PHA specimens with different chemical compositions were prepared (Table 3.1). The polymers used in the study were the homopolymer poly-3-hydroxybutyrate [P(3HB)] and three copolymers: poly(3-hydroxybutyrate/ 3-hydroxyvalerate) [P(3HB/3HV)], poly(3-hydroxybutyrate/4-hydroxybutyrate) [P(3HB/4HB)], and poly(3-hydroxybutyrate/3-hydroxyhexanoate) [P(3HB/3HHx)] with similar molar fractions of the second monomer (10–12%). Polymer was extracted from cells with chloroform, and the extracts were precipitated using hexane. The extracted polymers were re-dissolved and precipitated again 3–4 times to prepare homogeneous specimens.

Films were prepared from 2 % (w/v) polymer solutions in chloroform. The PHA solution was cast in Teflon-coated metal molds, and then solvent evaporation occurred, and they were placed into a vacuum drying cabinet (Labconco, U.S.) for 3–4 days, until complete solvent evaporation took place. The films were then weighed on the analytical balance. PHA films were weighed, placed into nylon mesh bags, and buried in the field soil (200 g in 250-cm^3 containers) at a depth of 2 cm. The agro-transformed soil was collected in the temperate zone of Siberia (56°4' N, 92°41' E) and placed into 250-cm^3 plastic containers (200 g soil per container). The soil was cryogenic-micellar agro-chernozem with high humus content in the 0–20-cm layer (7.9–9.6 %). The soil was weakly alkaline (pH 7.1–7.8), with high total exchangeable bases (40.0–45.2 mequiv/100 g). The soil contained nitrate nitrogen $N\text{-}NO_3$ – 6 mg/kg, and P_2O_5 – 6 and K_2O – 22 mg/100 g soil. The polymer films were incubated in soil for 35 days at a temperature of 21 or 28°C and soil moisture content of 50%. Every 7 days, the specimens (3 in a bag) were removed from the soil to determine their mass loss. Soil pH and moisture content were measured by using conventional physicochemical methods (Zvyagintsev et al., 2005). The specimens removed from the soil were thoroughly rinsed in distilled water, dried to constant weight in a thermostat at 40°C for 24 h, and weighed on the Discovery analytical balance

TABLE 3.1 PHA Chemical Composition and Properties of the Specimens Before and After Incubation in Soil for 28 Days (Volova et al., 2017)

PHA composition, mol.%	Mw, kDa	PD	Cx,%	Tmelt,°C	Tdegr, °C
	Initial/after 28 days				
P(3HB)	625/598	3.79/4.15	78/81	179.0/181.3	294.8/289.1
P(3HB-*co*–12.0 mol.%–3HV)	1120/1005	2.43/2.56	60/64	173.1/175.2	283.5/286.5
P(3HB-*co*–12.0 mol.%–3HHx)	356/310	3.92/3.88	56/60	173.8/176.4	246.2/251.1
P(3HB-*co*–10.0 mol.%–4HB)	597/317	3.87/9.87	50/63	153.9/157.9	286.4/283.8

(Ohaus, Switzerland). Parameters indicating PHA biodegradation were the mass loss of the specimens, changes in their molecular-weight parameters and degree of crystallinity, and surface morphology and properties.

The structure of the soil microbial community was analyzed by conventional methods of soil microbiology (Netrusov, 2005). The total number of organotrophic bacteria was determined on nutrient agar medium (NA, HiMedia); microscopic fungi were counted on malt extract agar (MEA, Sigma-Aldrich, U.S.). The ecological-trophic groups of microorganisms were identified by plating them onto diagnostic media. Ammonifying (copiotrophic) bacteria were identified on NA; microorganisms capable of utilizing mineral nitrogen (prototrophs) were identified on starch ammonium agar (SAA). Microorganisms involved in mineralization of humus substances (oligotrophs) were identified on soil extract agar (SEA) (Netrusov, 2005). Platings were performed in triplicate from 10^{3}–10^{7} dilutions of soil suspension. The plates were incubated for 3–7 days at a temperature of 30°C for counting bacteria and for 7–10 days at 25°C for counting fungi. Mineralization coefficient was determined as a ratio between microorganisms on SAA and on NA. Oligotrophy coefficient was determined as a ratio between microorganisms on SEA and on NA. To prepare pure cultures of microorganisms, 8–10 morphotypes of each isolate were transferred to the plates with the corresponding media. The morphology of bacterial cells was examined by Gram staining. Dominant microorganisms were isolated and identified by conventional methods, based on their cultural and morphological properties, and by using standard biochemical tests mentioned in identification keys (Garrity et al., 2006; Holt et al., 1997; Vos et al., 2011). Soil microscopic fungi were identified by their micro- and macro-morphological features (the structure and color of colonies and the structure of mycelium and spore-forming organs), which are objective parameters for identifying these microorganisms (Sutton et al., 2001; Watanabe, 2002).

To identify the microorganisms degrading four PHA types with different chemical compositions, polymer specimens were incubated in laboratory soil microecosystems. By examining the microbial communities that formed on the surface of PHA specimens and by plating the samples onto selective media (using the clear zone technique), we isolated and identified degraders specific for certain polymer types and microorganisms degrading all PHA types used in this study.

The soil had high mineralization and oligotrophy coefficients (1.52 and 11.74, respectively), indicating soil maturity and low contents of available nitrogen forms. The number of copiotrophic bacteria was 16.3±5.1 million

CFUs/g soil – 1.5 and 11.7 times lower than the number of prototrophic and oligotrophic bacteria, respectively, while the number of nitrogen-fixing bacteria was very high (26.1±4.7 million CFUs/g soil). Analysis of the taxonomic composition of soil microflora, which included morphological, biochemical, cultural, and molecular-genetic tests showed that the soil microbial community was dominated by Actinobacteria (37%), with *Streptomyces* reaching 24% (Figure 3.1a). Other major Gram-positive bacteria were represented by *Arthrobacter* (18%) and *Corynebacterium* (12%). Gram-negative bacilli were dominated by *Pseudoxanthomonas* (12%). The composition of the microscopic fungi was typical for the northern soils (Zvyagintsev et al., 2005), with the community dominated by *Penicillium* species (58–65% of microscopic fungi). *Trichoderma* and *Aspergillus* comprised 8% each. The counts of *Fusarium* species reached 3.2 thousand CFUs/g soil, constituting about 11% of the population of microscopic fungi (Figure 3.1b). The reason for such high counts of *Fusarium* species was that the soil was collected in the field where wheat had been grown; many of the *Fusarium* species are cereal pathogens (Bateman, Murray, 2001).

After the polymers had been incubated and degraded in the soil, the composition of the microbial community changed considerably, both quantitatively and qualitatively, including changes in the dominant species (Figure 3.1) and percentages of the ecological-trophic groups of microorganisms. The counts of ammonifying and nitrogen-fixing bacteria and microscopic fungi increased by a factor of 3 and the number of prototrophs increased by a factor of 1.8, while the number of oligotrophs dropped by a factor of 8.3. These changes suggested increased rates of soil organic matter transformation processes and stimulation of soil organotrophic microflora by the polymer as a supplementary carbon source. The percentages of Gram-negative bacilli increased (*Pseudomonas* to 44%, *Stenotrophomonas* to 28%, and *Variovorax* to 14%), while the percentage of actinobacteria decreased to 6%. The composition of microscopic fungi did not change considerably: *Penicillium* remained the major genus (46% of the total microscopic fungi in the samples) (Figure 3.1b).

To identify PHA-degrading microorganisms, PHA samples were aseptically removed from the bags that had been incubated in the soil. The specimens were rinsed with sterile water to remove unattached microorganisms. Microbial biofilm samples were scraped off the surface of the polymer. PHA-degrading microorganisms were detected using the clear zone technique (Mergaert et al., 1993). On the surface of the agar medium, clear zones were formed around colonies of microorganisms with PHA-depolymerase activity.

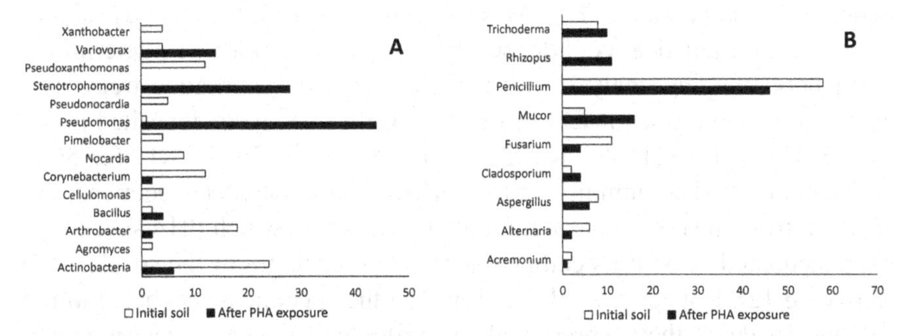

FIGURE 3.1 Major bacteria (A) and fungi (B) in laboratory soil microecosystems before and after incubation of degradable PHA specimens (% of the total counts). (Reprinted by permission from Springer Nature: Microbial Ecology, Microbial Degradation of Polyhydroxyalkanoates with Different Chemical Compositions and Their Biodegradability, Tatiana G. Volova, Svetlana V. Prudnikova, Olga N. Vinogradova et al. © 2016.)

In addition to the conventional examination, we also used molecular-genetic methods. DNA was extracted by using an AquaPure Genomic DNA Isolation reagent kit (Bio-Rad, U.S.), following the manufacturer's guidelines. The 16S rRNA gene of bacteria was amplified using universal primers 27F (5'-AGAGTTTGATCCTGGCTCAG–3') and 1492R (5'-GGTTACCTTGTTACGACTT–3'), corresponding to positions 8–27 and 1510–1492 of the *Escherichia coli* gene, respectively. The polymerase chain reaction (PCR) was performed on a Mastercycler Gradient amplifier (Eppendorf, Germany) according to the standard procedure.

High-quality sequences were obtained by cloning unpurified PCR products into vector pCR4-TOPO (Invitrogen, U.S.), used to transform TOP 10 *E. coli* cells. The resultant clones were tested by restriction analysis for the presence of the insert of the proper size in the vector. Plasmid DNA was extracted with a PureLink Quick Plasmid Miniprep kit (Invitrogen, U.S.) following the manufacturer's guidelines. Sequencing was performed in two directions in an ALF express II automatic DNA analyzer (Amersham Pharmacia Biotech Ltd, U.S.), using universal primers T3 and T7 and a Thermo Sequenase Cy5 Dye Terminator kit. The nucleotide sequences were compared with the sequences in the EMBL/DDBJ/GenBank using the BLAST program for searching highly homologous sequences of the NCBI web resource (http://www.ncbi.nlm.nih.gov/BLAST).

For phylogenetic analysis, the nucleotide sequences were aligned with homologous referential sequences from the database NCBI RefSeq using the Muscle package of MEGA software version 6. Phylogenetic analysis was carried out according to the 3-parameter model Tamura, using the neighbor-joining method in the MEGA version 6 software package (Tamura et al., 2013). Statistical confidence of the branching order was determined using bootstrapping analysis, by constructing 1000 alternative trees. The obtained nucleotide sequences of the 16S rRNA gene were deposited in the GenBank (No. KT321679 – KT321704 and No. KU052942 – KU052950).

The microbial community on the surface of the polymer specimens was different from the microbial community of the soil in which PHA specimens were incubated in both its composition and percentages of the species. We recovered 128 isolates from the biofilms on the specimens incubated in the soil for 35 days; they represented the following genera: *Achromobacter, Acidovorax, Alcaligenes, Arthrobacter, Bacillus, Cellulomonas, Chitinophaga, Corynebacterium, Cupriavidus, Delftia, Ensifer, Flavobacterium, Lysobacter, Microbacterium, Micrococcus, Mitsuaria, Mycobacterium, Nocardia, Pseudomonas, Pseudonocardia, Pseudoxanthomonas, Roseateles, Roseomonas, Stenotrophomonas, Streptomyces,* and *Variovorax*.

The subsequent plating of microorganisms isolated from the biofilms onto the selective solid media (mineral agar with the polymer as sole carbon source) showed that many of these species were not primary degraders of the polymers and were unable to metabolize the initial high-molecular-weight polymer. Thus, biofilms contained both primary degraders of PHAs, capable of metabolizing the initial high-molecular-weight polymer, and commensal organisms, which utilized oligomers, monomers, acetoacetate, and other products of degradation of high-molecular-weight polymers, which were present in the medium due to vital activities of primary true PHA degraders.

By plating the samples on solid media (using the clear zone technique, Figure 3.2), we, for the first time, isolated the major degraders of the four polymer types used in this study. Clear zone formation on the diagnostic medium with the polymer (depolymerase activity) was revealed for 35 isolates of bacteria, which represented 16 genera. (Table 3.2). *Fusarium solani* was the only PHA-degrading microscopic fungus. It was shown that each of the PHAs studied was degraded by a different set of species (Figure 3.3). P(3HB) was degraded by *Mitsuaria, Chitinophaga,* and *Acidovorax* species, which were not identified among the degraders of the other PHA types, and also by *Nocardia sp., Streptomyces albolongus, Variovorax sp.* and *Achromobacter sp.* In addition to the specific degraders of P(3HB/4HB) – *Roseateles depolymerans and Cupriavidus sp.* – we also identified strains of the genera *Nocardia, Streptomyces, and Variovorax. Roseomonas massiliae* and *Delftia acidovorans* were specific for P(3HB/3HV), but this polymer was also degraded by *Lysobacter gummosus, Streptomyces omiyaensis, Variovorax paradoxus,* and *Achromobacter sp.* Finally, P(3HB/3HHx) was degraded by *Streptomyces albolongus* and by its specific degraders: *Pseudoxanthomonas, Pseudomonas fluorescens, Ensifer adhaerens,* and *Bacillus pumilus*.

TABLE 3.2 Identification of the Primary PHA-Degrading Microorganisms

PHA type	PHA-degrading strains, access no.	The most similar strains from the GenBank database, degrading ability	Ident
P(3HB/HHx)	KT321684	*Bacillus pumilus* strain PRE14 (EU880532.1)	100%
	KT321685	*Bacillus cereus* strain FJM4 (KR493006.1)	100%
	KT321683	*Ensifer adhaerens* (*Sinorhizobium morelense*) strain BB2 (KP314234.1)	100%
	KT321681	*Pseudomonas fluorescens* strain JK15 (KF148637.1)	99%
	KT321679	*Pseudoxanthomonas mexicana* strain HX-N01 (KF501482.1)	99%
		Cellulose-degrading bacteria	
	KT321680	*Pseudoxanthomonas sp.* 11_4K (EF540482.1)	100%
	KT321682		
	KU052942	*Streptomyces albolongus* strain CSSP414 (NR_115372.1)	99%
P(3HB/3HV)	KU052943	*Achromobacter insuavis* strain G.W-CD.23 (KP645387)	99%
		Cellulose-degrading bacteria	
	KT321690	Candidatus *Roseomonas massiliae* 1461A (AF531769.1)	100%
		Roseomonas sp. BZ44 (HQ588850.1) isolated from hydrocarbon-contaminated soil	99%
	KT321691	*Delftia acidovorans* strain B201 (KJ781877.1)	100%
	KT321686	*Lysobacter gummosus* strain KCTC 12132 (NR_041005.1)	100%
	KT321688	*Lysobacter sp.* HX1 (EF601813.1)	100%
	KT321687	*Streptomyces omiyaensis* strain NBRC 13449 (NR_112403.1)	100%
	KT321689	*Variovorax paradoxus* DSM 30162 (AB622223.1)	100%
		Xenobiotic compounds degrading bacteria	
		Variovorax sp. GG6b (GQ337855.2)	
		Parathion-degrading bacteria	

TABLE 3.2 *(Continued)*

PHA type	PHA-degrading strains, access no.	The most similar strains from the GenBank database, degrading ability	Ident
P(3HB)	KT321695	*Achromobacter spanius* strain JN52 (KJ794193.1)	100%
		Salt-tolerant diesel-degrading bacteria	
	KU052944	*Achromobacter sp.* M23 (GU086442)	99%
	KU052950	*Achromobacter insuavis* strain G.W-CD.23 (KP645387)	99%
		Cellulose-degrading bacteria	
	KT321698	*Acidovorax sp.* THG-LW112	99%
	KT321694	Uncultured *Chitinophaga sp.* clone WCD76 (KJ123800.1)	99%
	KT321692	*Mitsuaria sp.* CR 3–06 (KM252964.1)	100%
	KT321693		
	KU052945	*Nocardia sp.* SXS5 (GQ357987)	99%
	KU052946	Uncultured bacterium clone B07–26-BAC (GQ340117)	100%
		Nocardia sp. OAct 132 (JX047071)	99%
	KT321696	*Streptomyces albolongus* strain CSSP414 (NR_115372.1)	99%
	KU052949	*Streptomyces sp.* HP5 (GQ867025)	100%
	KT321697	*Variovorax sp.* i7s (AB974276.1)	100%
	KT321699	*Variovorax sp.* c24 (AB167202.1)	99%
		Phenol and trichloroethylene degrading bacteria	
P(3HB/4HB)	KU052948	*Nocardia fluminea* strain 173590 (EU593589)	99%
	KT321704	*Ralstonia sp.* 80 (AY238507.1)	100%
		Populations associated with 2,4-D degradation	
	KT321702	*Roseateles depolymerans* HG-P (AB495143.1)	99%
	KT321700	*Streptomyces sp.* QZGY-A23 (JQ812080.1)	100%
	KT321701	*Streptomyces gardneri* NMCZ8 (JN999892.1)	100%
	KU052947	*Streptomyces sp.* GLY-P1 (KF917528)	97%
		4-hydroxybenzoic acid-degrading actinomycetes	
	KT321703	*Variovorax sp.* c24 (AB167202.1)	99%
		Phenol and trichloroethylene degrading bacteria	

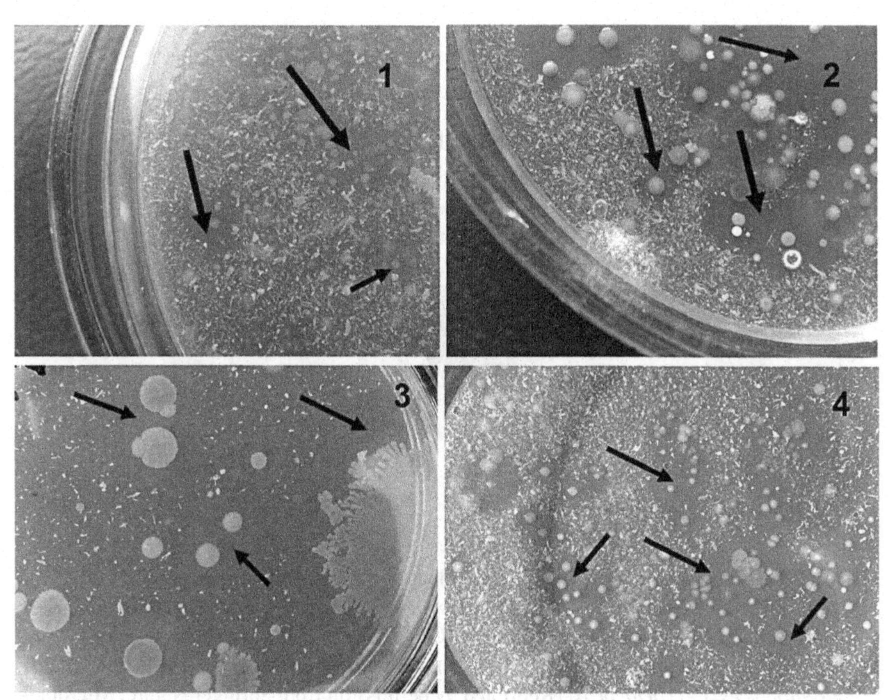

FIGURE 3.2 Determination of PHA-degrading microorganisms. PHA-depolymerase activity displayed as clear zone formation on the solid diagnostic medium with powdered PHAs of different chemical compositions as a sole carbon source; 1 – P(3HB), 2 – P(3HB/4HB), 3 – P(3HB/3HHx), 4 – P(3HB/3HV). (Reprinted by permission from Springer Nature: Microbial Ecology, Microbial Degradation of Polyhydroxyalkanoates with Different Chemical Compositions and Their Biodegradability, Tatiana G. Volova, Svetlana V. Prudnikova, Olga N. Vinogradova et al. © 2016).

Nocardia, Streptomyces, and *Variovorax* species were common degraders for P(3HB) and P(3HB/4HB). *Streptomyces* and *Variovorax* species were common degraders for three PHA types: P(3HB), P(3HB/4HB), and P(3HB/3HV). *Streptomyces* species were common degraders for all four PHA types.

Phylogenetic analysis showed that the community of degraders consisted of four clusters, which united representatives of the phyla *Proteobacteria, Bacteroidetes, Actinobacteria,* and *Firmicutes* (Figure 3.4). Most of the strains (11 genera) isolated from different polymer types were *Proteobacteria. Actinobacteria* (2 genera) also included representatives isolated from all four PHA types. The *Firmicutes* included two *Bacillus* strains – degraders of P(3HB/3HHx) – and one P(3HB)-degrading strain of *Chitinophaga sp.*, affiliated with the phylum *Bacteroidetes*.

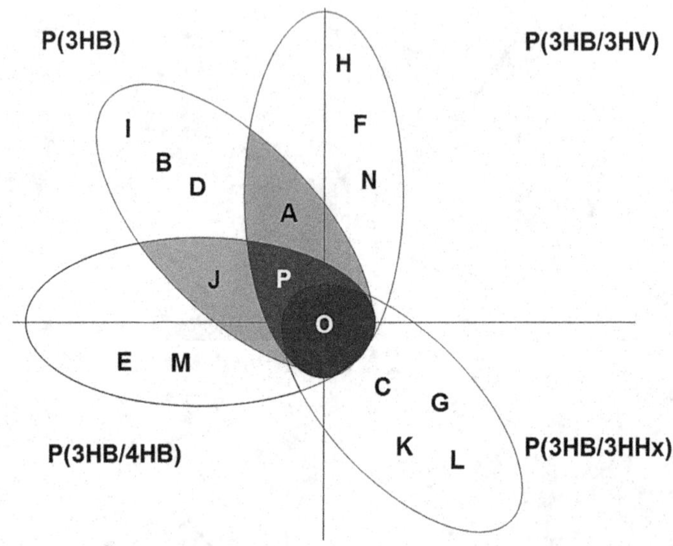

FIGURE 3.3 Occurrence of primary PHA-degrading microorganisms on PHA films of different chemical composition; A – *Achromobacter*, B – *Acidovorax*, C – *Bacillus*, D – *Chitinophaga*, E – *Cupriavidus*, F – *Delftia*, G – *Ensifer*, H – *Lysobacter*, I – *Mitsuaria*, J – *Nocardia*, K – *Pseudoxanthomonas*, L – *Pseudomonas*, M – *Roseateles*, N – *Roseomonas*, O – *Streptomyces*, P – *Variovorax*. (Reprinted by permission from Springer Nature: Microbial Ecology, Microbial Degradation of Polyhydroxyalkanoates with Different Chemical Compositions and Their Biodegradability, Tatiana G. Volova, Svetlana V. Prudnikova, Olga N. Vinogradova et al. © 2016).

Comparative analysis of the 16S rRNA gene nucleotide sequences of PHA-degrading strains with the sequences from the GenBank database revealed 99–100% homology to some of the strains capable of degrading other complex molecules. *Pseudoxanthomonas sp.* 3HH–1 showed homology to cellulose-degrading *Pseudoxanthomonas mexicana* strain HX-N01 (KF501482). *Achromobacter sp.* 3HV–2 and 3HB–2o had high homology to cellulolytic bacterium *Achromobacter insuavis* strain G.W-CD23 (KP645387). Another strain of *Achromobacter sp.*, 3HB–6p, showed 100% homology to a salt-tolerant diesel-degrading bacterium *Achromobacter spanius* strain JN52 (KJ794193). *Variovorax* species are among the most active PHA degraders (Mergaert et al., 1993). In the present study, we isolated four strains of this genus, which showed close similarity to *Variovorax paradoxus* strains capable of degrading xenobiotic compounds (AB622223), parathion (GQ337855), phenol, and trichloroethylene (AB167202) (Choi et al., 2009; Futamata et al., 2005). Some of the species isolated in this study had never been described as PHA degraders,

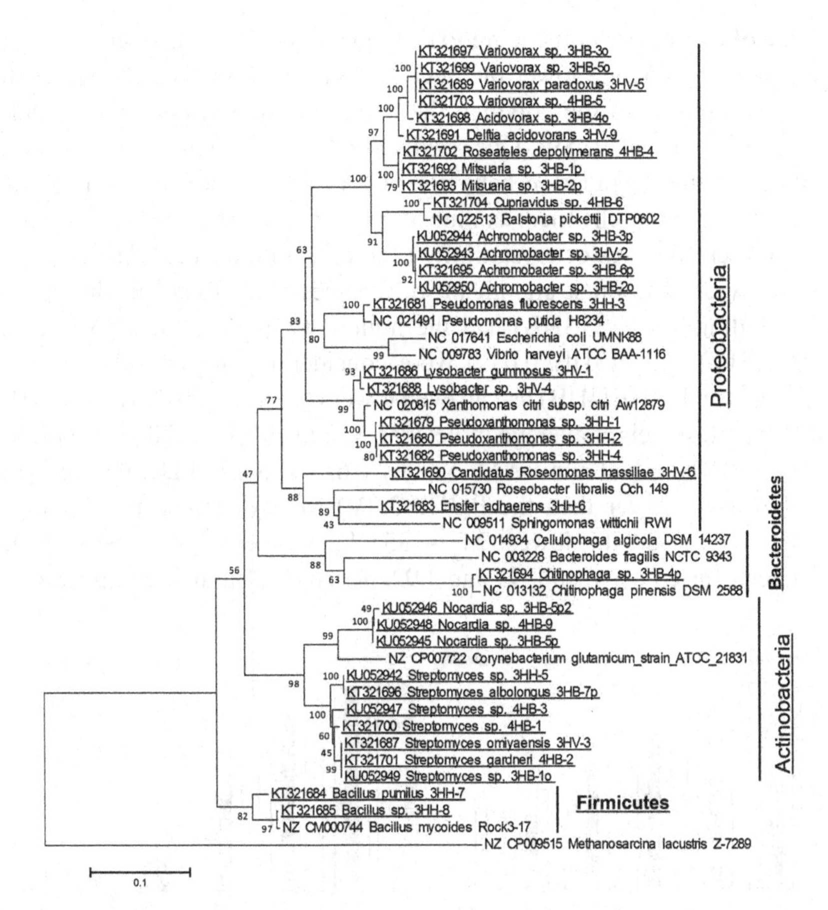

FIGURE 3.4 4 A rootless phylogenetic tree based on the analysis of 35 16S rRNA gene sequences of PHA-degrading bacteria. The scale shows the evolutionary distance corresponding to 10 nucleotide substitutions per each 100 nucleotides. The numbers indicate the confidence of branching determined by bootstrap analysis (values greater than 50 are accepted significant). The strains of degraders are underlined. (Reprinted by permission from Springer Nature: Microbial Ecology, Microbial Degradation of Polyhydroxyalkanoates with Different Chemical Compositions and Their Biodegradability, Tatiana G. Volova, Svetlana V. Prudnikova, Olga N. Vinogradova et al. © 2016).

but they degraded other chemical compounds. For instance, *Roseateles depolymerans* 4HB–4 showed 99% homology to HG-P strain (AB495143.1) characterized by the ability to utilize carboxymethyl-cellulose (Tani et al., 2011). Suyama et al. (1998) reported that *Roseateles depolymerans* were able to degrade aliphatic polycarbonates, including poly(tetramethylene succinate), poly(hexamethylene carbonate), and poly(ε-caprolactone), which are known as biodegradable plastics.

One of the factors influencing PHA biodegradation is the degree of crystallinity of a PHA: the lower the degree of crystallinity, i.e., the larger the amorphous region of the polymer, the higher the rate of polymer degradation (Kunioka et al., 1989). Our results are consistent with this reasoning. Analysis of the degradation behavior of four PHA types in soil is shown in Figure 3.5. The chemical composition and properties of the polymers with different degrees of crystallinity (Table 3.1) influenced PHA degradation behavior. The most significant differences were found in the degree of crystallinity (C_x). P(3HB) had the highest C_x (78%); the C_x values of P(3HB/3HV) and P(3HB/3HHx) were considerably lower (60 and 56%, respectively); P(3HB/4HB) had an even lower C_x value – 50%. P(3HB/3HV) had the highest weight average molecular weight (M_w) – 1120 kDa; P(3HB) and P(3HB/4HB) had similar M_w values – 625 and 597 kDa, respectively, and they were lower than the P(3HB/3HV) M_w by almost a factor of 2; P(3HB/3HHx) had the lowest M_w – 356 kDa. The polydispersity of the specimens varied between 2.43 and 3.92, depending on their composition.

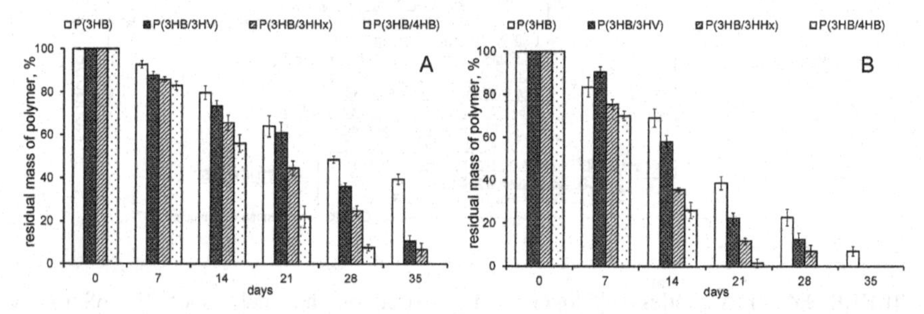

FIGURE 3.5 Degradation behavior of PHAs with different chemical compositions in the laboratory soil microecosystem at 21°C (A) and 28°C (B). (Reprinted by permission from Springer Nature: Microbial Ecology, Microbial Degradation of Polyhydroxyalkanoates with Different Chemical Compositions and Their Biodegradability, Tatiana G. Volova, Svetlana V. Prudnikova, Olga N. Vinogradova et al. © 2016).

Based on their degradation rates in the soil, the PHAs tested in this study can be ranked as follows: P(3HB/4HB) > P(3HB/3HHx) > P(3HB/3HV) > P(3HB). At a temperature of 28°C (the temperature known to be optimal for PHA degradation (Volova et al., 1998), for the first seven days, the mass loss of all PHA films was not substantial: the mass loss of P(3HB/4HB) films was the most significant – 30%, and the P(3HB) films lost 10% of their initial mass. After that, the degradation rates of all PHAs increased. For 21 days, P(3HB/4HB) films had been 97% degraded, but P(3HB) specimens had been only 60% degraded. P(3HB) degradation reached 93% as late as Day 35. P(3HB/3HHx) and P(3HB/3HV) were degraded with similar rates, which were slower than P(3HB/4HB) degradation rates but faster than P(3HB) ones (Figure 3.6b). At a soil temperature of 21°C, degradation rates of all specimens slowed down (Figure 3.6A). The average degradation rates of PHAs at 21°C were 0.61 mg/d for P(3HB), 0.89 mg/d for P(3HB/3HV), 0.93 mg/d for P(3HB/HHx), and 1.15 mg/d for P(3HB/4HB). At 28°C, degradation rates of PHA films were 20–35% higher, the lowest exhibited by P(3HB) (0.93 mg/d) and the highest by P(3HB/4HB) (1.63 mg/d).

Figure 3.7 shows photographs and SEM images of the films prepared from the four PHA types during their degradation. As the films were degraded, holes, and cracks developing on their surfaces grew larger. Then, the defects extended deeply into the specimens, and the specimens broke into fragments. This process was more noticeable in the films with higher degradation rates. After 35 days of incubation in the soil, the degree of crystallinity of P(3HB) films, which were the least degraded ones, remained almost unchanged, suggesting that both crystalline and amorphous regions of this polymer were degraded at similar rates. The degrees of crystallinity of P(3HB/3HV) and P(3HB/3HHx) specimens increased, and the most considerable increase was observed at 21°C: from the initial C_x values of 60 and 56% to 64 and 60%, respectively; thus, the amorphous regions of these polymers were degraded with higher rates. An even greater increase in the degree of crystallinity was recorded for P(3HB/4HB) specimens: from the initial 50% to 61% at 28°C and 63% at 21°C (Table 3.1). We did not find any direct relationship between the degradation rates of PHAs and their initial molecular weights. However, there were differences in the changes occurring during degradation in the molecular weights of the films prepared from different PHAs. The weight average molecular weights (M_w) of the slowly degraded homopolymer P(3HB) and the copolymers P(3HB/3HV) and (P3HB/3HHx) remained almost unchanged, despite the considerable mass loss. At Day 7, however, we recorded a considerable decrease in the molecular weight and an increase in a polydispersity (Đ) of the most rapidly degraded P(3HB/4HB); then the molecular weight declined insignificantly. Over the first week, the number average (M_n) and weight average (M_w) molecular weights of P(3HB/4HB) films incubated at 21°C decreased by a factor of 4.5 and 1.88, respectively, and their polydispersity increased by a factor of 2.2. At the end of the experiment, these parameters were equal to 33 and 317 kDa, respectively (Table 3.1). At 28°C, molecular weight decreased by a factor of 2 and 1.5, respectively, and Đ increased by a factor of 1.4, reaching 5.32, over the first week.

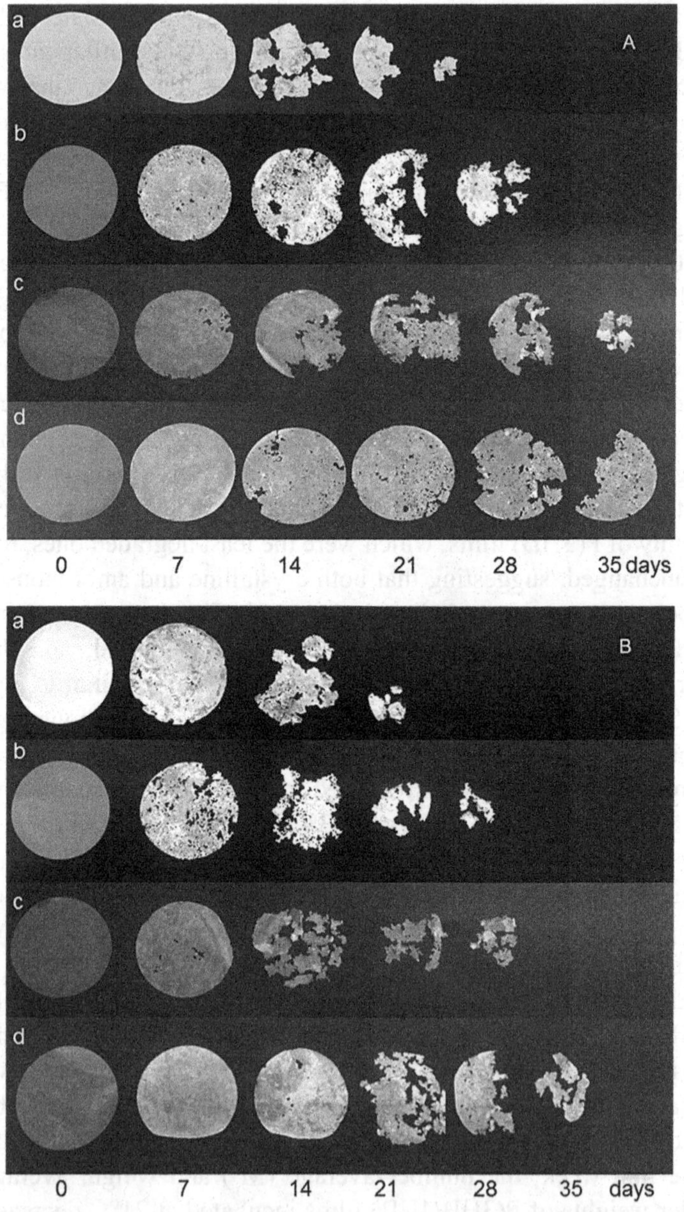

FIGURE 3.6 Photographs of the films of PHAs with different chemical compositions during their degradation in soil at 21°C (A) and 28°C (B): a – P(3HB/4HB); b – P(3HB/3HV); c – P(3HB/3HHx), d – P(3HB). (Reprinted by permission from Springer Nature: Microbial Ecology, Microbial Degradation of Polyhydroxyalkanoates with Different Chemical Compositions and Their Biodegradability, Tatiana G. Volova, Svetlana V. Prudnikova, Olga N. Vinogradova et al. © 2016.

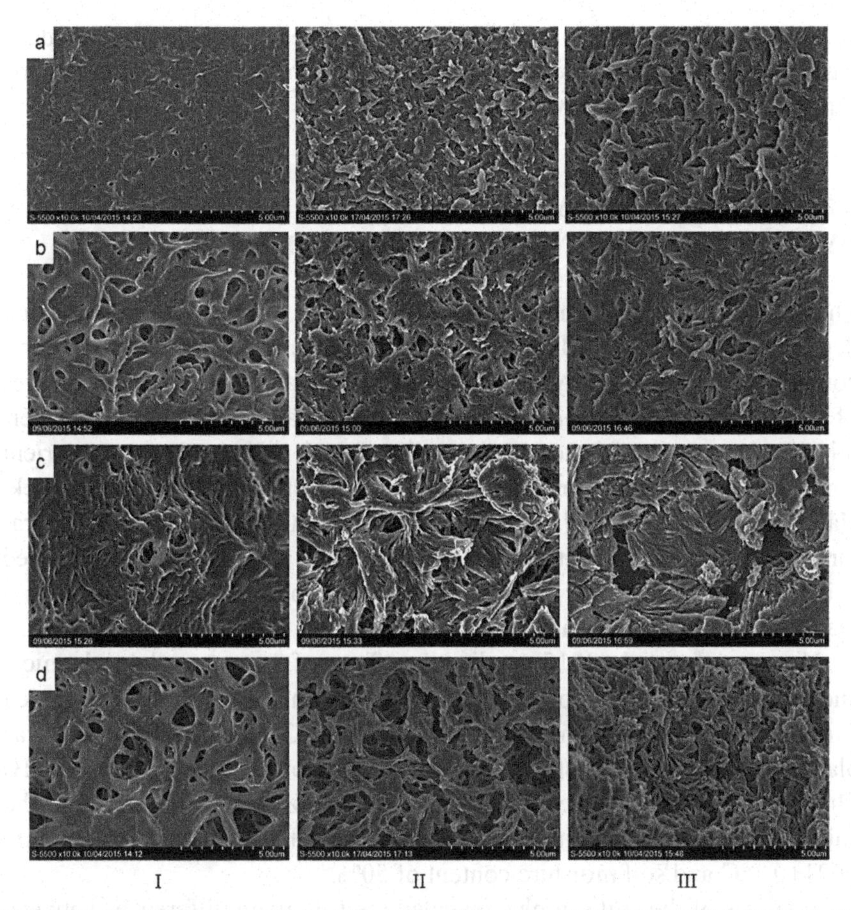

FIGURE 3.7 SEM images of initial films (I) and films degraded in soil at 21°C (II) and 28°C (III): a – P(3HB), b – P(3HB/3HV), c – P(3HB/3HHx), d – P(3HB/4HB) during degradation in soil. Bar = 5 μm (10 000 ×) (Reprinted by permission from Springer Nature: Microbial Ecology, Microbial Degradation of Polyhydroxyalkanoates with Different Chemical Compositions and Their Biodegradability, Tatiana G. Volova, Svetlana V. Prudnikova, Olga N. Vinogradova et al. © 2016.)

The degradation of four PHA types with different chemical structure was studied in the soil with the known composition of the microbial community. Based on their degradation rates, the PHAs were ranked as follows: P(3HB/4HB) > P(3HB/3HHx) > P(3HB/3HV) > P(3HB). PHA degradation influenced the total counts of microorganisms and composition of soil microflora. The microbial community that formed on the polymer surface and the soil microbial community differed in the composition and percentages of the species. By employing the clear zone technique, we, for the first time, showed that each of the PHA types studied had specific degraders. Determined the primary degraders both specific for each PHA and common for all the polymers (Volova et al., 2017).

3.3 MICROBIOLOGICAL DEGRADATION OF POLY-3-HYDROXYBUTYRATE IN AGRO-TRANSFORMED SOILS

The microbial component of soil is an active agent of hydrolytic and degradation processes. Many soil microorganisms utilize various natural compounds, including polyhydroxyalkanoates (Wang et al., 2005). The rates of degradation of PHA polymer products in natural environments are determined by a number of factors such as climate, pH, moisture content, nutrient concentrations, temperature, and activity of microorganisms (Jendrossek, Handrick, 2002; Yew et al., 2006; Philip et al., 2007). A search of the literature revealed very few data on PHA degradation rates in agro-transformed soils, although such studies should provide the basis for using PHAs as a matrix for embedding agricultural chemicals.

Vinogradova et al. (2015) studied the effects of the type and the chemical and microbiological composition of the agro-transformed field and garden soil specimens on biodegradation behavior of P(3HB) films. The soil was placed into 250-mm^3 plastic containers (200 g soil per container). P(3HB) films were weighed, placed in close-meshed gauze bags, and buried in the soil. The containers were incubated in a thermostat at a constant temperature of 21±0.1°C and soil moisture content of 50%.

Analysis of the soil samples revealed the following differences between them. The field soil (the village of Minino, the Krasnoyarskii Krai) was cryogenic-micellar agro-chernozem with high humus content in the 0–20-cm layer (7.9–9.6 %). The soil was weakly alkaline (pH 7.1–7.8), with high total exchangeable bases (40.0–45.2 mequiv/100 g). The soil contained nitrate nitrogen N-NO_3 – 6 mg/kg and P_2O_5 – 6 and K_2O – 22 mg/100 g soil (according to Machigin). The garden soil was highly agro-transformed soil (the village of Subbotino, the Krasnoyarskii Krai), with pH 6.6 (close to neutral). It contained high amounts of humus (17.4%), nitrate nitrogen N-NO_3 (122.0 mg/kg), and available phosphorus and potassium: P_2O_5 – 151.2 mg/100 g soil; K_2O – 80 mg/100 g soil (according to Machigin).

Microbiological analysis of the initial soil samples showed considerable differences between their eco-trophic groups of microorganisms (Figure 3.8). In the field soil, the number of copiotrophic bacteria was lower than the number of prototrophic and oligotrophic ones (Table 3.3); by contrast, the number of nitrogen-fixing bacteria was high. The field soil had high mineralization and oligotrophy coefficients, suggesting active mineralization processes, low contents of available nitrogen forms, and no input of fresh organics.

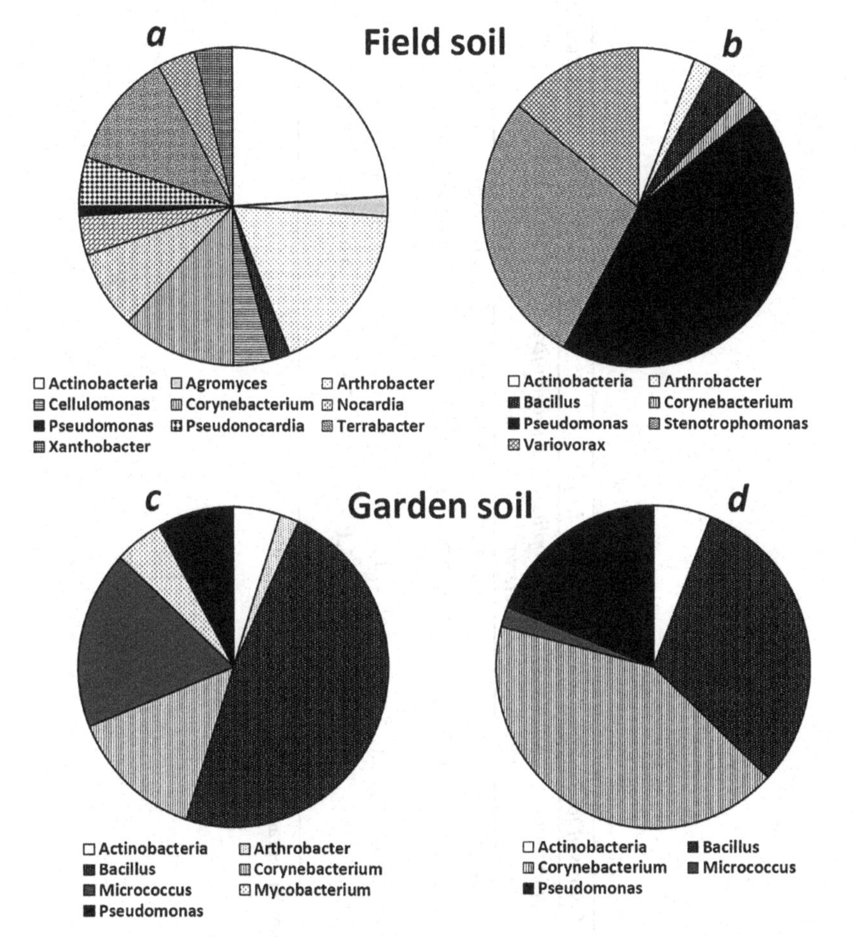

FIGURE 3.8 Dominant genera of bacteria isolated from samples of the field and garden soils before the experiment (*a*) and after incubation of P(3HB) (*b*) (Reprinted with permission from Vinogradova et al., 2015).

Analysis of the garden soil showed different results. The number of copiotrophic bacteria was four times higher than their number in the field soil samples. The high abundance of copiotrophic bacteria and very high concentrations of nutrients (N, P, and K) suggested input of fresh organics

TABLE 3.3 Microbiological Characterization of Soil Samples

Soil samples	***Counts of microorganisms, million CFU in 1 g***					*Mineralization coefficient*	*Oligotrophy coefficient*
	Copiotrophic bacteria	*Prototrophic bacteria*	*Oligotrophic bacteria*	*Nitrogen-fixing bacteria*	*Microscopic fungi*		
Field soil (initial)	16.3±5.1	24.7±7.1	190.9±70.7	26.1±4.7	0.03±0.01	1.52	11.74
Field soil (after incubation with P(3HB))	48.8±5.7	44.9±4.5	22.9±7.7	79.1±5.9	0.09±0.02	0.92	0.47
Garden soil (initial)	66.3±29.5	5.0±2.01	30.9±4.4	3.82±1.4	0.25±0.06	0.07	0.46
Garden soil (after incubation with P(3HB))	38.1±5.5	15.1±3.8	23.85±5.1	16.1±7.3	0.01±0.01	0.40	0.63

(Reprinted with permission from Vinogradova et al., 2015)

and active transformation processes. Low mineralization and oligotrophy coefficients confirmed this suggestion. The presence of available nitrogen forms decreased the number of nitrogen-fixing bacteria, which was lower than the number of these bacteria in the field soil by a factor of 6.8. These are the most favorable conditions for rapid biodegradation of organic matter. In the field soil, the total abundance of bacteria was lower, but the community was more diverse (Figure 3.8). Actinobacteria and representatives of *Arthrobacter* and *Corynebacterium* dominated. The microbial community of the garden soil was dominated by *Bacillus*, *Micrococcus*, *Corynebacterium*, and *Pseudomonas*.

The differences in the total counts of microorganisms and proportions of their eco-trophic groups between the soils influenced the degradation behavior of P(3HB) films.

Rates of P(3HB) film biodegradation differed depending on the type of soil (Figure 3.9). The curve of polymer mass loss in the field soil was flatter. Over the first seven days of the experiment, the mass of the films in the field and garden soils decreased by 7 and 11%, respectively, of their initial mass.

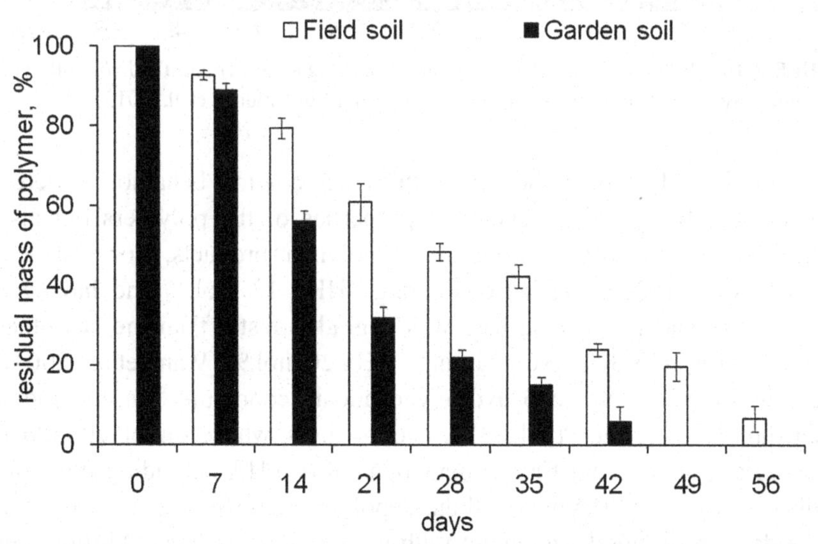

FIGURE 3.9 Changes in the mass of P(3HB) films incubated in the garden and field soils in laboratory experiments (Reprinted with permission from Vinogradova et al., 2015).

Then, in the garden soil, polymer degradation rates increased, and after 42 days, degradation of P(3HB) films reached 94%, while in the field soil, the residual mass of the polymer was about 30% at the same time point. Only after 56 days, P(3HB) degradation in the field soil reached 93.4%. Thus, P(3HB) was degraded 1.5–1.7 times faster in the garden soil.

Changes in the films are shown in Figure 3.10. As the films were degraded and their mass decreased, their surface became rougher and more uneven, and the films became thinner. Formation of small pores was observed, and the number of the pores and their size increased throughout the experiment. As the number of pores increased, the films disintegrated and broke into pieces, which gradually disappeared. The films incubated in the field soil retained their mass and shape for longer periods of time.

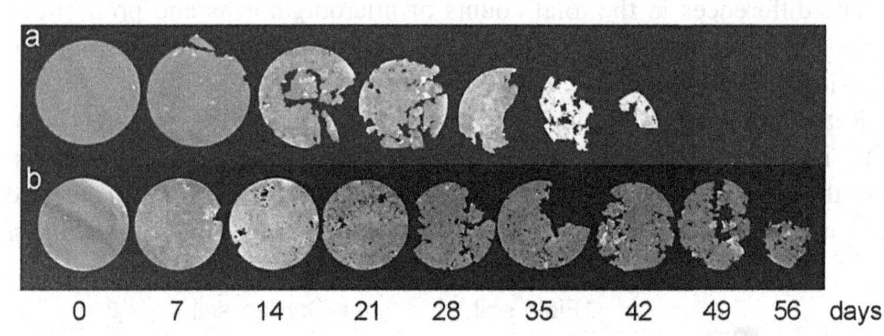

FIGURE 3.10 P(3HB) films during incubation in the garden (*a*) and field (*b*) soils in a laboratory experiment (Reprinted with permission from Vinogradova et al., 2015).

A number of studies show that PHA degradation behavior is determined not only by physicochemical properties of the polymers but also by the shape and surface structure of polymer products. For instance, the copolymer P(3HB/3HHx) containing 3HHx 12 mol.% and having a more porous and rougher surface was degraded faster than the smoother specimen of P(3HB/3HHx) containing 3HHx 20 mol.% (Wang et al., 2004). The 3-hydroxybutyrate-co–3-hydroxyhexanoate copolymer was degraded faster than P(3HB) and P(3HB/3HV) copolymers, whose smoother surface hindered attachment and further development of PHA-degrading microorganisms possessing PHA extracellular depolymerases (Sridewi et al., 2006). These data are in good agreement with the previous findings (Molitoris et al., 1996; Tsuji, Suzuyoshi, 2002), which suggested that rates of degradation of PHA products were determined by their surface morphology. Molitoris et al. (1996) showed that hydrolysis started at the surface and at lesions and proceeded to the inner part of the polymer.

Electron microscopy revealed differences in the surface microstructure of the initial specimens and changes that occurred during their degradation

(Figure 3.11). The surface of the initial specimens had an indistinct pattern, without defects or fractures, with few 2- to 4-μm micropores (Figure 3.11, 1a and 1b). After 28 days, as the films were degraded, the surface became more uneven, with chaotically positioned sharply defined structures varying in shape and pores larger than 2 μm in size, probably as a result of faster leaching of the polymer amorphous phase; the number of micropores increased too (Figures 3.11, 2a and 2b).

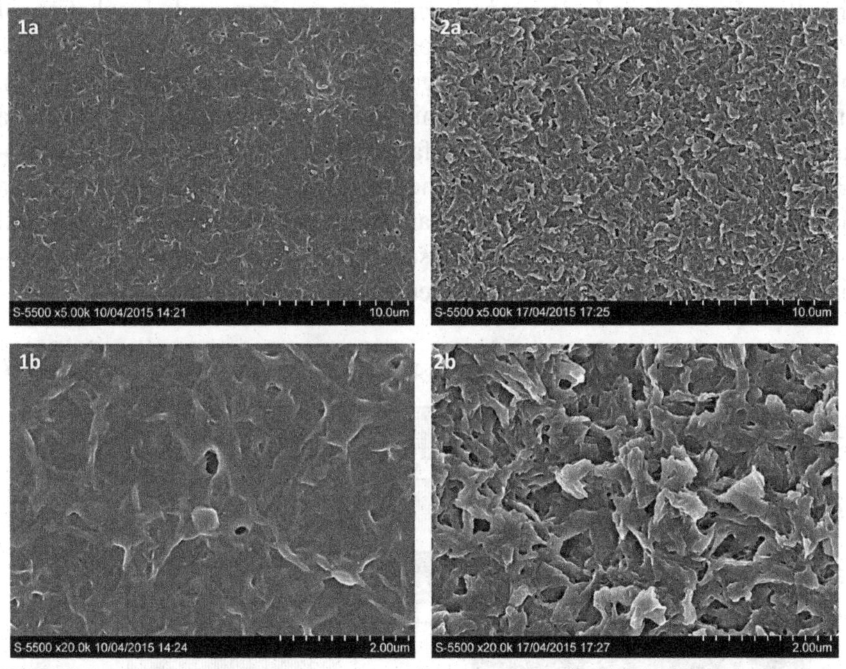

FIGURE 3.11 SEM images of surface microstructure of P(3HB) films: 1a – pristine P(3HB) film, before degradation, bar = 10 μm; 1b – bar = 2 μm; 2a – the same film after 28-day incubation in the field soil, bar = 10 μm; 2b – bar = 2 μm (Reprinted with permission from Vinogradova et al., 2015).

After P(3HB) films had been incubated and degraded in soil, the composition of the microbial community changed considerably, both quantitatively and qualitatively. In the field soil, the counts of ammonifying bacteria increased by a factor of 3, the number of Gram-negative bacteria increased, while the number of actinobacteria decreased. The community was dominated by *Pseudomonas*, *Stenotrophomonas*, and *Variovorax* species. In the garden soil, no significant changes in the abundance of organotrophic bacteria were found, although their number tended to decrease. These changes could be associated with the incubation of polymer films as a supplementary carbon source, which could counter the excess of nitrogen in the soil and accelerate mineralization of organics in the closed space. This was confirmed by higher mineralization and oligotrophy coefficients of the garden soil after incubation of P(3HB) films.

After P(3HB) incubation, microbial community of the garden soil was dominated by different species: the abundance of the spore-forming bacteria and Gram-positive cocci decreased, while the counts of bacteria representing the *Pseudomonas* and *Corynebacterium* genera increased. Incubation of P(3HB) films in the soil was accompanied by changes in the percentages of the eco-trophic groups of microorganisms in soil samples (Figure 3.12). In the field soil, the percentage of copiotrophic and nitrogen-fixing bacteria and microscopic fungi increased by a factor of 3 and the percentage of prototrophic bacteria by a factor of 1.8; the abundance of oligotrophic bacteria decreased by a factor of 8.3. These changes suggested increased rates of soil organic matter transformation processes and stimulation of soil organotrophic microflora by P(3HB) as a supplementary carbon source.

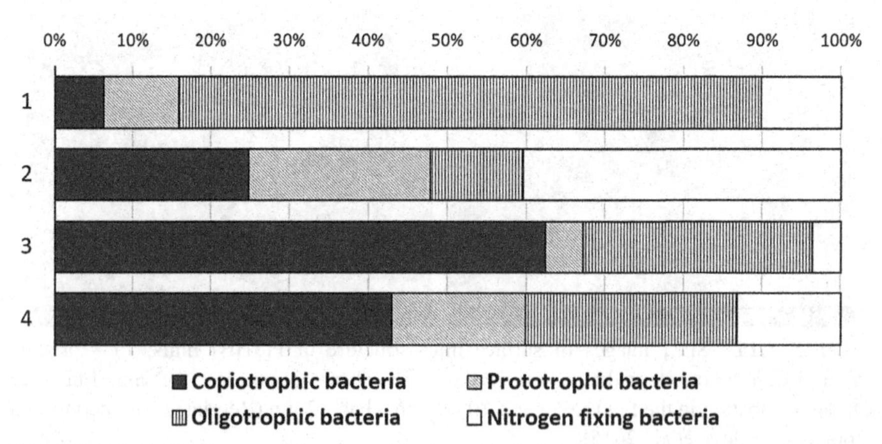

FIGURE 3.12 Percentages of eco-trophic groups of microorganisms under incubation of P(3HB) films in different soil samples: 1 – field soil, initial sample; 2 – field soil after 42-day incubation of P(3HB); 3 – garden soil, initial sample; 4 – garden soil after 21-day incubation of P(3HB) (Reprinted with permission from Vinogradova et al., 2015).

In the garden soil, the percentage of copiotrophic bacteria decreased, suggesting high rates of mineralization of organic matter and polymer biodegradation, which supplied supplementary substrate to soil microflora. Under conditions of unlimited supply of nitrogen-containing organics, fast polymer degradation and utilization of organic substrate by bacteria led to a stable abundance of organotrophic microflora in soil followed by an increase in the percentage of prototrophic and nitrogen-fixing bacteria.

The present study showed that different amounts of nutrients and different total abundances of microorganisms and diverse structures of microbial communities in the two soil types substantially influenced the degradation behavior of P(3HB) films. In the garden soil, which contained significantly higher amounts of organotrophic bacteria (biopolymer-degrading bacteria), P(3HB) degradation rates were 1.5–1.7 times higher than in the field soil. During incubation of P(3HB) films in the soil, the taxonomic composition of the dominant microorganisms changed as follows: in both field and garden soils, the percentages of Gram-negative bacteria of the *Pseudomonas*, *Stenotrophomonas*, and *Variovorax* genera increased.

3.4 BIODEGRADATION OF PHAS BY SOIL MICROFLORA IN NATURAL ENVIRONMENTS

The soil is the natural environment with the greatest capacity for PHA degradation. However, most of the studies addressing PHA degradation in soil were carried out in laboratory (Bonartseva et al., 2003; Mergaert et al., 1993; Erkske et al., 2006; Suyama et al., 1998; Woolnough et al., 2008) and some of them used isolated cultures of PHA degrading microorganisms (Mokeeva et al., 2002; Bhatt et al., 2008; Colak, Güner, 2004; Nishida and Tokiwa, 1993). Samples of soil suspension were used as laboratory microcosms to study degradation of polymer films based on two PHA types (poly(3-hydroxybutyrate) and copolymers of 3-hydroxybutyrate and 3-hydroxyvalerate) under stable temperature and moisture conditions (Volova et al., 1992; 1996). Degradation of both types of PHA was strongly influenced by the temperature, but no pH effect was observed. Copolymer samples were degraded with a higher rate than homogenous P(3HB), and as the molar fraction of 3HV increased, the difference in degradation rates became more significant.

There are very few published data on PHA biodegradation in soil under field conditions. One of the first studies that addressed PHA degradation under natural conditions showed (Mukai, Doi, 1993) that a golf tee made of the polymer was almost completely degraded in soil within four weeks; unfortunately, the authors of this study did not describe either the exact composition of the PHA or the soil characteristics. There are data, however, suggesting that the type of soil is an essential factor affecting PHA degradation. Mergaert and coauthors (1994) studied biodegradation of P(3HB) and

P(3HB/3HV)s with different molar fractions of 3-hydroxyvalerate (10 and 20%) in household compost heaps and showed that significant mass loss was only recorded in the P(3HB/3HV) specimens with a high 3HB percent (20 mol.%). Lim et al. (2005) studied degradation of the polymer consisting of 3-hydroxyhexanoate, 3-hydroxyoctanoate, 3-hydroxydecanoate, 3-hydroxydodecanoate, 3-hydroxytetradecanoate, and 3-hydroxyhexadecanoate in forest and mangrove soils. After 112 d of burial, there was 16.7% reduction in gross weight of the films buried in acidic forest soil, 3.0% in the ones buried in alkaline forest soil, and 4.5% in those buried in mangrove soil. Sridewi and coauthors reported (2006) that the weight of P(3HB) and P(3HB/3HHx) films was reduced in 7 days after burial in the mangrove soil. All PHA types were degrading with similar rates. The half-life was 42 days for all PHA samples on the soil surface and 28 days for the samples buried 20 cm deep in the soil. Yew and co-authors showed (2006) that PHA degradation rate in the garden soil was influenced by the burial depth and the density of microbial populations. The mass of P(3HB) films decreased by 55% and 25% in the soils with microorganism concentrations 1.0×10^8 and 3.2×10^6 CFU/g soil, respectively, within 43 days. The degradation rate of the films placed on the soil surface was 50% slower. The density of soil microbial populations affected PHA degradation in garden soils.

Wang et al. (2005) studied degradation of (3HB/3HV) films in natural media and reported the highest degradation rate in activated sludge (residual mass 10.87% of the initial mall at day 480 of the experiment) and lower rates in river water (80.44%), seawater (83.96%), and farm soil (84.32); and the lowest degradation rate was recorded in the infertile garden soil (98.88%). Rapid degradation of (3HB/3HV) and its blends with atactic P(3HB) was observed in compost containing activated sludge (almost 100% for 6 weeks) (Rutkowska et al., 2008). The available data on the effect of PHA properties such as a molar mass on its degradation are inconsistent. PHA properties, however, are related to the molar mass of the polymer and, specifically, on its Mw (weight average molecular weight). Several authors reported a direct relationship between the M_w of a polymer and its degradation (Quinteros et al., 1999; Mokeeva et al., 2002; Bonartseva et al., 2003; Bonartsev et al., 2009). The only way to correctly determine Mw is to use high-performance gel permeation chromatography. Degradation of polymer samples (biomass loss, changes in the surface and structure) by both mixed populations of microorganisms (Woolnough et al., 2008) and single-species isolates (Reddy et al., 2008; Bhatt et al., 2008) has been described. Fungi degrade PHAs more actively than bacteria, due to the higher mobility of fungal

PHA depolymerases (Reddy et al., 2008). PHA depolymerases of some soil microorganisms have been isolated, and their properties have been studied (Colak, Güner, 2004; Reddy et al., 2003).

In order to gain insight into PHA biodegradation patterns and mechanisms of this process, it is important to isolate and identify PHA degrading microorganisms. Among PHA degraders that have been described in the literature are representatives of various genera: *Bacillus, Pseudomonas, Alcaligenes, Comamonas, Rhodococcus, Rhodocyclus, Syntrophomonas, Ilyobacter* (Jendrossek et al., 1996), *Terrabacter, Terracoccus, Brevibacillus, Agrobacterium, Duganella, Ralstonia, Matsuebacter, Rhodoferax, Variovorax, Acinetobacter, Pseudomonas, Bacillus, Azospirillum, Mycobacterium, Streptomyces* etc. (Suyama et al., 1998; Jendrossek, Handrick, 2002; Mergaert, Swings, 1996; Bonartsev et al., 2009; Volova et al.,2006). Fungi are considered to be the most efficient PHA degraders: *Ascomycetes, Basidiomycetes, Deuteromycetes, Zygomycetes* (Matavulj, Molitoris, 1992) and *Mixomycetes, Mastigiomycetes, Penicillium, Fusarium* (Brucato, Wong, 1991; Mokeeva et al., 2002; Kim, Rhee, 2003). The higher degradation capacity of fungi is accounted for by the fact that fungal PHA-depolymerases are more mobile than PHA-depolymerases secreted by bacteria (Reddy et al., 2008). Most soil PHA degraders are believed to be capable of degrading short-chain PHAs, i.e., ones that consist of monomers containing not more than 5 carbon atoms. Only a few of them can degrade medium-chain PHAs, and this is accounted for by the substrate specificity of PHA extracellular depolymerases (Kim et al., 2003, 2007). Authors of earlier studies showed that under their study conditions, most PHA degrading soil microorganisms degraded short-chain PHAs, but the portion of degraders of medium-chain PHAs was rather small: 0.8% to 18% of all PHA degraders (Nishida and Tokiwa, 1993; Suyama et al., 1998).

It is noteworthy that isolation of PHA degraders is often performed by analyzing the media (soil, compost, water) in which polymer specimens have been maintained and microorganisms isolated from biofilms on the surface of polymers, by inoculating them onto standard microbiological media. Among the microorganisms isolated, there may be commensal organisms, which utilize monomers and other degradation products of high-molecular-mass PHAs and which exist in the medium due to the vital activity of primary and true PHA degraders. A reliable way to isolate true PHA degraders is to use the clear-zone technique (Mergaert et al., 1993), which involves inoculation of the isolates onto mineral agar that contains PHA as a sole carbon source. Clear zones are formed around colonies of microorganisms with

PHA-depolymerase activity on the surface of the agar medium, as a result of polymer degradation.

The discrepancies between the data reported by different authors at different times must be due to dissimilarities in PHA specimens used: they were synthesized by different producers on different media, the amounts of residual impurities (such as lipids) in the samples were not equal, the polymers were processed by various techniques, exposure conditions were not the same, and, finally, the effects of degradation were determined using different methods. The degree of crystallinity and PHA molecular weight can be determined using X-ray structure analysis and high-performance liquid chromatography, but the data of some authors are based on indirect evidence, obtained by differential thermal analysis and viscometry.

Extensive pioneering research into PHA biodegradation behavior in natural soil ecosystems was performed at the Siberian Federal University and Institute of Biophysics SB RAS. So, an integrated approach to the investigation of PHA degradation by soil microorganisms under natural conditions was employed in a series of papers by Boyandin et al. (2012a, 2012b, 2013). The authors studied this process in different climates and soils, taking into account the diversity of soil microbial communities, shapes of polymer samples and methods of their preparation, and the chemical composition of the PHAs tested.

One part of the studies addressed degradation of PHAs with different chemical structures in the form of film discs and pressed pellets by soil microorganisms inhabiting the rhizosphere of coniferous and broadleaved trees under varying soil temperature conditions. Experiments were performed under natural conditions, in the arboretum at the V.N. Sukachev Institute of Forest SB RAS (Krasnoyarsk) during two field seasons, which differed in temperature conditions. The first field season lasted from 2 July to 19 October 2007; the second – from 7 June to 7 September 2010. The second field season was preceded by severe winter, with the average winter temperature –22.1°C (in the second half of January and the first part of February, the temperature fell to –40°C), which was 13.5°C lower than the average temperature of the winter preceding the first field season (–8.6°C). In addition to that, the spring temperature rise occurred later in the year of the second experiment; the arboretum had been covered with snow until late April. The soddy-carbonate soil of the arboretum consists of a 10–15 to 30–40-cm thick humus layer and the underlying carbonate rock.

A polymer of 3-hydroxybutyric acid [P(3HB)] [crystallinity (C_x) 61%, molecular weight (M_w) 710.0±1.5 kDa] and a copolymer of 3-hydroxybutyric and 3-hydroxyvaleric acids [P(3HB/3HV)] containing 10 mol.% of hydroxyvalerate [C_x 50%; M_w 799.0±3.1 kDa (2007); M_w 680.0±1.1 kDa (2010)] were used in experiments. PHA film discs (60.0±6.5 mg, 30 mm in diameter, 80±7 μm thick) were weighed and placed in close-meshed gauze jackets, which were then buried 5 cm deep in the rhizosphere of the larch and the birch.

During the course of the experiment, polymer specimens were extracted from the soil, polymer discs were taken out of the jackets and cleaned: the remaining soil was removed using mechanical and enzymatic techniques, rinsed in distilled water, and dried for 24 h at 40°C. Soil temperature, pH, and moisture content were measured at the same time points, using conventional physicochemical methods. The mass loss dynamics of the polymer specimens is shown in Figure 3.13. PHA degradation was definitely influenced by the chemical composition of the polymer and the place where the specimens were buried. During the first field season, soil temperature rose significantly (from 18°C to 28°C) in July and gradually decreased to 8–10°C by the end of the experiment (October 2007). Neutral or low-alkalinity pH was recorded under the larch (7.1–8.1), with a short-term decrease at the end of July (to 6.7). The soil under the birch was weakly acidic (6.1–7.1), with a brief decrease to 4.5 at the end of July.

Moisture content of the soil was higher under the larch than under the birch: 18–20% and 11–23%, respectively, during the period between late July and late September. During the 2010 field season, the parameters of the soils were different: during the first month of the field experiment, the temperatures of the soils were lower, 14–18°C, gradually rising to 26°C and then falling to 13–18°C at the end of the experiment. The levels of pH (6.5–7.2 in the soil under the birch and 7.2–8.0 under the larch) remained almost unchanged throughout the field season. The moisture content of the soil was generally higher than in 2007, but it was considerably lower under the birch than under the larch.

The initial microbial communities in the soils under the two tree species differed in both their total counts (Table 3.4) and their compositions. At the beginning of June 2007, the total count of aerobic microflora was higher under the larch than under the birch, amounting to $(1.47 \pm 0.08) \times 10^9$ of colony-forming units per gram (CFU/g) of soil. After 3 months, the total counts of bacteria in the soil were $(5.11 \pm 0.42) \times 10^9$ CFU/g soil under the larch and $(2.21 \pm 0.24) \times 10^9$ CFU/g soil under the birch. The counts of prototrophs and oligotrophs were also higher in soil samples collected under the larch, in which oligotrophic index (PA/FPA) reached 0.75, indicating high rates of nutrient uptake and assimilation.

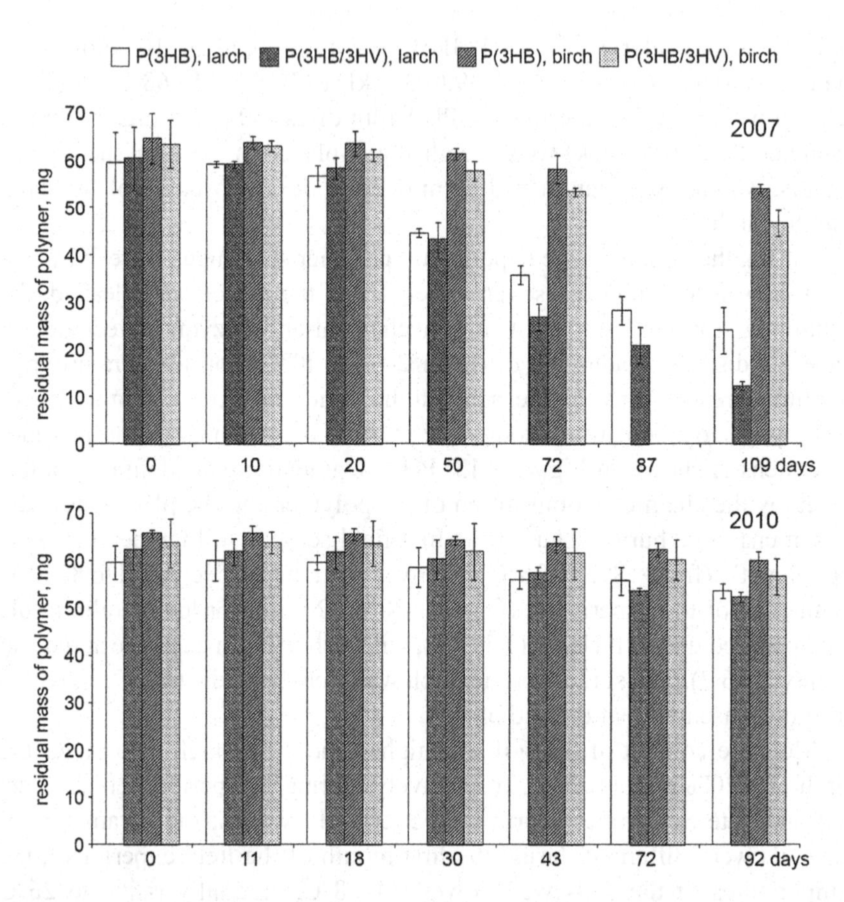

FIGURE 3.13 Mass loss dynamics of PHA specimens in the soil of the arboretum (Krasnoyarsk) in the larch and the birch rhizosphere in the experiments of 2007 and 2010 (Reprinted by permission from Springer Nature: Applied Biochemistry and Microbiology, Biodegradation of polyhydroxyalkanoates by soil microbial communities of different structures and detection of PHA degrading microorganisms, A. N. Boyandin, S. V. Prudnikova, M. L. Filipenko et al., © 2012.)

The bacterial component of the soil microbial community in the larch rhizosphere was represented by the following dominant bacteria: *Alcaligenes* (25.0%), *Aureobacterium* (15.9%), *Pseudomonas* (4.5%), *Cellulomonas* (52.3%), and *Acinetobacter* (2.3%). The microbial community in the soil under the birch was more diverse. The dominant species were *Pimelobacter* (48.4%), *Actinomyces* (16.1%), and *Micrococcus* (9.7%); *Flavimonas, Mycobacterium, Corynebacterium, and Arthrobacter* were present in small numbers. In the soil under the larch, fungi were represented by *Acremonium* (1.8% of the total count), *Mucor* (5.5%), *Verticillium* (18.2%), and the dominant species,

Penicillium (74.5%). In the soil under the birch, fungal flora was more diverse: in addition to the species inhabiting the soil under the larch (*Penicillium* and *Verticillium*), we identified *Cladosporium* (1.7%), *Hyphoderma* (1.7%), *Pytium* (3.3%), *Cephalosporium* (8.3%), and *Beltrania* (31.7%); *Beltrania* and *Penicillium* were the dominant species. The total counts of the fungi inhabiting soils under the larch and under the birch were similar.

In 2010, a colder year, the composition of soil microbial communities was similar to that in 2007, but the initial total counts of soil microorganisms were lower (Table 3.4). At the beginning of July 2010, the total counts of aerobic microflora were $(3.16\pm0.24)\times10^6$ CFU/g soil under the larch and $(2.67\pm0.16)\times10^5$ CFU/g soil under the birch. At the end of the experiment, the total counts of bacteria were two orders of magnitude lower than in 2007: $(5.28\pm1.76)\times10^7$ CFU/g soil under the larch and $(6.77\pm4.51)\times10^7$ CFU/g soil under the birch.

TABLE 3.4 Total Counts of Bacteria in Soil Samples From the Rhizospheres of the Larch and the Birch.

Soil sample		Total microbial count, CFU per 1 g soil	
		2007	**2010**
L. sibirica	Rhizosphere at the time of PHA placement	(1.47±0.08)×109	(3.16±0.24)×106
	Rhizosphere at the end of the season	(5.11±0.42)×109	(5.28±1.76)×107
	PHA surface	(1.60±0.04)×1011	(1.35±0.11)×108
B. pendula	Rhizosphere at the time of PHA placement	(1.33±0.47)×108	(2.67±0.16)×105
	Rhizosphere at the end of the season	(2.21±0.24)×109	(6.77±4.51)×107
	PHA surface	(1.29±0.08)×109	(1.96±0.05)×108

(Reprinted by permission from Springer Nature: Applied Biochemistry and Microbiology, Biodegradation of polyhydroxyalkanoates by soil microbial communities of different structures and detection of PHA degrading microorganisms, A. N. Boyandin, S. V. Prudnikova, M. L. Filipenko et al., © 2012.)

The species composition of microflora from the control soil samples is given in Table 3.5.

Gram-positive bacteria were dominated by *Micrococcus sp.* Other isolated species were *Bacillus* spore-forming rods and arthrobacteria. Gram-negative microflora was represented by *Acinetobacter, Flavobacterium, and Pseudomonas.* In soil samples removed from the surface of polymer specimens, proportions of microorganisms were different. The community of the soil in the larch rhizosphere was dominated by *Agrobacterium* species followed by *Cellulomonas. Aureobacterium, Acinetobacter, Pseudomonas,* and *Arthrobacter* were present in small numbers. The soil from the film surface of the polymer from the birch rhizosphere was mainly inhabited by Bacillus species, with some representatives of *Arthrobacter, Micrococcus, Nocardia, Actinomyces, Pimelobacter,* and *Alcaligenes.*

TABLE 3.5 Species Composition of the Microflora in the Rhizosphere of the Larch and the Birch (September 20, 2007) (Volova's data)

Soil sample	Fungi		Bacteria	
	Species	% of total count	Species	% of total count
L. sibirica– control	*Aureobasidium pullulans*	38.9	*Micrococcus varians*	41.4
	Penicillium madriti	16.7	*Micrococcus roseus*	35.1
	Penicillium decumbens	16.7	*Acinetobacter sp.*	11.7
	Penicillium atrovenetum	11.1	*Arthrobacter sp.*	3.6
	Penicillium adametzioides	5.6	*Bacillus alvei*	2.7
	Mycelia sterilia	5.6		
	Paecilomyces lilacinus	5.6		
L. sibirica – polymer surface	*Paecilomyces lilacinus*	81.5	*Agrobacterium sp.*	24.3
	Aureobasidium pullulans	11.1	*Cellulomonas sp.*	17.5
	Acremonium butyri	3.7	*Alcaligenes sp.*	6.6
	Zygosporium masonii	1.9	*Aureobacterium terregens*	4.5
	Penicillium novae-caledoniae	1.9	*Acinetobacter sp.*	1.9
			Pseudomonas sp.	1.9
			Arthrobacter sp.	1.3
B. pendula – control	*Penicillium fusco-flavum*	50.0	*Micrococcus roseus*	49.2
	Penicillium steckii	13.6	*Micrococcus varians*	24.6
	Aureobasidium pullulans	13.6	*Actinomyces sp.*	6.2
	Verticillium lateritium v, beticola	9.1	*Micrococcus luteus*	3.1
	Trichoderma piluliferum	4.5	*Arthrobacter sp.*	3.1
	Trichoderma sp,	4.5	*Bacillus fastidiosus*	1.5
	Penicillium aurantio-violaceum	4.5	*Pseudomonas sp.*	1.5
			Flavobacterium sp.	1.5

TABLE 3.5 *(Continued)*

Soil sample	Fungi		Bacteria	
	Species	**% of total count**	**Species**	**% of total count**
B. pendula – polymer surface	*Penicillium canescens*	47.6	*Bacillus fastidiosus*	21.1
	Penicillium corylophyloides	33.3	*Arthrobacter sp.*	5.3
	Aureobasidium pullulans	7.1	*Micrococcus luteus*	5.3
	Paecilomyces lilacinus	7.1	*Nocardia sp.*	5.3
	Verticillium lateritium v, beticola	2.4	*Actinomyces sp.*	5.3
	Nigrospora gallarum	2.4	*Pimelobacter*	5.3
			Bacillus brevis	5.3
			Alcaligenes sp.	5.3

Note: bold type denotes PHA degrading fungi.

PHA degradation behavior was influenced by properties of the soils under the trees and characteristics of the microbial communities (Figure 3.13). In 2007, in the soil under the larch, which was moister and housed more microorganisms, PHA degradation rates were higher than those recorded under the birch. By the end of the experiment, the residual mass of P(3HB) specimens had decreased to 45% of their initial mass, and the residual mass of P(3HB/3HV) specimens – to 22%; the half-lives of these polymers were 83 d and 68.5 d and their average mass losses for the field season 0.325 and 0.44 mg/d, respectively.

In the soil of the birch rhizosphere, degradation rates of both PHA types were lower, in spite of the great variety of the fungi present in this soil. At day 109 of the exposure, the residual masses of P(3HB) and P(3HB/3HV) specimens amounted to 84% and 74% of their initial masses, respectively, with the mass losses of the homopolymer and the copolymer 0.097 and 0.15 mg/d. The degradation rate of the copolymer was generally higher than that of the high-crystallinity P(3HB). These results are in good agreement with the data reported in papers by other authors and in our previous studies, which show that PHA copolymer specimens are degraded in biological media faster than the P(3HB) homopolymer (Mergaert et al., 1993, 1994; Woolnough et al., 2008; Volova et al., 1992). In 2010, PHA degradation rates were lower than during the 2007 field season. At the end of the field experiment, the residual mass of the specimens amounted to 89.9% and 74% for P(3PHB) and P(3HB/3HV) specimens buried under the larch and to 91.4%

and 89% for the specimens buried under the birch. As in 2010, the mass loss was so small, and we failed to find any reliable differences in the degradation of the two PHAs used in this study.

Thus, at temperate latitudes (Siberia, Krasnoyarsk) with a markedly continental climate, in the soddy-carbonate soil of the arboretum, during the warmer summer season, P(3HB/3HV) copolymer films were degraded faster than the higher-crystallinity P(3HB) specimens. These results are in good agreement with the data reported by other authors and those obtained in our previous studies, which show that PHA copolymer specimens are degraded in biological media faster than the homopolymer of 3-hydroxybutyrate (Volova et al., 1992; 1996; Mergaert et al., 1993; 1994; Woolnough et al., 2008), but contradict the data reported by Rosa and coauthors (2003). Degradation rates of P(3HB) recorded by these authors were higher than P(3HB/3HV) degradation rates, and they explained their results as being due to specific surface structure and properties of their specimens.

PHA biodegradability is influenced not only by the chemical composition of the polymer and the temperature of the environment, but also by the polymer stereoconfiguration, crystallinity, and molecular weight (Nishida, Tokiwa, 1993; Jendrossek, Handrick, 2002). PHA specimens used in this study had different degrees of crystallinity (Table 3.6). X-ray structure analysis performed at the end of the field experiment showed increases in the degrees of crystallinity of both PHAs, suggesting preferential disintegration of the amorphous phase of both PHAs in the soil under the study conditions, which resulted in a higher degree of crystallinity of the specimens. This result is consistent with the data reported by a number of authors (Abe et al., 1998; Sridewi et al., 2006). Briese et al. (1998) also showed that in the pure culture of *Alcaligenes faecalis* degradation rates of P(3HB) were higher than those of the samples prepared from P(3HB/3HV). Sridewi et al. (2006) incubated PHA films in the media containing marine sediment suspension in laboratory experiments and found a greater mass loss for P(3HB) films than for P(3HB/3HV) ones.

SEM images clearly showed the effects of degradation and larger defects, which are more pronounced in the films incubated in the rhizosphere of the pine, where degradation occurred at higher rates than under the birch. In contrast to some other degradable polymers (polysaccharides, polylactides), PHAs undergo true biological degradation, which occurs via the cellular and the humoral pathways and is affected by phagocytes and PHA-depolymerizing enzymes secreted by microflora (Kim et al., 2007). The total counts and composition of control soil microbial communities examined at the end of the field experiments were considerably different from those of microbial communities of soil samples removed from the surface of polymer specimens.

TABLE 3.6 Changes in PHA Crystallinity and Molecular Weight During Biodegradation.

Parameter	Year	Initial value	90 days of exposure in soil under the birch	90 days of exposure in soil under the larch
Crystallinity, Cx,%, P(3HB)	2007	61	69	65
Crystallinity, Cx %, P(3HB/3HV)	2007	50	54	68
Mw, kDa, P(3HB)	2007	710.0 ± 1.5 %	691 ± 3 %	577.0 ± 4.6 %
Polydispersity, P(3HB)	2007	2.05 ± 2.10 %	2.22 ± 6.50 %	2.39 ± 4.10 %
Mw, kDa, P(3HB/3HV)	2007	799.0 ± 3.1 %	633.0 ± 5.4 %	660.0 ± 2.6 %
Polydispersity, P(3HB/3HV)	2007	1.77 ± 4.40 %	2.23 ± 8.60 %	2.41 ± 2.60 %
Mw, kDa, P(3HB)	2010	710.0 ± 1.5 %	704.0 ± 2.3 %	692.0 ± 2.9 %
Polydispersity, P(3HB)	2010	2.05 ± 2.10 %	2.32 ± 3.20 %	2.25 ± 3.40 %
Mw, kDa, P(3HB/3HV)	2010	680.0 ± 1.1 %	628.0 ± 5.0 %	66.0 ± 2.6 %
Polydispersity, P(3HB/3HV)	2010	2.32 ± 0.30 %	2.35 ± 2.50 %	2.41 ± 2.60 %

The counts of prototrophs increased by about 3 times compared to the initial counts in the soils under both tree species. The fact that the counts of oligotrophs in the soil under the birch increased to (3.81±0.71) ×10^8 CFO/g soil suggested the greater activity of this group of microorganisms. By contrast, the counts of oligotrophs and oligotrophic index in the soil under the larch decreased. The reason may be that high-density populations of hydrolytic bacteria (primary degraders of organic matter) in the soil under the larch could suppress the development of oligotrophs. True PHA degraders were isolated by inoculating samples onto diagnostic agar that contained 0.25% PHA powder as a sole carbon source. Growth of microorganisms with PHA-depolymerase activity was accompanied by the formation of clear zones around colonies of microorganisms (Figure 3.14). In addition to the conventional morphological and biochemical examination, PHA degrading bacteria were identified by 16S rRNA gene sequence analysis. This approach enabled identifying true PHA degraders: 16 bacterial and 5 fungal isolates (Boyandin et al., 2012a). Inoculation of the samples onto the diagnostic medium confirmed PHA-depolymerase activity of species belonging to the genera *Penicillium, Paecilomyces, Acremonium, Verticillium, and Zygosporium.* Mokeeva and coauthors in their recent study (2002) described a wider range of fungi degrading PHAs, which included representatives of

Penicillium, Aspergillus, Paecilomyces, Acremonium, Verticillium, Cephalosporium, Trichoderma, Chaetomium, and *Aureobasidium,* but the authors of that study did not inoculate samples onto the diagnostic medium, i.e., their data were not based on the use of the clear-zone technique.

PHA degraders in the soil of the larch rhizosphere were mainly represented by *Paecilomyces lilacinus,* amounting to 81.5%. This species was also described as a polymer degrader by Sang and coauthors (2002). The fungi localized on polymer surface in the birch rhizosphere were dominated by *Penicillium sp.* BP–1 and *Penicillium sp.* BP–2, totally amounting to 81%. Lopez-Llorca and coauthors (1993) also emphasized that *Penicillium* species were the major PHA degraders (up to 88% of the isolates).

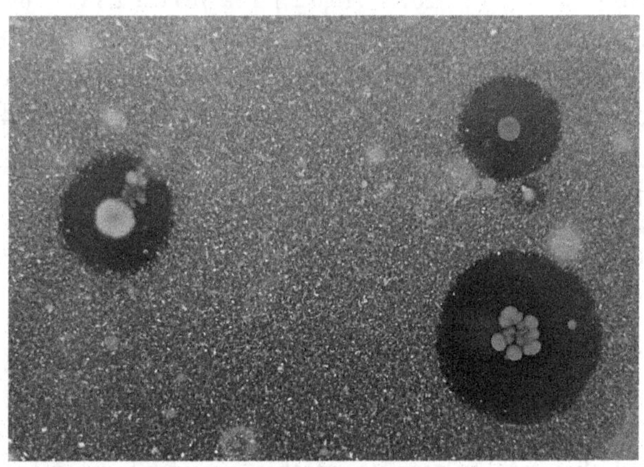

FIGURE 3.14 Manifestation of depolymerase activity of PHA degrading bacteria (formation of clear zones on diagnostic medium) (Reprinted by permission from Volova, T. G., Boyandin, A. N., Vasiliev, A. D., Karpov, V. A., Prudnikova, S. V., Mishukova, O. V., et al., Biodegradation of polyhydroxyalkanoates (PHAs) in tropical coastal waters and identification of PHA-degrading bacteria. Polym. Degrad. Stab., 95, 2350–2359. © 2010 Elsevier.)

The counts (Tables 3.4 and 3.5) and the species compositions of microorganisms (Figure 3.15 and 3.16) isolated from the soil on PHA surface were considerably different from the control soil samples. The total counts of microorganisms on the surface of polymer specimens were two orders of magnitude higher than the counts in the control soil samples in 2007 and one order of magnitude higher in 2010. In the control soil samples collected under the larch, the dominant species belonged to the genus *Micrococcus.* Other isolated species were Bacillus spore-forming rods and arthrobacteria. Gram-negative microflora was represented by *Acinetobacter, Flavobacterium,* and *Pseudomonas.* In soil samples removed from the surface of polymer specimens, proportions of microorganisms were different. The bacterial community of the soil in the larch rhizosphere was dominated by *Agrobacterium* species followed by *Cellulomonas. Alcaligenes, Aureobacterium, Acinetobacter, Pseudomonas,* and *Arthrobacter* were present in small numbers. Analysis of the control soil samples collected in the birch rhizosphere and those removed from the surface of the polymer specimens buried under the birch yielded similar results. The bacterial community of the soil in the birch rhizosphere was dominated by *Micrococcus* species. The soil from the polymer film surface was mainly inhabited by *Bacillus* species, with some representatives of *Arthrobacter, Micrococcus, Nocardia, Actinomyces, Pimelobacter,* and *Alcaligenes.* Not all isolated fungi exhibited P(3HB) depolymerase activity. The occurrence of clear zones around colonies on the diagnostic medium was recorded for *Penicillium, Paecilomyces, Acremonium, Verticillium,* and *Zygosporium* species. At the end of the growth season, control samples contained representatives of the genera *Penicillium, Paecilomyces, Aureobasidium,* and *Verticillium*, with *Penicillium* fungi remaining the dominant genus (Figure 3.16).

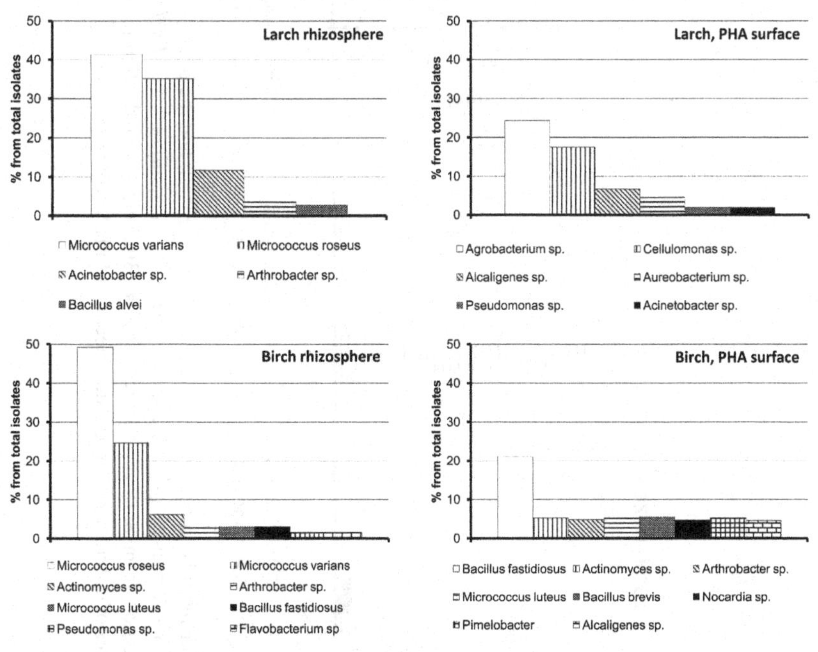

FIGURE 3.15 Species dominating bacterial populations in the soil in the larch rhizosphere (a) and in the soil removed from the surface of polymer films (b); in the soil in the birch rhizosphere (c) and in the soil removed from the surface of polymer films (d) (September 20, 2007) (Reprinted by permission from Springer Nature: Applied Biochemistry and Microbiology, Biodegradation of polyhydroxyalkanoates by soil microbial communities of different structures and detection of PHA degrading microorganisms, A. N. Boyandin, S. V. Prudnikova, M. L. Filipenko et al., © 2012.)

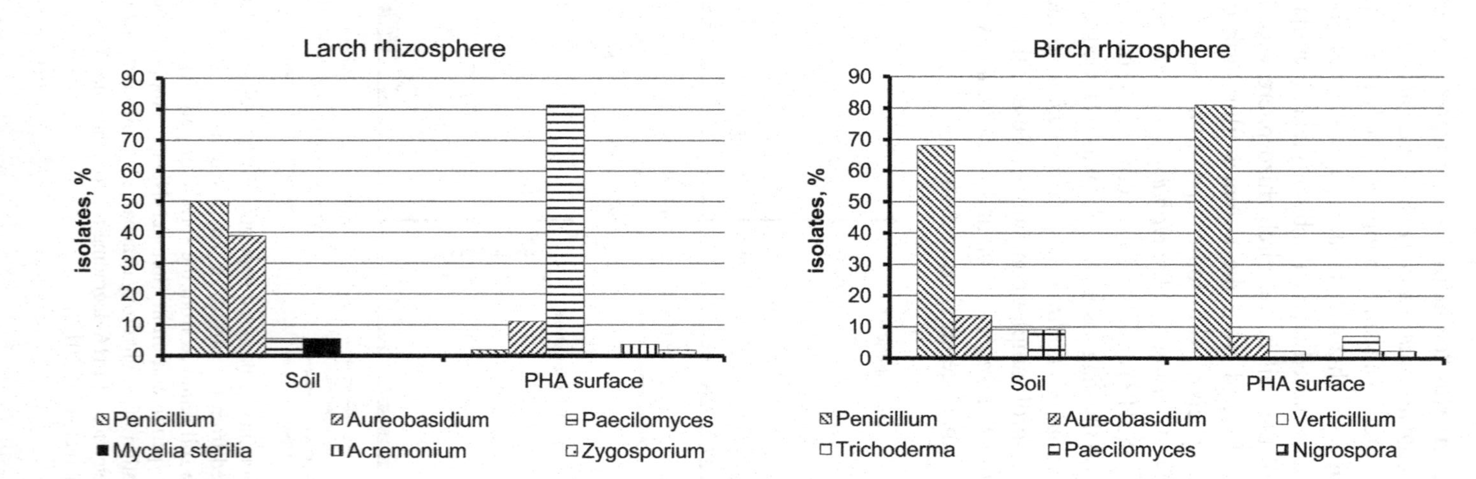

FIGURE 3.16 Major fungi isolated from the rhizospheres of the larch and the birch: 1 – *Penicillium*;* 2 – *Aureobasidium;* 3 – *Paecilomyces*;* 4 – *Mycelia sterilia;* 5 – Acremonium*; 6 – *Zygosporium*;* 7 – *Verticillium*;* 8 – *Trichoderma,* 9 – *Nigrospora* (* denotes true PHA degraders) (Reprinted by permission from Springer Nature: Applied Biochemistry and Microbiology, Biodegradation of polyhydroxyalkanoates by soil microbial communities of different structures and detection of PHA degrading microorganisms, A. N. Boyandin, S. V. Prudnikova, M. L. Filipenko et al., © 2012.)

Based on similarities of morphological types, 16 strains of bacteria capable of PHA biodegradation were selected, which were subsequently identified based on the combination of morphological, cultural, biochemical, and molecular-genetic properties. Strains IBP-SB5, IBP-SL5, IBP-SL9, and IBP-SL10 were identified as *Variovorax* species; strains IBP-SB4, IBP-SB7, IBP-SB9, and IBP-SL14 as *Stenotrophomonas* species; strains IBP-SB14, IBP-SL6, IBP-SL7, and IBP-SL13 as *Acinetobacter* species; strains IBP-SB6 and IBP-SB8 as *Pseudomonas* species; and strains IBP-SL8 and IBP-SL11 as *Bacillus* species (Figure 3.17) (Boyandin et al., 2012a).

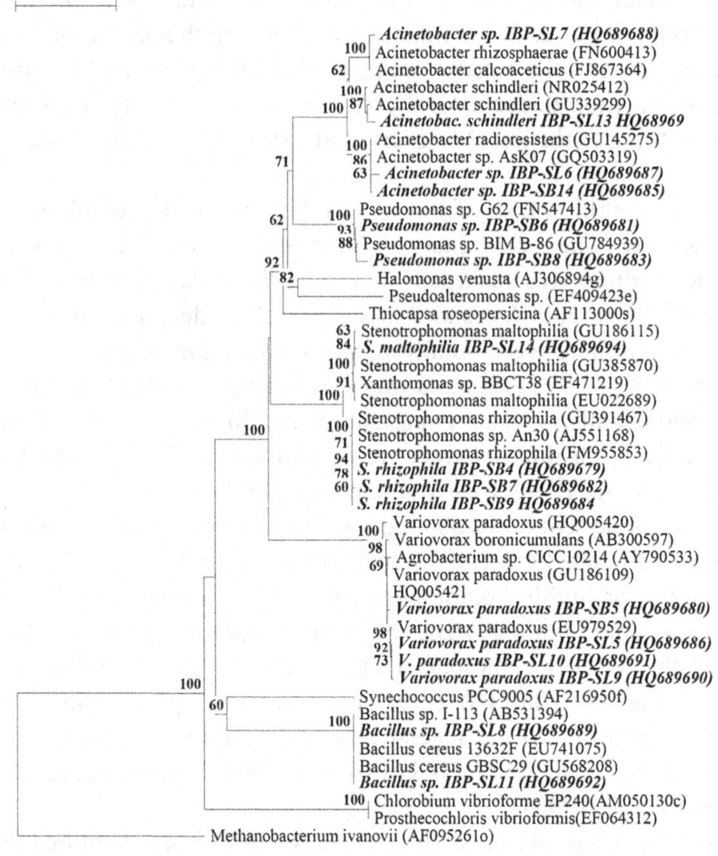

FIGURE 3.17 Phylogenetic positions of PHA degrading bacterial strains (italics) based on a comparison of nucleotide sequences of the 16S rRNA gene using the neighbor-joining method. The scale corresponds to 1 nucleotide substitution for every 10 sequences. Numbers show bootstrap indices 60% and more (Reprinted by permission from Springer Nature: Applied Biochemistry and Microbiology, Biodegradation of polyhydroxyalkanoates by soil microbial communities of different structures and detection of PHA degrading microorganisms, A. N. Boyandin, S. V. Prudnikova, M. L. Filipenko et al., © 2012.)

The integrated study of biodegradation of two PHA types by soil microbial communities of different structures performed during two field seasons, which differed in their weather conditions, showed that PHA degradation is influenced by both polymer chemical composition and soil parameters: temperature, moisture content, and composition of the microbial community. Among the microorganisms that grew on the surface of polymer specimens, there were both true PHA degraders displaying PHA-depolymerase activity and concomitant bacteria, which thrived on PHA hydrolysis products. This is consistent with the literature data on *Penicillium predominance* among micromycetes of northern soils (Egorova, 1986). Soil samples removed from the surface of polymer films contained 36.5 and 4 times higher total counts of micromycetes in the rhizosphere of the birch and that of the larch, respectively, than in the control samples. Thus, fungi are actively involved in PHA degradation under natural conditions, and there are literature data confirming this conclusion (Mergaert et al., 1993, Sang et al., 2002, Lee et al., 2005).

Of the 8 identified taxa of fungi, only 5 were capable of PHA-depolymerase activity. The occurrence of clear zones around colonies on the diagnostic medium was recorded for *Penicillium, Paecilomyces, Acremonium, Verticillium,* and *Zygosporium* species. PHA degraders in the soil of the larch rhizosphere were mainly represented by *Paecilomyces lilacinus,* amounting to 81.5%. This species was also described as a polymer degrader by Sang and coauthors (Sang et al., 2002). The fungi localized on polymer surface in the birch rhizosphere were dominated by *Penicillium sp.* BP–1 and *Penicillium sp.* BP–2, totally amounting to 81%.

The nucleotide sequences obtained were compared with the sequences in the databases of GenBank, EMBL, and DDBJ, using BLAST – the program for searching for highly homologous sequences – of NCBI web resource (http://www.ncbi.nlm.nih.gov/BLAST/). The nucleotide sequences were aligned with the most homologous sequences of cultured strains from the databases using the ClastalW (2.08) program. Phylogenetic analysis was carried out according to the Jukes-Cantor one-parameter model, using the neighbor-joining method in the TREECON (1.3b) software package (Van de Peer, De Wachter, 1993).

Twenty bacterial and eight fungal species have been identified as true PHA degraders. Under the study conditions, representatives of the bacterial genera *Variovorax, Stenotrophomonas, Acinetobacter, Pseudomonas,* and *Bacillus* have been identified as major PHA degraders. Populations of soil micromycetes that grew on polymer specimens were dominated by such

PHA degraders as *Penicillium sp.* BP–1, *Penicillium sp.* BP–2, and *Paecilomyces lilacinus,* which amounted to 80%.

A study of PHA degradation was also performed in tropical soil. Biodegradation of PHAs of two types – poly(3-hydroxybutyrate) [P(3HB)] and poly(3-hydroxybutyrate-co–3-hydroxyvalerate) [P(3HB/3HV)] – was analyzed in soils at field laboratories in the environs of Hanoi (Vietnam) and Nha Trang (Vietnam) (Boyandin et al., 2013). The methods of experiments were similar to those described for the arboretum in Krasnoyarsk. The air and soil temperatures and humidity in both study sites were similar throughout the study season. Precipitation at Hanoi was, however, almost an order of magnitude higher than in Nha Trang (Table 3.7).

Mass loss dynamics of PHA specimens is shown in Figure 3.18, which demonstrates that in Nha Trang PHAs were degraded at lower rates because of lower precipitation amounts in this area in summer. PHAs of all types were degraded at higher rates in the soil of the study site at Hanoi.

TABLE 3.7 Climate Parameters at Field Laboratories at Hanoi and in Nha Trang .

Month	Average monthly air temperature, °C		Average monthly humidity, %		Precipitation, mm	
	Hanoi	Nha Trang	Hanoi	Nha Trang	Hanoi	Nha Trang
May	29	29.5	81	80.6	150	27.4
June	31	29.4	74	78.2	175	7.2
July	31	29.0	74	78.8	280	40.2
August	29	28.4	82	81.4	274	82.0
September	29	27.9	79	82.9	172	138.8
October	26	26.9	70	84.0	25	325.7

PHA films were more prone to degradation than pressed pellets. At the end of the experiment (after 184 days of soil exposure), degradation of P(3HB) films reached more than 97%, and P(3HB/3HV) films were 33% degraded, while the pressed pellets were 42 and 23% degraded, respectively. In the more arid area (Nha Trang), the mass loss of P(3HB) and P(3HB/3HV) films was 16 and 7% and that of the pressed pellets – 18 and 3%. P(3HB) specimens were degraded faster than P(3HB/3HV) films and pellets. This is consistent with the data reported by other authors, showing that P(3HB) was degraded with higher rates than P(3HB/3HV) by most isolates of PHA degrading bacteria (Manna,

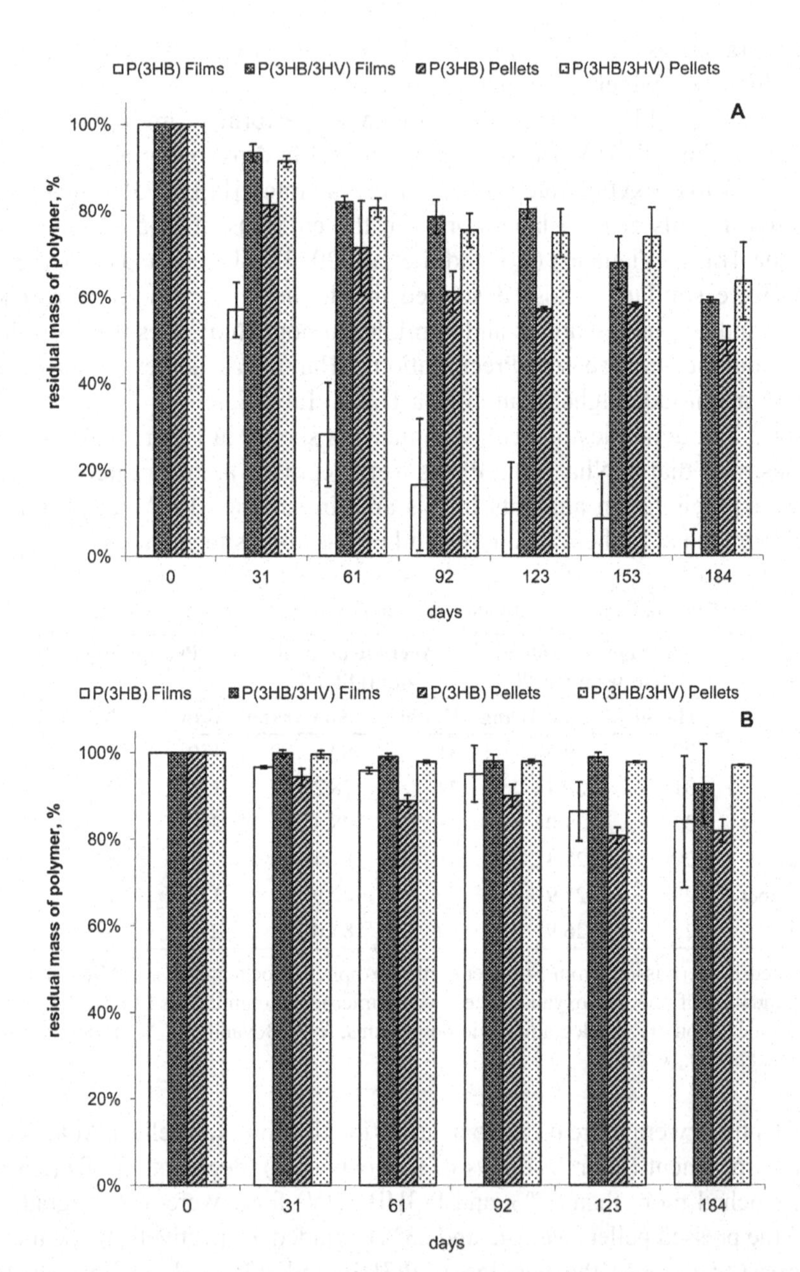

FIGURE 3.18 Mass loss of PHA specimens incubated in the tropical soil at Hanoi (A) and in Nha Trang (B) . (Reprinted by permission from Volova, T. G., Boyandin, A. N., Vasiliev, A. D., Karpov, V. A., Prudnikova, S. V., Mishukova, O. V., et al., Biodegradation of polyhydroxyalkanoates (PHAs) in tropical coastal waters and identification of PHA-degrading bacteria. Polym. Degrad. Stab., 95, 2350–2359. © 2010 Elsevier.)

Paul, 2000), actinomycetes (Manna et al., 1999), and fungi (Sanyal et al., 2006; McLellan, Halling, 1988). Rosa et al. (2003) also reported higher biodegradation rates of P(3HB) granules compared with P(3HB/3HV) and polycaprolactone, and they explained their results as being due to specific surface structure and properties of their specimens. Feng et al. (2004) showed that as a 3HV fraction of P(3HB/3HV) increased from 8 to 98 mol%, enzymatic hydrolysis of the copolymer by the depolymerase from Ralstonia pickettii occurred with higher rates. The copolymer containing more than 80% 3HV was not degraded by the depolymerase from *Acidovorax sp.*

Several authors reported faster degradation of P(3HB/3HHx) compared with P(3HB) and P(3HB/3HV), suggesting that this difference was caused by dissimilar structures of the polymers (Wang et al., 2004; Sridewi et al., 2006). Other authors reported faster degradation of copolymers compared with P(3HB) (Volova et al., 1992; 1996; Mergaert et al., 1992; 1993; 1994; Ya-Wu Wang et al., 2004). This can be accounted for by the diversity of depolymerases with different substrate specificity and by dissimilarities in polymer crystallinities (Manna, Paul, 2000).

Microbial communities of the soils in the two study areas in Vietnam were significantly different. Microbial populations of the Hanoi soil were dominated by *Acinetobacter calcoaceticus, Arthrobacter artocyaneus, Bacillus aerophilus, Bacillus megaterium, Bacillus sp., Brevibacillus agri, Brevibacillus invocatus, Chromobacterium violaceum, Cupriavidus gilardii, Mycobacterium fortuitum, Ochrobactrum anthropi, Staphylococcus arlettae, Staphylococcus haemolyticus, Staphylococcus pasteuri, Pseudomonas acephalitica,* and *Rhodococcus equi;* while the major species in the Nha Trang soil were *Bacillus cereus, Bacillus megaterium, Bacillus mycoides, Brevibacillus agri, Gordonia terrari,* and *Microbacterium paraoxydans.* The total counts of bacteria from the biofilm on the surface of polymer specimens showed that their concentration was one or two orders of magnitude higher than in the control soil. Analysis of fungi on Saburo medium showed that the counts of fungi on the surface of all PHA specimens were higher than in the control soil; the difference was more pronounced in the experiment in Nha Trang, reaching 2 or 3 orders of magnitude. Thus, fungi actively degrade polymers. Examination of control soil samples and PHA surface biofilms proved that the soils of the study areas contained PHA degrading microorganisms and that they were more numerous on polymer surfaces.

The compositions of microbial communities in the two study sites differed significantly. PHA is degrading bacteria dominated in the soils in Nha Trang, while PHA degrading fungi were major PHA degraders in the soil at Hanoi. This may be accounted for by differences in soil parameters such as pH: at Hanoi, the soil was weakly acidic (pH = 5.48), which is a favorable condition for the development of fungi, while in Nha Trang, soil pH was close to neutral (6.63). The clear-zone technique was used to identify and examine PHA degrading soil microorganisms; 62 isolates of heterotrophic bacteria, 23 isolates of actinomycetes, and 74 isolates of microscopic fungi were selected. Colonies of different morphological types were quantitatively differentiated. Eight to ten isolates for each type of the colony were cultured, and their morphological and cultural parameters were compared. Bacteria were additionally analyzed for their physiological and biochemical properties, using conventional tests (catalyze, oxidase, protease, and amylase activities, fermentation of carbohydrates: glucose, sucrose, lactose, maltose, and mannitol). Bacteria and fungi were identified by DNA extraction, amplification, and determination of nucleotide sequences of the sites encoding the 16S and 28S rRNA genes. The nucleotide sequences obtained were compared with the sequences in the GenBank, EMBL, and DDBJ databases, using the BLAST tool for the search for sequences with high homology, of the NCBI Web site (http://www.ncbi.nlm.nih.gov/BLAST/). PHA degrading microorganisms were identified based on their cultural, morphological, biochemical, and molecular-genetic parameters.

Determination of the species composition of microbial communities showed that PHA degrading bacteria were dominated by Gram-negative rods of *Burkholderia sp.;* they were isolated from the samples of both study areas. Actinobacteria of the genus *Streptomyces* were also present in the samples of both areas. Other PHA degraders isolated from the samples at Nha Trang were *Bacillus, Cupriavidus, and Mycobacterium spp.; Nocardiopsis* actinobacteria were isolated from the soil at Hanoi. *Gongronella butleri* and *Penicillium sp.* were the fungi found in both study areas. *Acremonium recifei, Paecilomyces lilacinus,* and *Trichoderma pseudokoningii* were only isolated from the soil at Hanoi. The major PHA degrading soil microorganisms were additionally identified using sequencing of the 16S rRNA gene. Having compared nucleotide sequences of the 16S rRNA gene segment of the isolated strains – true PHA degraders – with the sequences in the GenBank, we revealed high homology with the sequences of some previously identified strains of prokaryotes (Figure 3.19) and fungi (Figure 3.20).

Thus, polymer biodegradation in soils – both Siberian and tropical – is performed by bacteria and fungi. Bacteria inhabiting these soils belonged to different genera (except Bacillus), while most of the fungi in both Siberian and Vietnamese soils were represented by *Penicillium, Paecilomyces,* and *Acremonium.*

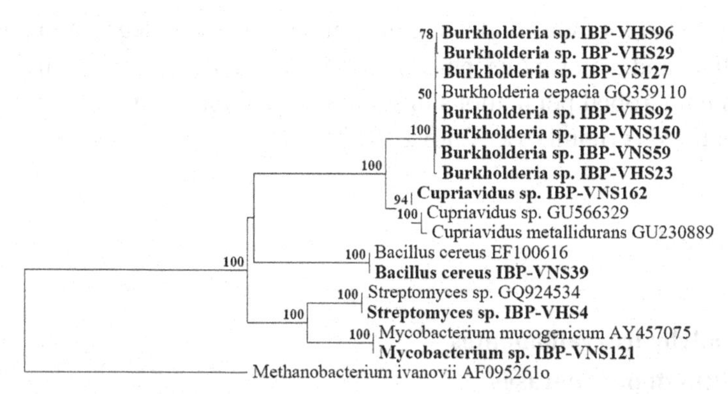

FIGURE 3.19 Phylogenetic positions of PHA degrading bacterial strains (bold type) based on a comparison of nucleotide sequences of the 16S rRNA gene using the neighbor-joining method. Numbers show bootstrap indices equal to or higher than 50%. (Reprinted by permission from Volova, T. G., Boyandin, A. N., Vasiliev, A. D., Karpov, V. A., Prudnikova, S. V., Mishukova, O. V., et al., Biodegradation of polyhydroxyalkanoates (PHAs) in tropical coastal waters and identification of PHA-degrading bacteria. Polym. Degrad. Stab., 95, 2350–2359. © 2010 Elsevier.)

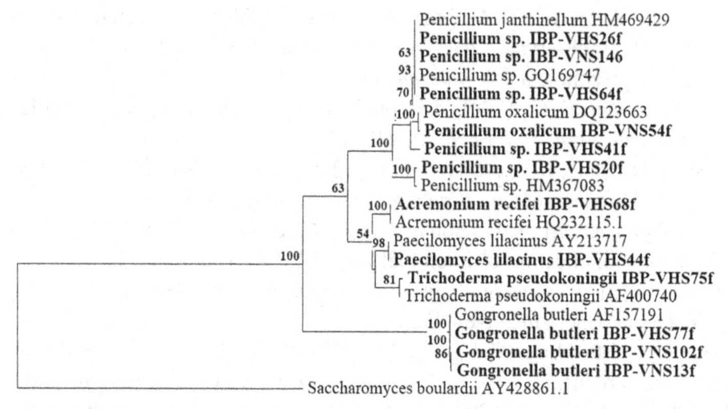

FIGURE 3.20 Phylogenetic positions of PHA degrading fungi (bold type) based on a comparison of nucleotide sequences of the 28S rRNA gene using the neighbor-joining method. Numbers show bootstrap indices equal to or higher than 60%. (Reprinted by permission from Volova, T. G., Boyandin, A. N., Vasiliev, A. D., Karpov, V. A., Prudnikova, S. V., Mishukova, O. V., et al., Biodegradation of polyhydroxyalkanoates (PHAs) in tropical coastal waters and identification of PHA-degrading bacteria. Polym. Degrad. Stab., 95, 2350–2359. © 2010 Elsevier.)

Studies of PHA degradation in different natural environments showed that PHA biodegradation is influenced by the chemical structure of the polymer, its geometry and the technique used to process it; climate and weather, the type of the natural ecosystem and its microbial component in particular, as the factor determining the mechanism of PHA biodegradation: preferential attack of the amorphous regions of the polymer or equal degradation of both crystalline and amorphous phases. PHA degrading microorganisms that dominate microbial populations in some soil ecosystems have been isolated and identified (Prudnikova, Volova 2012).

KEYWORDS

- **polyhydroxyalkanoates**
- **PHA depolymerases**
- **PHA biodegradation**

REFERENCES

Abe, H., Doi, Y., Aoki, H., & Akehata, T., (1998). Solid-state structures and enzymatic degradabilities for melt-crystallized films of copolymers of (R)-3-hydroxybutyric acid with different hydroxyalkanoic acids. *Macromolecules*, *31*, 1791–1797.

Bateman, G. L., & Murray, G., (2001). Seasonal variations in populations of Fusarium species in wheat-field soil. *Appl. Soil. Ecol.*, *18*, 117–128.

Bhatt, R., Shah, D., Patel, K. C., & Trivedi, U., (2008). PHA-rubber blends: Synthesis, characterization and biodegradation. *Bioresource Technology*, *99*, 4615–4620.

Bonartsev, A. P., Iordanskii, A. L., Bonartseva, G. A., & Zaikov, G. E., (2009). Biodegradation and medical application of microbial poly(3-hydroxybutyrate). In: Zaikov, G. E., & Krylova, L. P., (eds.), *Biotechnology, Biodegradation Water and Foodstuffs* (pp. 1–35). Nova Science Publishers Inc.: New York.

Bonartseva, G. A., Myshkina, V. L., Nikolaeva, D. A., Kevbrina, M. V., Kallistova, A. Y., Gerasin, V. A., Iordanskii, A. L., & Nozhevnikova, A. N., (2003). Aerobic and anaerobic microbial degradation of poly-β-hydroxybutyrate produced by Azotobacter chroococcum. *Appl. Biochem. Biotechnol.*, *109*, 285–301.

Boyandin, A. N., Prudnikova, S. V., Filipenko, M. L., Khrapov, E. A., Vasilev, A. D., Volova, T. G., (2012a). Biodegradation of polyhydroxyalkanoates by soil microbial communities of different structures and detection of PHA degrading microorganisms. *Appl. Biochem. Microbiol.*, *48*, 28–36.

Boyandin, A. N., Prudnikova, S. V., Karpov, V. A., Ivonin, V. N., Đỗ, N. L., Nguyễn, T. H., et al., (2013). Microbial degradation of polyhydroxyalkanoates in tropical soils. *Int. Biodeterior. Biodegrad.*, *83*, 77–84.

Boyandin, A. N., Rudnev, V. P., Ivonin, V. N., Prudnikova, S. V., Korobikhina, K. I., Filipenko, M. L., et al., (2012b). Biodegradation of polyhydroxyalkanoate films in natural environments. *Macromol. Symposia.*, *320*, 38–42.

Brandl, H., & Püchner, P., (1991). Biodegradation of plastic bottles made from 'Biopol' in an aquatic ecosystem under in situ conditions. *Biodegradation*, *2*, 237–243.

Briese, B. H., Jendrossek, D., & Schlegel, H. G., (1994). Degradation of poly(3-hydroxybutyrate-co–3-hydroxyvalerate) by aerobic sewage sludge. *FEMS Microbiol. Lett.*, *117*, 107–111.

Brucato, C. L., & Wong, S. S., (1991). Extracellular poly(3-hydroxybutyrate) depolymerase from *Penicillium funiculosum*: General characteristics and active site studies. *Arch. Biochem. Biophys.*, *290*, 497–502.

Choi, M. K., Kim, K. D., Ahn, K. M., Shin, D. H., Hwang, J. H., Seong, C. N., & Ka, J. O., (2009). Genetic and phenotypic diversity of parathion-degrading bacteria isolated from rice paddy soils. *J. Microbiol. Biotechnol.*, *19*, 1679–1687.

Chowdhury, A. A., (1963). Poly-β-hydroxybutyric acid-degrading bacteria and exoenzyme (Poly-ß-hydroxybuttersäure abbauende bakterien und exoenzym). *Arch. Microbiol.*, 47, 167–200 (in German).

Colak, A., & Güner, S., (2004). Polyhydroxyalkanoate degrading hydrolase-like activities by *Pseudomonas sp.* isolated from soil. *Int. Biodeterior. Biodegrad.*, *53*, 103–109.

Delafield, F. P., Doudoroff, M., Palleroni, N. J., Lusty, C. J., & Contopoulos, R., (1965). Decomposition of poly-β-hydroxybutyrate by pseudomonads. *J. Bacteriol.*, *90*, 1455–1466.

Doi, Y., Kanesawa, Y., Tanahashi, N., & Kumagai, Y., (1992). Biodegradation of microbial polyesters in the marine environment. *Polym. Degrad. Stab.*, *36*, 173–177.

Egorova, L. N., (1986). *Soil Fungi of the Far East. Hyphomycetes* (192 p.). Nauka: Leningrad (in Russian).

Erkske, D., Viskere, I., Dzene, A., Tupureina, V., & Savenkova, L., (2006). Biobased polymer composites for films and coatings. *Proceedings of the Estonian Academy of Sciences, Chemistry*, *55*, 70–77.

Feng, L., Wang, Y., Inagawa, Y., Kasuya, K., Saito, T., Doi, Y., & Inoue, Y., (2004). Enzymatic degradation behavior of comonomer compositionally fractionated bacterial poly(3-hydroxybutyrate-co–3-hydroxyvalerate)s by poly(3-hydroxyalkanoate) depolymerases isolated from *Ralstonia pickettii* T1 and Acidovorax sp. TP4. *Polym. Degrad. Stab.*, *84*, 95–104.

Futamata, H., Nagano, Y., Watanabe, K., & Hiraishi, A., (2005). Unique kinetic properties of phenol-degrading Variovorax strains responsible for efficient trichloroethylene degradation in a chemostat enrichment culture. *Appl. Environ. Microbiol.*, *71*, 904–911.

Garrity, G.M., Brenner, D.J., Krieg, N.R., & Staley, J. T., (2006). *Bergey's Manual® of Systematic Bacteriology. Vol. 2: The Proteobacteria* (2791 p.). Springer Science & Business Media. ISBN 0-387-95040-0 eBook ISBN 978-0-387-29299-1.

Hocking, P. J., Marchessault, R. H., Timmins, M. R., Lenz, R. W., & Fuller, R. C., (1996). Enzymatic degradation of single crystals of bacterial and synthetic poly(β-hydroxybutyrate). *Macromolecules*, *29*, 2472–2478.

Holt, J., Krieg, N., & Snit, P., (1997). *Bergey's Manual® of Systematic Bacteriology.* In 2 Volumes (1232 p.). Mir: Moscow (in Russian). ISBN: 5-03-003112-X.

Imam, S. H., Gordon, S. H., Shogren, R. L., Tosteson, T. R., Govind, N. S., & Greene, R. V., (1999). Degradation of starch-poly (β-hydroxybutyrate-co-β-hydroxyvalerate) bioplastic in tropical coastal waters. *Appl. Environ. Microbiol.*, *65*, 431–437.

Jendrossek, D., & Handrick, R., (2002). Microbial degradation of polyhydroxyalkanoates. *Ann. Rev. Microbiol.*, *56*, 403–432.

Jendrossek, D., Schirmer, A., & Schlegel, H., (1996). Biodegradation of polyhydroxyalkanoic acids. *Appl. Microbiol. Biot.*, *46*, 451–463.

Kijchavengkul, T., & Auras, R., (2008). Compostability of polymers. *Polym. Int.*, *57*, 793–804.

Kim, D. Y., & Rhee, Y. H., (2003). Biodergadation of microbial and synthetic polyesters by fungi. *Appl. Microbiol. Biot.*, *61*, 300–308.
Kim, D. Y., Kim, H. W., Chung, M. G., & Rhee, Y. H., (2007). Biosynthesis, modification, and biodegradation of bacterial medium-chain-length polyhydroxyalkanoates. *Journal of Microbiology*, *45*, 87–97.
Kim, H. J., Kim, D. Y., Nam, J. S., Bae, K. S., & Rhee, Y. H., (2003). Characterization of an extracellular medium-chain-length poly(3-hydroxyalkanoate) depolymerase from Streptomyces sp. KJ–72. *Antonie Van Leeuwenhoek.*, *83*, 183–189.
Knoll, M., Hamm, T. M., Wagner, F., Martinez, V., & Pleiss, J., (2009). The PHA depolymerase engineering database: A systematic analysis tool for the diverse family of polyhydroxyalkanoate (PHA) depolymerases. *BMC Bioinformatics*, *10*, 89–97.
Kozlovsky, A. G., Zhelifonova, V. P., Vinokourova, N. G., Antipova, T. V., & Ivanushkina, N. E., (1999). The study of biodegradation of poly-β-hydroxybutyrate by microscopic fungi. *Mikrobiologiya (Microbiology)*, *68*, 340–346 (In Russian).
Kumagai, Y., Kanesawa, Y., & Doi, Y., (1992). Enzymatic degradation of microbial poly(3-hydroxybutyrate) films. *Macromol. Chem. Phys.*, *193*, 53–57.
Kunioka, M., Kawaguchi, Y., & Doi, Y., (1989). Production of biodegradable copolyesters of 3-hydroxybutyrate and 4-hydroxybutyrate by *Alcaligenes eutrophus*. *Appl. Microbiol. Biot.*, *30*, 569–573.
Kusaka, S., Iwata, T., & Do, Y., (1999). Properties and biodegradability of ultra-high-molecular-weight poly [(R)–3-hydroxybutyrate] produced by a recombinant *Escherichia coli*. *Int. J. Biol. Macromol.*, *25*, 87–94.
Lee, K. M., Gimore, D. F., & Huss, M. J., (2005). Fungal degradation of the bioplastic PHB (Poly-3-hydroxy-butyric acid). *J. Polym. Environ.*, *13*, 213–219.
Lim, S. P., Gan, S. N., & Tan, I., (2005). Degradation of medium-chain-length polyhydroxyalkanoates in tropical forest and mangrove soils. *Appl. Biochem. Biotechnol.*, *126*, 23–32.
Lopez-Llorca, L. V., Colom, V. M. F., & Gascon, A., (1993). A study of biodegradation of poly-β-hydroxyalkanoate (PHA) films in soil using scanning electron microscopy. *Micron.*, *24*, 23–29.
Madden, L. A., Anderson, A. J., & Asrar, J., (1998). Synthesis and characterization of poly (3-hydroxybutyrate) and poly (3-hydroxybutyrate-co–3-hydroxyvalerate) polymer mixtures produced in high-density fed-batch cultures of *Ralstonia eutropha* (*Alcaligenes eutrophus*). *Macromolecules*, *31*, 5660–5667.
Manna, A., & Paul, A. K., (2000). Degradation of microbial polyester poly(3-hydroxybutyrate) in environmental samples and in culture. *Biodegradation*, *11*, 323–329.
Manna, A., Giri, P., & Paul, A. K., (1999). Degradation of poly(3-hydroxybutyrate) by soil streptomycetes. *World J. Microbiol. Biotechnol.*, *15*, 705–709.
Matavulj, M., & Molitoris, H. P., (1992). Fungal degradation of polyhydroxyalkanoates and a semiquantitative assay for screening their degradation by terrestrial fungi. *FEMS Microbiol. Lett.*, *103*, 323–331.
McLellan, D. W., & Halling, P. J., (1988). Preparation and chromatographic analysis of poly(3-hydroxybutyrate) hydrolysis products. *J. Chromatogr.*, *445*, 251–257.
Mergaert, J., & Swings, J., (1996). Biodiversity of microorganisms that degrade bacterial and synthetic polyesters. *J. Ind. Microbiol.*, *17*, 463–469.
Mergaert, J., Anderson, C., Wouters, A., & Swings, J., (1994). Microbial degradation of poly (3-hydroxybutyrate) and poly (3-hydroxybutyrate-co–3-hydroxyvalerate) in compost. *J. Environ. Polym. Degrad.*, *2*, 177–183.

Mergaert, J., Glorieux, G., Hauben, L., Storms, V., Mau, M., & Swings, J., (1996). Biodegradation of poly(3-hydroxyalkanoates) in anaerobic sludge and characterization of a poly(3-hydroxyalkanoates) degrading anaerobic bacterium. *Syst. Appl. Microbiol.*, *19*, 407–413.

Mergaert, J., Webb, A., Anderson, C., Wouters, A., & Swings, J., (1993). Microbial degradation of poly (3-hydroxybutyrate) and poly (3-hydroxybutyrate-co–3-hydroxyvalerate) in soils. *Appl. Environ. Microbiol.*, *59*, 3233–3238.

Mergaert, J., Wouters, A., Swings, J., & Anderson, C., (1995). *In situ* biodegradation of poly (3-hydroxybutyrate) and poly (3-hydroxybutyrate-co–3-hydroxyvalerate) in natural waters. *Canadian J. Microbiol.*, *41*, 154–159.

Mergaert, J., Wouters, A., Swings, J., & Kersters, K., (1992). Microbial flora involved in the biodegradation of polyhydroxylakanoates. In: Vert, M., Feijen, J., Albertsson, A., Scott, G., & Chiellini, E., (eds.), *Biodegradable Polymers and Plastics* (pp. 267–270), Royal Society of Chemistry: London.

Mokeeva, V. L., Chekunova, L. N., Myshkina, V. L., Nikolaeva, D. A., Gerasin, V. A., & Bonartseva, G. A., (2002). Poly-β-hydroxybutyrate biodestruction by micromycetes: Resistance and fungicidity tests. *Mikologiya i Fitopatologiya (Mycology and Phytopathology)*, *36*, 59–63 (in Russian).

Molitoris, H. P., Moss, S. T., De Koning, G. J. M., & Jendrossek, D., (1996). Scanning electron microscopy of polyhydroxyalkanoate degradation by bacteria. *Appl. Microbiol. Biotechnol.*, *46*, 570–579.

Morse, M. C., Liao, Q., Criddle, C. S., & Frank, C. W., (2011). Anaerobic biodegradation of the microbial copolymer poly (3-hydroxybutyrate-co–3-hydroxyhexanoate): Effects of comonomer content, processing history, and semi-crystalline morphology. *Polymer.*, *52*, 547–556.

Mukai, K., & Doi, Y., (1993). Microbial degradation of polyhydroxyalkanoates. *RIKEN Review*, *3*, 21–22.

Netrusov, A. I., (2005). *Practical Course in Microbiology.* Akademiya: Moscow, (in Russian).

Nishida, H., & Tokiwa, Y., (1993). Effects of higher-order structure of poly(3-hydroxybutyrate) on its biodegradation. II. Effects of crystal structure on microbial degradation. *J. Environ. Polym. Degrad.*, *1*, 65–80.

Nobes, G. A. R., Marchessault, R. H., Briese, B. H., & Jendrossek, D., (1998). Microscopic visualization of the enzymatic degradation of poly(3HB-co–3HV) and poly(3HV) single crystals by PHA depolymerases from *Pseudomonas lemoignei*. *J. Environ. Polym. Degrad.*, *6*, 99–107.

Philip, S., Keshavarz, T., & Roy, I., (2007). Polyhydroxyalkanoates: Biodegradable polymers with a range of applications. *J. Chem. Technol. Biotechnol.*, *82*, 233–247.

Prudnikova, S. V., & Volova, T. G., (2012). *Ecological Role of Polyhydroxyalkanoates – An Analog of Synthetic Plastics: Biodegradation Behavior in Natural Environments and Interaction with Microorganisms (Ekologicheskaya rol Poligidroksialkanoatov – Analoga Sinteticheskikh Plastmass: Zakonomernosti Biorazrusheniya v Prirodnoi Srede i Vzaimodeistviya s Mikroorganizmami)* (p. 183). Krasnoyarskii Pisatel: Krasnoyarsk, (in Russian).

Quinteros, R., Goodwin, S., Lenz, R. W., & Park, W. H., (1999). Extracellular degradation of medium chain length poly (β-hydroxyalkanoates) by *Comamonas* sp. *Int. J. Biol. Macromol.*, *25*, 135–143.

Reddy, C. S. K., Ghai, R., Rashmi, R., & Kalia, V. C., (2003). Polyhydroxyalkanoates: An overview. *Bioresour. Technol.*, *87*, 137–146.

Reddy, S. V., Thirumala, M., & Mahmood, S. K., (2008). Biodegradation of polyhydroxyalkanoates. *Internet J. Microbiol.*, *4*(2).

Rodriguez-Contreras, A., Calafell-Monfort, M., & Marqués-Calvo, M. S., (2012). Enzymatic degradation of poly (3-hydroxybutyrate-co-4-hydroxybutyrate) by commercial lipases. *Polym. Degrad. Stab.*, *97*, 597–604.

Rosa, D. S., Filho, R. P., Chui, Q. S. H., Calil, M. R., & Guedes, C. G. F., (2003). The biodegradation of poly-β-(hydroxybutyrate), poly-(β-hydroxybutyrate-co-β-valerate) and poly(ε-caprolactone) in compost derived from municipal solid waste. *Eur. Polym. J.*, *39*, 233–237.

Rutkowska, M., Krasowska, K., & Heimowska, A., (2008). Environmental degradation of blends of atactic poly[(R,S)-3-hydroxybutyrate] with natural PHBV in Baltic sea water and compost with activated sludge. *J. Polym. the Environ.*, *16*, 183–191.

Salim, Y. S., Sharon, A., Vigneswari, S., Ibrahim, M. M., & Amirul, A. A., (2012). Environmental degradation of microbial polyhydroxyalkanoates and oil palm-based composites. *Appl. Biochem. Biotechnol.*, *167*, 314–326.

Sang, B. I., Hori, K., Tanji, Y., & Unno, H., (2002). Fungal contribution to in situ biodegradation of poly(3-hydroxybutyrate-co–3-hydroxyvalerate) film in soil. *Appl. Microbiol. Biotechnol.*, *58*, 241–247.

Sanyal, P., Samaddar, P., & Paul, A., (2006). Degradation of poly(3-hydroxybutyrate) and poly(3-hydroxybutyrate-co–3-hydroxyvalerate) by some soil *Aspergillus spp. J. Polym. the Environ.*, *14*, 257–263.

Schöber, U., Thiel, C., & Jendrossek, D., (2000). Poly(3-Hydroxyvalerate) depolymerase of *pseudomonas lemoignei. Applied and Environmental Microbiology, 66*, 1385–1392.

Shah, A. A., Hasan, F., Hameed, A., & Ahmed, S., (2007). Isolation and characterization of poly (3-hydroxybutyrate-co–3-hydroxyvalerate) degrading bacteria and purification of PHBV depolymerase from newly isolated *Bacillus sp. AF3. Int. Biodeter. Biodegrad.*, *60*, 109–115.

Sridewi, N., Bhubalan, K., & Sudesh, K., (2006). Degradation of commercially important polyhydroxyalkanoates in tropical mangrove ecosystem. *Polym. Degrad. Stab.*, *91*, 2931–2940.

Sutton, D., Fotergill, A., & Rinaldi, M., (2001). *Determinant of Pathogenic and Conditionally Pathogenic Fungi* (p 486). Mir: Moscow, (In Russian).

Suyama, T., Tokiwa, Y., Ouichanpagdee, P., Kanagawa, T., & Kamagata, Y., (1998). Phylogenetic affiliation of soil bacteria that degrade aliphatic polyesters available commercially as biodegradable plastics. *Appl. Environ. Microbiol.*, *64*, 5008–5011.

Tamura, K., Stecher, G., Peterson, D., Filipski, A., & Kumar, S., (2013). MEGA6: Molecular evolutionary genetics analysis version 6.0. *Mol. Biol. Evol.*, *30*, 2725–2729.

Tani, A., Akita, M., Murase, H., & Kimbara, K., (2011). Cultural bacteria in hydroponic cultures of moss *Racomitrium japonicum* and their potential as biofertilizers for moss production. *J. Biosci. Bioeng.*, *112*, 32–39.

Tsuji, H., & Suzuyoshi, K., (2002). Environmental degradation of biodegradable polyesters 1. Poly(e-caprolactone), poly[(R)–3-hydroxybutyrate], and poly(L-lactide) films in controlled static seawater. *Polym. Degrad. Stab.*, *75*, 347–355.

Urmeneta, J., Mas-Castella, J., & Guerrero, R., (1995). Biodegradation of poly-(beta)-hydroxyalkanoates in a lake sediment sample increases bacterial sulfate reduction. *Appl. Environ. Microbiol.*, *61*, 2046–2048.

Van de Peer, Y., & De Wachter, R., (1993). TREECON: A software package for the construction and drawing of evolutionary trees. *Comput. Appl. Biosci.*, *9*, 177–182.

Vinogradova, O. N., Prudnikova, S. V., Zobova, N. V., & Kolesnikova, V. L., (2015). Microbial degradation of poly-3-hydroxybutyrate in samples of agrogenically changed soils. *SibFU Journal Biology*, *8*, 199–209.

Volova, T. G., Belyaeva, O. G., Plotnikov, V. F., & Puzyr, A. P., (1996). Study of biodegradation of microbial polyhydroxylalkanoates. *Doklady Biological Sciences*, *350*, 504–508.

Volova, T. G., Belyaeva, O. G., Plotnikov, V. F., & Puzyr, A. P., (1998). Studies of biodegradation of microbial polyhydroxyalkanoates. *Appl. Biochem. Microbiol.*, *34*, 488–492.

Volova, T. G., Boyandin, A. N., Vasiliev, A. D., Karpov, V. A., Prudnikova, S. V., Mishukova, O. V., et al., (2010). Biodegradation of polyhydroxyalkanoates (PHAs) in tropical coastal waters and identification of PHA-degrading bacteria. *Polym. Degrad. Stab.*, *95*, 2350–2359.

Volova, T. G., Gladyshev, M. I., Trusova, M. Y., & Zhila, N. O., (2006). Degradation of polyhydroxyalkanoates and the composition of microbial destructors under natural conditions. *Microbiology*, *75*, 593–598.

Volova, T. G., Gladyshev, M. I., Trusova, M. Y., & Zhila, N. O., (2007). Degradation of polyhydroxyalkanoates in eutrophic reservoir. *Polym. Degrad. Stab.*, *92*, 580–586.

Volova, T. G., Lukovenko, S. G., & Vasiliev, A. D., (1992). Production of microbial polyhydroxyalkanoates and investigation of their physicochemical properties. *Biotekhnologiya (Biotechnology)*, *1*, 19–22 (in Russian).

Volova, T. G., Prudnikova, S. V., Vinogradova, O. N., Syrvacheva, D. A., & Shishatskaya, E. I., (2017). Microbial degradation of polyhydroxyalkanoates with different chemical compositions and their biodegradability. *Microb. Ecol.*, *73*, 353–367.

Vos, P., Garrity, G., Jones, D., Krieg, N.R., Ludwig, W., Rainey, F.A., Schleifer, K.-H., Whitman, W. (2009). *Bergey's Manual of Systematic Bacteriology, Vol. 3: The Firmicutes* (1422 p.). Springer Science & Business Media. ISBN 0-387-95041-9; eBook ISBN 978-0-387-68489-5; DOI: 10.1007/b92997.

Wang, S., Song, C., Mizuno, W., Sano, M., Maki, M., Yang, C., Zhang, B., & Takeuchi, S., (2005). Estimation on biodegradability of poly(3-hydroxybutyrate-co–3-hydroxyvalerate) (PHB/V) and numbers of aerobic PHB/V degrading microorganisms in different natural environments. *J. Polym. Environ.*, *13*, 39–45.

Wang, Y. W., Mo, W., Yao, H., Wu, Q., Chen, J., & Chen, G. Q., (2004). Biodegradation studies of poly (3-hydroxybutyrate-co–3-hydroxyhexanoate). *Polym. Degrad. Stab.*, *85*, 815–821.

Watanabe, T., (2002). *Pictorial Atlas of Soil and Seed Fungi: Morphologies of Cultured Fungi and Key to Species.* CRC Press LLC: Florida.

Weng, Y. X., Wang, L., Zhang, M., Wang, X. L., & Wang, Y. Z., (2013). Biodegradation behavior of P(3HB/4HB)/PLA blends in real soil environments. *Polym. Testing.*, *32*, 60–70.

Weng, Y. X., Wang, X. L., & Wang, Y. Z., (2011). Biodegradation behavior of PHAs with different chemical structures under controlled composting conditions. *Polym. Testing.*, *30*, 372–380.

Woolnough, C. A., Charlton, T., Yee, L. H., Sarris, M., & Foster, J. R., (2008). Surface changes in polyhydroxyalkanoate films during biodegradation and biofouling. *Polym. Int.*, *57*, 1042–1051.

Yew, S. P., Tang, H. Y., & Sudesh, K., (2006). Photocatalytic activity and biodegradation of polyhydroxybutyrate films containing titanium dioxide. *Polym. Degrad. Stab.*, *91*, 1800–1807.

Zvyagintsev, D. G., Babeva, I. P., & Zenova, G. M., (2005). *Soil Biology.* The Moscow State University Press, Moscow, (In Russian).

CHAPTER 4

EXPERIMENTAL FORMULATIONS OF HERBICIDES AND EVALUATION OF THEIR EFFICACY

Herbicides (from Latin: *herba* "herb" + *caedo* "killing; killer") are chemicals that destroy unwanted plants. Herbicides represent the largest group of agrochemicals, whose scale of the application constitutes 40–50% of all pesticides and the diversity of the types produced is about 40%. The reason for this is that weeds, which are represented by thousands of species, because of the greatest damage to agriculture. Most of them are annuals, which germinate from the seeds and live only one season. Perennial weeds are able to reproduce every season. Weeds compete with crops for light, water, and nutrients. Therefore, farmers have to control them.

Herbicide-based weed control is one of the major components of modern, efficient agriculture. Herbicide application considerably enhances crop yields and, thus, contributes to food production for the growing human population. Moreover, as herbicides remove unwanted plants, they increase the efficiency of harvesters and dramatically decrease labor inputs. Thus, herbicides used in combination with agrotechnical measures remove weeds from the fields and contribute to the achievement of high crop yields (Shcheglov, 1961; Fisyunov, 1984; Shabaev, 2000).

Herbicides can be non-selective or selective. Non-selective herbicides kill all plants while selective ones affect certain species, without causing any harm to other plants. By the mode of action and application method, herbicides can be classified as contact herbicides, systemic herbicides, and ones only affecting roots or germinating seeds. Contact herbicides are applied directly to the aboveground plant parts, impairing their functions and eventually killing them. Systemic herbicides are able to move in the plant vascular system. They are absorbed by foliage or roots and translocated to other parts of the plant, killing it. Systemic herbicides are particularly effective against perennial weeds, which have an extensive root system. Herbicides applied to soil to kill the seeds and roots of weeds belong to the third group.

The choice of the method of application of herbicides depends on their mode of action. Herbicides can be used in different periods of field seasons: before sowing of crops, before the emergence of weeds or crops, after the emergence of crops, and in different periods of plant growth.

According to their origin and mode of action, herbicides are classified as:

- inhibitors of photosynthesis, which penetrate into chloroplasts and disrupt processes of electron capture by ferredoxin and reduction of the coenzyme NADP in Photosystem I and which prevent electrons from being transferred to plastoquinone in Photosystem II;
- inhibitors of cell division – N-aryl carbamates and dinitroanilines, which are used to inhibit seed germination and root growth;
- herbicides affecting plant respiration by interrupting oxidative phosphorylation chain and inhibiting ATP formation – dinitrophenols and halogen phenols; and
- herbicides regulating plant growth, or "synthetic auxins," which accelerate plant growth, thus causing their malnutrition and death.
- To make herbicide treatment sufficiently effective, the following rules should be observed:
- specific treatment should be performed for each weed species and crop species; time and rate of application should be correctly determined;
- the choice of the herbicide should be based on the active ingredient it contains, as there are many preparations with the same composition but different names;
- for herbicide treatment to be effective, the soil must contain sufficient moisture while the crop should not be wet because of rain or dew;
- herbicide treatment is more effective when weed plants are fresh and are rapidly growing;
- treatment should not be performed at night temperatures below 6°C and day temperatures above 25°C;
- herbicides should not be applied to protect damaged crops; and
- the same herbicide should not be applied for a long time; the alternation of herbicides is more effective.

Farming practice shows that successful weed control can be only achieved by combining agrotechnical measures and rational herbicide application. An inevitable side effect of using herbicides, as well as other pesticides, is an accumulation of high concentrations of chemical compounds in soils. This not only poses a hazard to human health but also causes the development of species resistant to herbicides, threatens the stability of agroecosystems, and endangers the long-term soil fertility. Moreover, herbicides remove the herb layer, leading to increased soil erosion. Another unwanted consequence of herbicide application is the emergence of other weeds, more resistant to herbicides. Also, a more sensitive crop planted after the herbicide-treated crop in the succession of crops may be inhibited or even killed by the herbicides accumulated in the soil. Soil microorganisms metabolize between 10 and 70% of pesticides, but some of the decomposition products, which are more toxic than the initial formulation, accumulate in the environment.

Since herbicides play an important part in agriculture and are used in great quantities, construction, and use of new-generation environmentally safe slow-release herbicide formulations is a pressing problem for agrochemists, biotechnologists, and crop farmers.

4.1 NEW-GENERATION HERBICIDE FORMULATIONS IN WEED CONTROL

Modern herbicides include sulfonylurea formulations for different regions and with different persistence and phenoxyphenoxy propionic and phenoxybenzoic acid derivatives, which are effective against a wide range of weeds, including monocotyledons. Important broad-spectrum herbicides are glyphosate and glyphosinate, which decompose in the soil to CO_2, H_2O, and phosphoric acid. Application of aryl-hydroxyphenoxy propionic acids manufactured as individual optic isomers reduces the rate of herbicide application considerably. A new pyridine insecticide – imidacloprid – is used to control pests resistant to other types of insecticides.

Some relatively recent studies describe the encapsulation of herbicides in polymer materials. However, there are few published data on this subject. For instance, formulations of herbicides alachlor (Fernandez-Urrusuno et al., 2000) and norflurazon (Sopeña et al., 2005) were prepared by encapsulating them in ethyl cellulose. Sopeña et al. (2008) described the technique of producing polymeric microspheres of ethyl cellulose with encapsulated alachlor. The authors showed that the rate of release of the active ingredient could be regulated by varying the initial alachlor content, thus controlling weed growth.

Microencapsulated pesticides can be used as aqueous dispersions to spray plant leaves (RF Patent No. 2407288). Such formulations are, however, harmful to the environment, and special safety measures should be taken to decrease their toxicity.

The crucial part of constructing environmentally friendly preparations is the availability of appropriate materials with the following properties: biodegradability; safety for living and non-living nature; long-term (weeks and months) presence in the natural environment and controlled degradation followed by formation of non-toxic products; processability by available methods; and compatibility with the chemicals embedded in them.

It is important to construct and use formulations of chemicals that would be safe for humans, the environment, and beneficial biota, which would be simple to fabricate and convenient to use and which would enable slow and smooth release of the chemical to soil throughout the plant growing period.

The use of the encapsulated haloacetanilide herbicide alachlor (U.S. Patents: No. 4280833; No. 4417916 IPC) is more effective than the use of free alachlor at a high rate (0.55 kg/ha). However, application of herbicides at high rates may adversely affect beneficial biota. The limitation of this formulation is that at rates below 0.55 kg/ha, its activity decreases in two weeks after application. A technique has been developed (U.S. Patent No. 4285720) to prepare microencapsulated herbicidal, organophosphorus insecticidal, and thiocarbamate herbicidal compositions in capsules of polyurea, which are rather quickly dissolved in moist soil, enabling quick release of the chemicals. The technique of producing these capsules is very complicated, consisting of several phases. Another microencapsulated herbicidal composition with controlled release of the active ingredient from polyurea capsules (RU Patent No. 2108036) contains acetochlor mixed with the antidote dichloroacetamide, to alleviate the adverse effect of the active ingredient on the beneficial biota. However, this formulation has a very complex composition; it is difficult to fabricate; it is used to spray plant leaves in one week after sowing; and the toxic herbicide is dispersed in the air.

The pressed pesticide in a unit-of-use packaging (RU Patent No. 2147179) contains the active ingredient based on polyvinylpyrrolidone as well as a dispersing agent, a wetting agent, and additives. The following chemicals have been proposed as active ingredients: herbicides – atrazine, simazine, cyanazine, tributhylazine, diuron, chlorsulfuron, metsulfuron; insecticides – deltamethrin, lindane, carbaryl, endosylfan, etc. This formulation is safer than liquids or powders. Its disadvantages are manufacturing complexity and complex composition.

Some soil-applied herbicide formulations contain post-emergence herbicides (e.g., RU Patent No. 2261596). The herbicide formulation contains one or several post-emergence herbicides and the carrier material – Fuller's earth, aerogel, high-molecular-weight polyglycols, and polymers based on acrylic acid, methacrylic acid, and copolymers thereof. These are foliar herbicides – 4-(hydroxy(methyl)phosphinoyl)-L-homoalanyl-L-alanyl-L-alanine (bilanofos), 1,1′-ethylene–2,2′-bipyridyldiylium dibromide (diquat), ammonium-DL-homoalanine–4-yl(methyl)-phosphinate (glufosinate-ammonium), N-(phosphonomethyl)glycine (glyphosate), and 1,1′-dimethyl–4,4′-bipyridinium dichloride (paraquat). This formulation can also be used as a pre-emergence herbicide.

Thus, the literature data suggest an increasing research effort in the construction of novel herbicidal formulations. The main purpose of that research is to produce less toxic and more selective formulations and to decrease application rates.

S–1,3,5-triazines are commonly used broad-spectrum selective herbicides, which do not persist for a very long time in soil. Metribuzin (sencor) is a pre-emergence and post-emergence herbicide based on the derivative of 1,2,4-triazine, which is used to treat various crops. Metribuzin (MET) is an asymmetric triazine herbicide, which has very high biological efficacy in different climate zones. MET is used to control such weeds as *Phalaris minor*, *Cynodon dactylon*, *Chenopodium album*, *Cyperus spp*., etc. This herbicide readily dissolves in water, and it is weakly sorbed in soil; when leached from the lower horizons, it contaminates groundwater (Fedtke, 1981).

The first studies describing the preparation of slow-release MET formulations appeared in the late 1980s – early 1990s. They reported MET encapsulation in polymer matrices based on 2-methyl–4-chlorophenoxyacetic acid and pentachlorophenol, a maleic anhydride/methyl methacrylate copolymer, and chitin, and MET binding with β-cyclodextrin (McCormick, 1985a; Ikladious and Messiha, 1984; McCormick and Anderson, 1984; Dailey et al., 1990). MET has attracted the attention of researchers as a model for developing slow-release herbicide formulations based on various synthetic and natural materials such as polyvinylchloride, carboxymethyl cellulose (Kumar et al., 2010), acrylamide (Sahoo et al., 2014), methacrylic acid combined with ethylene glycol and dimethacrylate (Zhang et al., 2009), sepiolite (Maqueda et al., 2008), alginate (Flores-Céspedes et al., 2013), phosphatidylcholine (Undabeytia et al., 2011), kraft lignin (Chowdhury, 2014), lignin/polyethylene glycol (PEG) blends (Fernández-Pérez et al., 2011; 2015), chitin, cellulose, starch (Fernández-Pérez et al., 2010; Rehab et al., 2002), bentonite, activated carbon (McCormick,1985b), etc.

Release kinetics of MET embedded in different materials was studied in laboratory systems (sterile water, soil) (McCormick, 1985b; Rehab et al., 2002; Maqueda et al., 2008; Zhang et al., 2009; Kumar et al., 2010;

Flores-Céspedes et al., 2013; Sahoo et al., 2014; Fernández-Pérez et al., 2010, 2015). The duration of MET release varied within a very wide range, between several tens of hours and several tens of days, suggesting the possibility of constructing systems for controlled release of this herbicide.

Some of the authors observed the fast release of MET. Sepiolite-based granular formulations loaded at 16.7 and 28.6% of MET showed a fast release of MET (50 to 80% of the encapsulated herbicide) in the first few hours (Maqueda et al., 2008). McCormick (1985b) reported a study of MET release from the granules prepared from the gelling sodium alginate combined with bentonite, acid-treated bentonite, anthracite, and activated carbon loaded at 12% of MET. Only 20% of the initially encapsulated MET was released from the alginate/activated carbon granules after 70 h, while the release of the herbicide from other formulations reached 100% after 40 h. MET release from the polymer prepared using methacrylic acid and ethylene glycol was studied in soil (Zhang et al., 2009). During 45 days, MET concentration in soil gradually decreased from 0.81 to 0.05 mg/kg. Kumar et al. (2010) studied release kinetics of MET loaded in polyvinyl chloride, carboxymethyl cellulose, and carboxymethyl cellulose-kaolinite composite (CMC-KAO) in water in comparison with the commercial formulation (75DF). After three days, MET was completely released from the commercial formulation, while its release rate from the experimental formulations was slower. Fernández-Pérez et al. (2015) studied MET release from lignin-PEG granules for 500 h in a glass reactor filled with silicone oil. MET release from the granules reached 90% over 48 h, while free MET was dissolved in oil in 30 min. MET loaded into lignin-based granules was released into the water in 13 days; the increase in the size of the granules from 0.5 to 2–3 mm slowed down the process.

MET formulations that show slower release rates have also been described. A study of acrylamide-bentonite composites as a matrix for MET encapsulation (Sahoo et al., 2014) showed that 50% of the encapsulated MET was released after 25 to 51 days, while it took only 14 days for MET to be completely released from the commercial formulation. Flores Céspedes et al. (2013) studied MET release from alginate-based formulations, with bentonite and anthracite used as modifying agents, in the soil for 60 days, at a temperature of 25°C in a thermostat incubator. About 80% MET was released from the MET-alginate-bentonite or MET-alginate-anthracite granules in the first 6 days, i.e., the release of MET was much slower from these formulations than from the commercial formulation (80% MET was released in 2 days). Fernández-Pérez et al. (2010) studied MET release into the buffer solution for 3–4 months at different pH values (5, 7, and 9) from the poly(N,N-diacryloyl)/MET and N, N-diacryloyl-methyl methacrylate/MET systems. The copolymers contained MET via an imide linkage and were prepared by the free-radical polymerization of MET monomer with N, N-diacryloyl or methyl methacrylate. MET release was observed during the first 5–10 days, at pH 7. The release rate was growing as pH of the medium was elevated, as at high pH values, the degrees of ionization and swelling of the copolymers were increased. At the same time, the amount of MET hydrolyzed in different buffer solutions after 4 months ranged from 0.15 to 0.45 mg/L. The N-diacryloyl-methyl methacrylate/MET system showed a higher release rate – up to 4.5 mg for the first 3–5 days. MET release kinetics was influenced not only by pH of the medium but also by the hydrophilic properties of the polymer matrix. Rehab et al. (2002) studied MET release in buffered solutions from natural polymers (chitin, dextran, cellulose, and starch) and synthetic acrylic polymers with amide bonds to MET. In most cases, MET was released after 40 days; this process was influenced by the hydrophilicity of the polymer. Cellulose and chitin formulations showed lower levels of release than dextran and polyvinyl alcohol (PVA).

Analysis of the literature shows that release kinetics of MET can be varied widely by loading it into different materials, suggesting the possibility of constructing controlled-release herbicide formulations.

4.2 POLYHYDROXYALKANOATES AS DEGRADABLE MATRIX FOR CONSTRUCTING SLOW-RELEASE HERBICIDE FORMULATIONS

It is important that materials used as matrices for pesticides should both enable slow-release and targeted and effective delivery and be readily processable and reasonably priced. The cost of the product is one of the major factors determining whether a given line of research should be developed and whether the novel agrochemical formulations will be used in practical agriculture.

Analysis of the literature suggests lively interest in synthesis and investigation of polymers based on derivatives of carbonic acids. In addition to polylactides and polyglycolides, special attention is given to such biodegradable polyesters as polyhydroxyalkanoates (PHAs) – microbial polyesters that have many useful properties. PHAs are promising materials for various applications. PHAs are thermoplastic and have good physical and mechanical properties, like synthetic polyolefins such as polypropylene and polyethylene, but, in addition to this, they are also biocompatible and biodegradable. These polymers are degraded in biological media (soil, rivers, lakes, seas) by natural microflora. In contrast to polylactides and polyglycolides, PHAs do not undergo rapid chemical hydrolysis; they decompose via truly biological degradation, and, thus, it takes months for them to be fully degraded in biological media, which is very important for the construction of long-term formulations. The ability of PHAs to be degraded in soil is a basis for constructing pre-emergence herbicide formulations, which can be buried in soil together with seeds of cultivated plants and thus prevent the growth of undesirable plants.

The first studies addressing the use of P(3HB) and P(3HB/3HV) copolymers as matrices for embedding pesticides Ronilan, Sumilex, and α-hexachlorocyclohexane, lindane, were reported by Savenkova (Savenkova et al., 2002) and Volova et al (Voinova et al., 2008; Volova et al., 2009) respectively. Some authors reported encapsulation of pesticides ametrine, atrazine, and malathion in microspheres prepared from P(3HB) and P(3HB/3HV) (Grillo et al., 2010; 2011; Suave et al., 2010; Lobo et al., 2011). In a more recent study, Prudnikova, and co-authors described embedding of herbicide Zellek Super in P(3HB/3HV) granules and films to prepare slow-release formulations (Prudnikova et al., 2013).

In the Institute of Biophysics, Krasnoyarsk, Russia (Volova et al., 2016a) for the first time, the natural polymer poly–3-hydroxybutyrate (P(3HB)) was tested as a degradable matrix for embedding herbicide in order to prolong its release into the surrounding medium. The purpose of this study was to construct slow-release herbicide (MET) formulations by using a natural degradable polymer, P(3HB). As MET exhibits rather high physiological activity towards various weeds and is convenient to use, it has attracted the attention of researchers as a model for developing slow-release herbicide formulations.

Chemically pure MET ($C_8H_{14}N_4OS$) (99.7% pure) was used (State Standard Sample 7713–99 – the state standard accepted in Russia (Blok–1, Moscow). MET has a systemic effect against many undesirable plants in vegetable and grain crop fields, and it both has a foliar action and can penetrate into plants through their roots; this herbicide inhibits plant photosynthesis. The structural formula of MET is shown in Figure 4.1. It has the following main physicochemical properties: colorless crystals; molecular weight 214.3 g mol^{-1}; melting point 126.2°C; solubility at 20°C (g/L) in water – 1.2, in chloroform – 850, and in acetone – 820. Log K_{ow} 1.60. pK_a – 7.1.

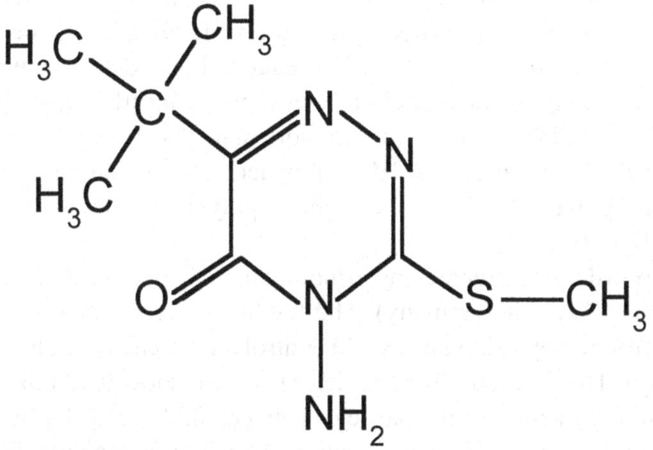

FIGURE 4.1 The structural formula of metribuzin.

Polymer poly(3-hydroxybutyrate) – P(3HB) – was used as a degradable polymer matrix for embedding the herbicide. The polymer was synthesized in the Laboratory of Chemoautotrophic Biosynthesis at the Institute of Biophysics SB RAS by using bacterium *Cupriavidus eutrophus* B10646. The inoculum was prepared by resuspending the museum culture maintained on agar medium. Polymer was extracted from cells with chloroform, and the extracts were precipitated using hexane. The extracted polymers were re-dissolved and precipitated again 3–4 times to prepare homogeneous specimens. P(3HB) had the following physicochemical parameters: weight average molecular weight (M_w) 920 kDa; polydispersity (D) 2.52; the degree of crystallinity 74%; melting point and thermal decomposition temperature 179.1 and 284.3°C, respectively (Volova et al., 2013; 2014).

Detection of MET was performed by using gas chromatography. Measurements were done on the gas chromatograph equipped with a mass spectrometer (7890/5975C, Agilent Technologies, U.S.), using a capillary column, under varied temperature. The chromatography conditions were as follows: an HP–5MS capillary column, 30 m long and 0.25 mm in diameter; carrier gas – helium, flow rate 1.2 mL/min; sample introduction temperature 220°C; initial temperature of chromatography – 150°C; temperature rise to 310°C at 10°C per min; transfer line temperature – 230°C, ion source temperature – 150°C, electron impact mode at 70 eV, fragment scan from m/z 50 to m/z 550 with a 0.5 second cycle time. The peak corresponding to MET was detected by the mass spectrometer. We used State Standard

Sample 7713–99 – the state standard accepted in Russia: 99.7% pure. A calibration curve was prepared by using a wide range of concentrations of MET in acetone (0.1–4.2 μg/μL). The range of linear detection was obtained for a wide variety of concentrations: between 0.1 μg/μL and 4.2 μg/μL. The standard error of the method was no more than 3%.

Various phase states of P(3HB) (powder, solution, emulsion) were used to prepare polymer/MET mixtures. Each polymer matrix was loaded with 25% (w/w) MET.

The powder was prepared by grinding the polymer in a ZM 200 ultracentrifugal mill (Retsch, Germany). The fractional composition of the powder was determined by using an AS 200 control analytical sieve shaker (Retsch, Germany). The fraction of the particles of sizes below 0.50 mm constituted 65%, and the fraction of the particles between 0.80 and 1.00 mm was 45%. P(3HB) and MET powder samples were weighed on the analytical balance and then homogenized with a laboratory stirrer for 2 min. P(3HB) solutions of different concentrations were prepared by adding a polymer sample to chloroform. The polymer sample was dissolved in chloroform, and a solution of MET in chloroform was added to the polymer solution. The polymer/MET solution was mixed for 2–3 h (until completely dissolved) by using an MR Hei-Standart magnetic stirrer (Heidolph, Germany) and heated to 35–40°C under reflux condenser or was left to stay at room temperature for 3–4 h. The polymer emulsion was prepared as follows. The oil phase, represented by a 1% P(3HB) solution, and MET (25% of the mixed solution) in chloroform were combined with the aqueous phase –PVA solution (Sigma, U.S., M_w 30 kDa) – and mixed for 24 h, until complete solvent evaporation took place.

The P(3HB)/MET mixtures (solutions, powders, and emulsions,) were used to construct MET-loaded films, granules, pellets, and microparticles. In every formulation, the content of MET was 25% of the polymer matrix (w/w). Polymer/MET solutions in chloroform were used to prepare MET formulations in the form of films and granules. Films were prepared as follows: the P(3HB)/MET solution was cast in glass or Teflon-coated metal molds, and then solvent evaporation occurred. We used 2 and 4% (w/v) polymer solutions in chloroform. The viscosity of the solutions was measured by using a HAAKE Höppler Falling Ball Viscometer C (Thermo Scientific, Germany). The following procedure was used to embed the active ingredient into the polymer matrix (25% w/w). The homogeneous polymer/MET solution was filtered and poured into the degreased mold under a bell-glass (to protect it from draught and dust). The films stayed under the bell-glass for 24 h at room temperature, and then they were placed into a vacuum drying cabinet (Labconco, U.S.) for 3–4 days, until complete solvent evaporation took place. The films were then weighed on the analytical balance. The film thickness was measured with a digital micrometer (LEGIONER EDM–25–0.001, Germany). Squares of 25 mm^2 in the area (5 mm × 5 mm) were then cut from the film.

MET-loaded polymer granules were prepared as follows: the polymer was precipitated from the solution into the sedimentation tank filled with the reagent in which P(3HB) did not dissolve (hexane), and, thus, crystallization occurred and polymer granules were formed. We used P(3HB) solutions in chloroform of three different concentrations: 8, 10, and 12% (w/v). A solution of MET was added to the polymer solution; the system was mixed to achieve homogeneity, by using a Silent Crusher high-speed homogenizer (Heidolph, Germany). A Pumpdrive 5001 peristaltic pump (Heidolph, Germany) was used to drop the polymer/MET solutions into the sedimentation tank that contained hexane, where the polymer was crystallized and granules formed. We varied polymer concentration in the solution, needle diameter, the rate at which the solution was fed to the sedimentation tank, and the thickness of the layer of the precipitating agent.

Pellets were prepared from the powdered polymer/MET mixture ground in a ZM 200 ultracentrifugal mill (Retsch, Germany). The fractional composition of the polymeric powder was determined by using an AS 200 control analytical sieve shaker (Retsch, Germany); apparent density of the fractions was determined with PT-TD 200 Touch (Retsch, Germany).

The powdered P(3HB) and MET were mixed mechanically. Samples of the two powders of different fraction compositions (0.10 mm to 1.0 mm) were weighed on the analytical balance and then homogenized with a laboratory stirrer for 2 min. The mixture was used to cold press pellets, 13 mm in diameter, by using a Carver Auto Pellet 3887 press (Carver, U.S.) under different pressing forces: 6,000, 14,000, and 24,000 F.

Microparticles were prepared by the emulsion technique. To produce large particles (10-μm diameter and more), we tested different production conditions. We varied the concentration of the polymer solution (between 1 and 4%); agitation speed (300, 750, 1000, 6000, and 16 000 rpm), and the type of the surfactant (PVA, M_w 30–50 and 150 kDa; sodium dodecyl sulfate (SDS), and polyoxyethylene–20-sorbitan-monooleate (Tween ® 80). To vary the conditions and speed of agitation of the emulsion, we used a Silent Crusher M high-speed homogenizer (Heidolph, Germany) (between 6,000 and 16,000 rpm) and an MR Hei-Standard magnetic stirrer (Heidolph,

Germany) – (between 300 and 1000 rpm). After solvent evaporation, microparticles were collected by centrifugation (Centrifuge 5810 R, 5417 R, Eppendorf, Germany, 10,000 rpm), rinsed, and freeze-dried (Alpha 1–2 LD plus, Christ®, Netherlands).

Initial substances in the form of powders (MET and poly–3-hydroxybutyrate); powdered P(3HB)/MET mixture, and MET formulations constructed as films, granules, pellets, and microparticles were examined by using state-of-the-art physicochemical methods. Thermal analysis was performed with a DSC–1 differential scanning calorimeter (Mettler Toledo). Samples of films, powders, granules, and pellets (4.0 ± 0.2 mg) were placed in aluminum crucibles and heated at 5°C per min. The melting point (T_{melt}) and thermal decomposition temperature (T_{degr}) were determined from exothermic peaks on thermograms, using the StarE software. X-ray structure analysis and determination of the degree of crystallinity (C_x,%) of films, powders, or pellets were performed using an X-ray spectrometer (D8 Advance, Bruker Corporation, Bremen, Germany) (graphite monochromator on a reflected beam) in a scan-step mode, with a 0.04°C step and exposure time 2 s, to measure intensity at point. The instrument was operating at 40 kV × 40 μA. Fourier transform infrared spectroscopy (FTIR) was conducted as follows. Infrared spectra of the films and powders were taken in the 500–4500 cm^{-1} range, using an INFRALUM FT–02 FTIR spectrometer (Lumex, Russia). Morphology of the microparticles and films was studied by electron microscopy, using an S–5500 scanning electron microscope (Hitachi, Japan). Samples of granules and pellets were examined under a TM 3000 electron microscope (Hitachi, Japan). Platinum sputter coating of the specimens was conducted in an Emitech K575XD Turbo Sputter Coater (Quorum Technologies Limited, U.K.). Granules were examined to determine their size, morphology, and active ingredient encapsulation efficiency (EE). Parameters of microparticles of sizes under 10 μm (size distribution and surface charge (ξ-potential) were investigated with a Zetasizer Nano ZS particle analyzer (Malvern, U.K.), employing dynamic light scattering, electrophoresis, and laser Doppler anemometry. The surface charge of microparticles was characterized by the value of ξ-potential, which was measured with a Zetasizer Nano ZS microparticle analyzer, using Henry's formula. Microparticles of sizes between 10 μm and 100 μm were measured with a Flow Cam system for quantitative and qualitative particle analysis (Fluid Imaging, U.S.). Triplicate measurements of each sample were performed. The EE of MET in microparticles and granules was calculated using the following formula (4.1):

$$EE = (M_{enc}/M_{init}) \times 100\% \quad (4.1)$$

where M_{enc} is the mass of the encapsulated MET in the polymer matrix (mg), and M_{init} is the mass of the initial amount of MET (mg).

P(3HB)/MET solutions, emulsions, and powders – were prepared from the preliminarily purified and thoroughly studied P(3HB) and crystalline MET powder. For the powdered P(3HB) and MET and for their mixture, thermograms were taken within a wide temperature range, including the polymer melting point and thermal decomposition temperature. The thermogram of the P(3HB)/MET mixture showed two peaks: one melting peak is at 126.5°C and the other at 168.9°C. The thermal decomposition temperature had one peak, at 280.1°C, as the thermal decomposition temperature of MET was in the same region. The enthalpies of melting of the mixture were 13.7 and 58.6 J/g and the enthalpy of its thermal decomposition was 610.6 J/g, and that was also lower than the enthalpies of the initial polymer, whose melting temperature was 179.1°C and thermal decomposition temperature 284.3°C: 89.1 and 731.8 J/g, respectively. A similar result – the presence of two melting peaks (127.2 and 168.2°C) – was obtained for the pellets.

The presence of two melting peaks suggested that in the dry mixture, the MET part did not interact physically with P(3HB), causing the mixture to split into layers during heating. However, the decrease in the melting temperature of P(3HB) in the mixture compared with the initial polymer was indicative of the weak physical interaction. The formulations prepared from solutions (microparticles, films, and granules) did not show a MET peak, suggesting that the system was a physical mixture of components. In all formulations, the temperatures of melting and thermal decomposition and enthalpy were lower than those of the initial polymer. The most significant decrease in the melting temperature was observed for films and pellets (161.1 and 168.2°C, respectively). We assume that during the preparation of formulations, the size of crystallites may have changed. MET molecules may have occupied the free space in the polymer and prevented P(3HB) crystals from growing. This may have led to the formation of small crystals, whose melting temperature was lower. The temperature properties of the formulations may have been influenced not only by the size of the crystals but also by the processes used to produce them, as they were prepared in different ways. Films were crystallized when the solvent was evaporated; crystallization of granules occurred during their precipitation in hexane; pellets were produced by high-pressure processing. These processes affected the structure of the crystals. By contrast, analysis of the number and shapes of endothermic peaks on the thermograms of films, granules, and microparticles showed the formation of a stable mixture of the polymer and MET, which was not separated under heating, as there was only one peak of melting and thermal decomposition. The T_m decrease suggested an increase in the viscosity of the melts and, thus, inhibition of polymer crystallization. The small MET crystals had greater interfacial energy and, therefore, they began to melt before the polymer did, and this lowered the melting temperature of the mixture and made the structure of P(3HB) more amorphous. This conclusion was also supported by a decrease in the enthalpy of melting, which was indicative of a decrease in the degree of crystallinity of the initial polymer, and certain smearing of the peaks, which is typical of melting of amorphous regions.

X-ray structure analysis showed that the loading of the P(3HB)/MET decreased the degree of crystallinity of the polymer. The C_x of the P(3HB)/MET powders was 61%. That was lower than the C_x values of P(3HB) (74%) and MET (90%). The C_x of MET formulations (granules, microparticles, and pellets) was also lower than that of the initial polymer (62–64%); the C_x of the films was even lower (51%). That was indicative of changes in the crystallization process in P(3HB)/MET, i.e., an increase in the proportion of the amorphous region.

Results of FTIR of MET, P(3HB), and P(3HB)/MET films and powders suggested that the most informative range of the wavenumbers was that between 1450 and 1700 1 cm^{-1}. The absorption peaks (bands) observed in the P(3HB)/MET films were those associated with the specific structural groups of MET. FTIR spectra of the mixture showed an absorption band in the 1520 cm^{-1} region, which was characteristic of pulsation vibrations of the carbon skeleton of four substituted pyridines. Enhanced intensity of absorption bands was observed in the 1400–1419 cm^{-1} region. They were characteristic of both P(3HB) and MET and were attributed to deformation vibrations of CH_3 groups, which were adjacent to C=O, and pendulum oscillations of CH3 groups in the 1130 cm^{-1} region. Similar intensity enhancement was observed in the 2874 cm^{-1} region for peaks of the bound OH groups. The absorption band in the 1632 cm^{-1} region was attributed to deformation vibrations of the NH_2 group of MET, and the absorption band of the mixture was shifted relative to that of pure MET (1617 cm^{-1}). Therefore, we assumed the possible involvement of the NH_2 group in the formation of hydrogen bonds. Peaks in the 3200–3400 cm^{-1} region were attributed to stretching vibrations of the bound OH group. No groups that would form due to a chemical interaction between P(3HB) and MET were detected. Peaks of the existing groups became higher because of the overlap of the P(3HB) and MET spectra, indicating that P(3HB)/MET was a physical mixture of the polymer and the herbicide.

Thus, results of DSC, X-Ray, and FTIR suggested that there were no chemical bonds between the pesticide and the polymer and that the mixtures were physical mixtures of components. The decrease in the melting temperature and enthalpy leads us to conclude that the active ingredient of MET plays the role of a filler of the polymer matrix.

Figures 4.2 and 4.3 are photographs and SEM images of the P(3HB)/MET formulations (films, pellets, granules, and microparticles) prepared in this study.

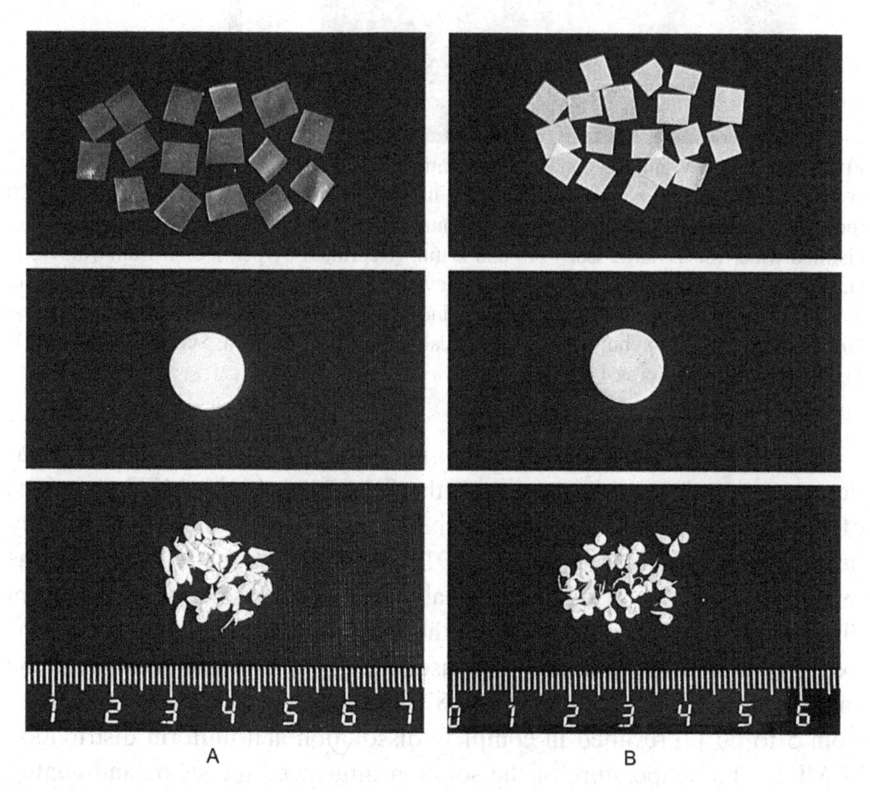

FIGURE 4.2 Photographs of the P(3HB)/metribuzin films, pellets, and granules: A – initial, B – after 49 days of incubation in water. (Reprinted by permission from Volova, T. G., Zhila, N. O., Vinogradova, O. N., Nikolaeva, E. D., Kiselev, E. G., Shumilova, A. A., et al., Constructing herbicide metribuzin sustained-release formulations based on the natural polymer poly–3-hydroxybutyrate as a degradable matrix. J. Environ. Sci. Health, P. B., 51, 113–125. © 2016 Taylor & Francis.)

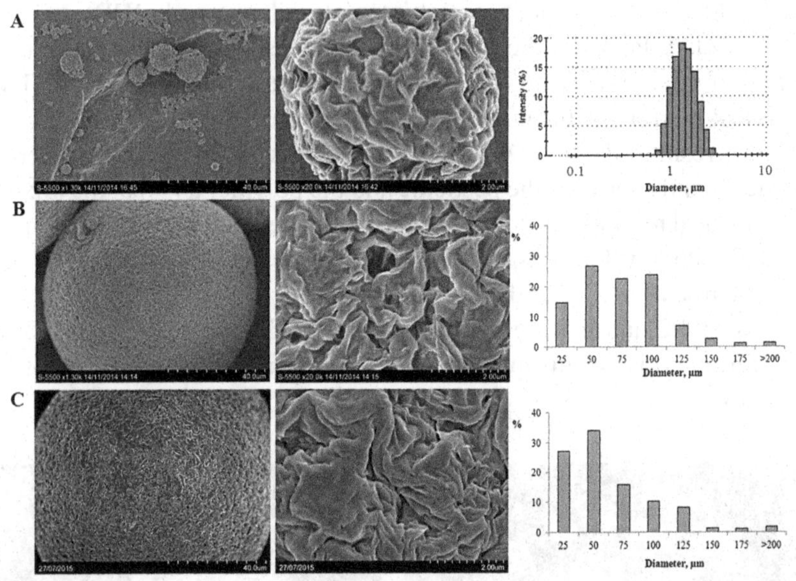

FIGURE 4.3 SEM images and size distribution of microparticles loaded with metribuzin at 25% of the polymer mass, prepared by using: A – a high-speed homogenizer (6000 rpm), B – magnetic stirring (750 rpm), C – magnetic stirring (750 rpm) after incubation in distilled water for 49 days. Bar = 40 and 2 µm. (Reprinted by permission from Volova, T. G., Zhila, N. O., Vinogradova, O. N., Nikolaeva, E. D., Kiselev, E. G., Shumilova, A. A., et al., Constructing herbicide metribuzin sustained-release formulations based on the natural polymer poly-3-hydroxybutyrate as a degradable matrix. J. Environ. Sci. Health, P. B., 51, 113–125. © 2016 Taylor & Francis.)

In the work (Volova et al., 2016a) studied the effect of polymer solution concentration on the quality of the films loaded with MET. Solutions of polymer in chloroform of concentrations 4% or higher (the viscosity of the solution decreased from 236.91 to 87.30 cP as the temperature was increased from 5 to 60°C) did not enable complete dissolution and uniform distribution of MET in the solution; the films prepared from these solutions exhibited nonuniform surfaces. The use of 2% solutions (the viscosity of the solution decreased from 26.79 to 12.82 cP as the temperature was increased from 5 to 60°C) resulted in complete dissolution and uniform distribution of MET. The temperature of the solution influenced the shape and quality of the films, too: at room temperature, films were deformed during solvent evaporation. As the solution and the surface of the glass mold were heated to 30–35°C, flexible films of uniform thickness (0.045 ± 0.005 mm) were produced. P(3HB) films loaded with MET had smooth, dense structure with a few small pores. X-ray spectral analysis showed that MET was uniformly distributed inside the polymer matrix and was present on the film surface as dispersed particles (smaller than 1 μm), which were also uniformly distributed on the surface. The EE of MET was 100%. We observed a few 2–3 μm pores on the film surface.

By varying P(3HB) solution density, hose diameters, and needle size, we determined the parameters that enabled the production of high-quality granules: 10% polymer concentration of the solution, needle size – 20 G, and the thickness of the layer of the precipitating agent – 200 mm. At polymer solution concentrations below 10% and with a smaller needle size, the granules were misshapen, and their formation was incomplete (some granules had voids or "tails") (Figure 4.2a). The use of solutions of higher concentrations was technologically infeasible. The size of MET -loaded granules prepared from 10% polymer solutions was 2.5–3 mm. We developed a method for producing polymer granules with the maximum active ingredient EE (95–100%). The surface of the granules was slightly rough, with easily visible MET crystals.

The method of contact cold pressing was used to prepare pellets of diameter 13 mm and mass 200 ± 0.15 mg, with MET constituting 25% of the pellet mass, from P(3HB)/MET powders (Figure 4.2a). The amount of the applied force influenced the structure of the pellets: when the force below 6000 F was applied, the resulting pellets had loose structure; at 24,000 F, partial surface sintering was observed. An optimal force for preparing high-quality pellets (with even surface and uniform distribution of chemicals) was 14,000 F. The surface of the pellets loaded with MET was generally dense, with slight lesions, probably where MET was located, as MET particles were larger than the polymer particles. Spectral analysis of elemental composition proved that 100% of MET was embedded in the polymer matrix and that its structure remained unchanged.

The study of the effect of microparticle preparation conditions on the size of microparticles showed that not all study parameters had favorable effects on this value. The main factors determining the size of microparticles were polymer emulsion concentration and agitation speed. As the polymer concentration of the emulsion was increased from 1% to 2% and to 4% (surfactant PVA 30 kDa), the average diameter of the microparticles increased from 7.5 to 17 μm. Agitation speed influenced the particle size. At a high agitation speed, 6000 and 16,000 rpm, (PVA 30 kDa, polymer concentration 1%), the size of the resulting particles was 2.1 and 1.2 μm, respectively. At lower agitation speeds (300, 750, 1000 rpm) (PVA 30 kDa, polymer concentration 1%), when the magnetic stirrer was used, the size of the particles was 17.3, 11.4, and 4.25 μm, respectively. Thus, as the agitation speed was increased, the size of the microparticles decreased. When PVA 30 kDa was replaced by PVA 150 kDa (polymer concentration 1%), the size of the particles increased from 1.2 to 1.5 μm. The use of SDS increased the size of the particles to 7.0 μm. The effect of Tween 80 and its concentration in the emulsion was studied at 16,000 rpm. At 1% Tween 80 in the emulsion, the size of the resulting particles was 3.0 μm; at 4%, it was somewhat greater – 3.9 μm. The quality and the yield of microparticles were influenced by the type of the surfactant used: microparticles prepared from emulsions containing PVA and SDS had regular spherical shapes, and their yield was 77.1 and 75.5%, respectively. The use of Tween 80 produced some misshapen particles, and their yield was no more than 60%. Variations in the density of polymer emulsion did not have a significant effect on the particle yield. A similar result was obtained when we varied the speed and type of agitation (particle yield varied between 68 and 73%).

The loading of particles with MET was studied at 750 and 6000 rpm. The average diameter of the MET-loaded particles prepared at the high agitation speed (6000 rpm) was 1.83 μm, but at 750 rpm it was much greater (54 μm). The surface of microparticles was wrinkled, regardless of their size (Figure 4.3A, B). The study of the ξ-potential as an indicator of the stability of MET-loaded particles, showed that at 750 rpm, ξ-potential was 26 mV and at 6000 rpm – 30 mV. Furthermore, the increase in the surfactant (PVA) concentration from 1 to 3% had almost no effect on the ξ-potential. An important parameter is EE of the substances in the polymer matrix of microparticles. By varying the agitation speed (750 and 6000 rpm) and the type of agitation, we prepared particles with similar values of metribuzin EE: 18% and 16.9%, respectively.

Thus, the best conditions for preparing MET -loaded microparticles were as follows: P(3HB) concentration 1%, PVA 30 kDa concentration 1%, and agitation speed 750 rpm, on a magnetic stirrer. The size of the resulting particles was 54 μm, the yield of the particles was 71.6%, the ξ-potential was 26 mV, and MET encapsulation efficiency was 18%.

Release kinetics of MET from the polymeric matrices was studied *in vitro* in laboratory systems: the specimens were sterilized and placed into 500-mL sterile conical flasks filled with sterile distilled water (100 mL). The number of granules, large microparticles (of the average size of 54 μm), films, or pellets placed in a flask was determined in such a way that the samples in each flask contained equal total amounts of the active ingredient (50 mg). The flasks were incubated at 25°C in an Innova 44 New Brunswick temperature controlled incubator shaker at 150 rpm. Samples (2 mL) for analysis were collected periodically, under aseptic conditions, and an aliquot of water was added to the flask to maintain a constant volume of liquid in it. MET was extracted with chloroform three times to determine its concentration. The chloroform extracts were passed through sodium sulfate. Chloroform was removed in a rotary vacuum evaporator. After chloroform was removed, we added 200 μL of acetone.

The amount of MET released (RA) was determined as a percentage of the MET encapsulated in the polymer matrix, using the following formula (4.2):

$$RA = r / EA * 100\%, \tag{4.2}$$

where EA is the encapsulated amount, mg; and r is the amount released, mg.

For describing herbicide release kinetics from different formulations, we used the Korsmeyer – Peppas model (formula 4.3):

$$M_t / M_\infty = Kt^n, \tag{4.3}$$

Here, M_t is the amount of the herbicide released at time t, M_∞ is the amount of the herbicide released over a very long time, which generally corresponds to the initial loading. K is a kinetic constant and n is the diffusional exponent.

For the case of cylindrical pellets, $0.45 \leq n$ corresponds to a Fickian diffusion mechanism, $0.45 < n < 0.89$ to non-Fickian transport, $n = 0.89$ to Case II (relaxation) transport, and $n > 0.89$ to super case II transport. To find the exponent n, the portion of the release curve where $M_t/M_\infty < 0.6$ should only be used (Ritger and Peppas, 1987; Peppas and Narasimhan, 2014). To analyze the stability of the formulations incubated in water for long periods of time, we used electron microscopy and monitored the temperature properties and degree of crystallinity of the samples.

Figure 4.4 shows MET release from P(3HB)/MET formulations: films, pellets, granules, and microspheres; Figures 4.2B, 4.3C are photographs and SEM image of the formulations after incubation in water. As poly–3-hydroxybutyrate is neither dissolved nor hydrolyzed in water, MET was passively released from the polymer matrix, diffusing through the pores and microcracks of the polymer matrix. The differences that can be clearly seen in the graph are determined by the construction of the carrier. MET was released at the highest rate from microspheres.

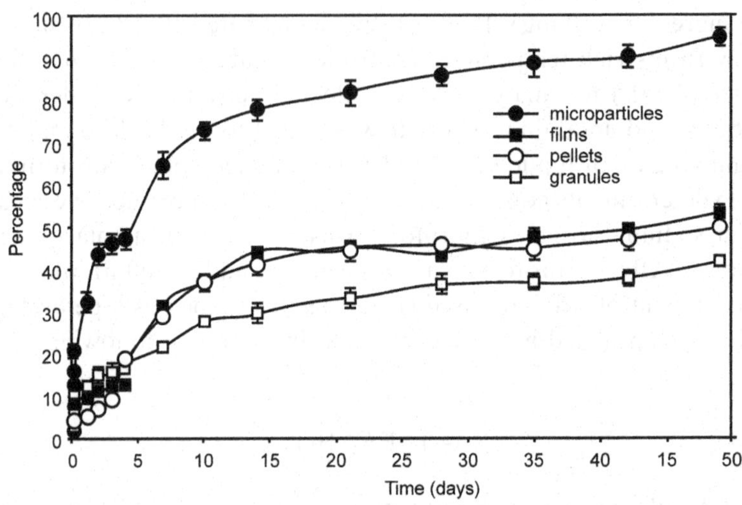

FIGURE 4.4 The release profile of metribuzin from: 1 – microparticles, 2 – films, 3 – pellets, 4 – granules (% of the initial amount). (Reprinted by permission from Volova, T. G., Zhila, N. O., Vinogradova, O. N., Nikolaeva, E. D., Kiselev, E. G., Shumilova, A. A., et al., Constructing herbicide metribuzin sustained-release formulations based on the natural polymer poly–3-hydroxybutyrate as a degradable matrix. J. Environ. Sci. Health, P. B., 51, 113–125. © 2016 Taylor & Francis.)

The reason for this was that microparticles were the smallest MET carriers and, thus, had the largest total particle/water interface area. In the first 3 days, the rate of MET release from microparticles was 7.7 mg/d; in the following 11 days, it dropped to 1.5 mg/d. Then, between Day 14 and the end of the experiment, the rate of MET release from microparticles was very low (0.24 mg/d), the herbicide release reaching 95% by the end of the experiment (49 days). MET release rates from films and pellets were much lower. For 14 days, they were about 1.5–1.6 mg/d, and, then, both curves exhibited plateaus, i.e., no more MET diffused through the pores of the polymer matrix. The total percentage of the encapsulated agent released from the polymer matrices of films and pellets in the experiment reached about 50–53%. MET encapsulated in granules showed a similar release behavior: in the first 10–14 days, the herbicide was released at a rate of 1.1 mg/d. Then, the curve exhibited a plateau, and the percentage of MET released from the granules was the lowest – 42%.

The main kinetic parameters of MET release from different formulations into the water were the following: exponent n varied between 0.40 and 0.55 for all formulations. For granules and microparticles, n was 0.4, suggesting diffusion of the herbicide through the polymer layers in accordance with Fick's law. For pellets and films, the exponent was equal to 0.51 and 0.55, respectively, indicating abnormal MET release behavior, which did not follow Fick's law. The highest value of constant kinetic K was shown by microspheres (0.081 h^{-1}) and the lowest by films and pellets (0.01 h^{-1}). The reason was that microparticles had the largest surface area for diffusion to occur. The lowest percentage of MET was released from the pellets. Taking into account that MET is a water-soluble substance, we assume that pellet production by compression blocked the pores of the polymer matrix, preventing water penetration and hindering herbicide diffusion. This is supported by the presence of the MET melting peak in the diagram of the melting of the pellet after incubation in water.

All specimens incubated in water were rather stable and retained their initial shapes (Figure 4.2B). Only the initially transparent films had become cloudy. Slight changes in the surface structure were captured by electron microscopy. The films became more porous, and the pores were of larger sizes (reaching 2–3 μm). Lesions developed in some regions of the pellets, most likely where MET was dissolved and released. The morphology of microparticles changed very little (Figure 4.3C). The ξ-potential and size of the particles did not change, reaching –25.7 mV and 55.1 μm, respectively. The physicochemical properties, such as temperature parameters and crystallinity, of all P(3HB)/MET formulations did not change considerably.

Thus, in the study (Volova et al., 2016a), for the first time, the natural polymer poly–3-hydroxybutyrate was tested as a degradable matrix for embedding MET in order to prolong its release into the surrounding medium. The biodegradable polymer poly (3-hydroxybutyrate) was used as a basis for constructing controlled release formulations of MET: microparticles, granules, pellets, and films. Analysis of the polymer/MET mixtures by DSC, X-ray, and FTIR methods showed that MET and polymer formed a stable physical mixture. MET release from the constructed formulations into water occurred gradually, and release rates depended on the geometry of the forms.

4.3 HERBICIDAL ACTIVITY OF EXPERIMENTAL METRIBUZIN FORMULATIONS IN MODEL WEED ECOSYSTEMS

Slow-release formulations of the herbicide MET embedded in the polymer matrix of degradable poly–3-hydroxybutyrate (P(3HB)) in the form of microparticles, films, microgranules, and pellets were developed and tested. The kinetics of polymer degradation, MET release and accumulation in soil were studied in laboratory soil microecosystems with higher plants (Volova et al., 2016b).

P(3HB) was used as the matrix for preparing films, pellets, microgranules, and microparticles. Each polymer matrix was loaded with 10, 25, 50% (w/w) MET. The choice of these loadings was based on MET concentrations in the formulations commonly used in agriculture (0.180–0.96 kg/ha). Three concentrations of Sencor Ultra added to the soil in control corresponded to MET concentrations in the experimental formulations, i.e., 3, 7.5, and 15 μg MET/g soil. The techniques employed to construct P(3HB)/MET blends and their properties are described in detail elsewhere.

Films were prepared as follows: a chloroform solution containing 2% (w/v) of P(3HB) was mixed with the MET solutions (the polymer/herbicide mass ratios in the film were 90:10, 75:25 and 50:50). The films were then weighed on an analytical balance. The film thickness was measured with an EDM–25–0.001 digital micrometer (LEGIONER, Germany). The films were 25 ± 0.3 μm thick. Squares of 25 mm^2 in the area (5 mm × 5 mm) were then cut from the film. Polymer microgranules loaded with MET were prepared from a solution of the herbicide and P(3HB) in chloroform. Three types of microcapsules containing different proportions of herbicide were prepared. In the first batch of microcapsules, the polymer/herbicide mass ratio was 90:10, in the second, the mass ratio was 75:25, and in the third, the mass ratio was 50:50. The average diameter of the granules with the MET encapsulation efficiency close to 100% was 2–3 mm. MET-loaded pellets were prepared as follows: the polymer was ground in a ZM 200 ultracentrifugal mill (Retsch, Germany). Samples of the P(3HB) and MET powders were weighed on an analytical balance, mixed at polymer/MET ratios of 90:10, 75:25, and 50:50, and then homogenized with a laboratory stirrer for 2 min. Pellets were prepared from the P3HB/MET powder by cold pressing, using a laboratory bench-top hand-operated screw press (Carl Zeiss Jena, Germany) under pressing force of 6 000 F. Pellets prepared from polymer powder and MET were 3 mm in diameter and 1 mm thick. Microparticles were prepared by the emulsion technique. The polymer emulsion was prepared as follows. The oil phase, represented by a 2% P(3HB) solution with different proportions of MET in chloroform, was combined with the aqueous phase –PVA solution (Sigma, U.S., M_w 30 kDa). Microparticles 10 μm or more in diameter were prepared with an MR Hei-Standard magnetic stirrer (Heidolph, Germany), taking into account the previously determined effects of the type of the emulsion and agitation speed on the particle diameter. After solvent evaporation, microparticles were collected by centrifugation (Centrifuge 5810 R, 5417 R, Eppendorf, Germany, 10000 rpm), rinsed, and freeze-dried (Alpha 1–2 LD plus, Christ®, Germany).

Formulations of MET embedded in the P(3HB) matrix, with different MET loadings, were prepare (Figures 4.5 and 4.6).

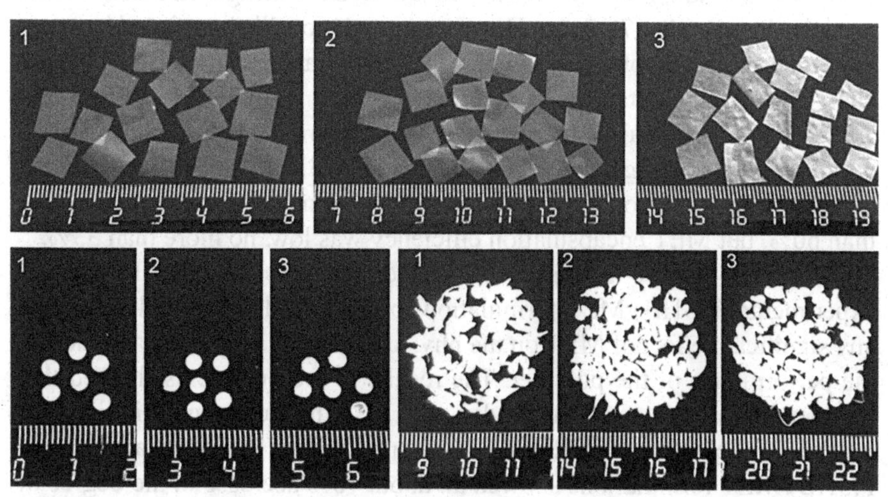

FIGURE 4.5 Photographs of the P(3HB)/metribuzin films, pellets, and granules loaded with 10 (1), 25 (2) and 50% (3) (w/w) metribuzin . (Reprinted by permission from Springer Nature: Environmental Science and Pollution Research: Poly(3-hydroxybutyrate)/metribuzin formulations: characterization, controlled release, Volova, T., Zhila, N., Kiselev, E., Prudnikova, S., Vinogradova, O., Nikolaeva, E., et al. © 2016.)

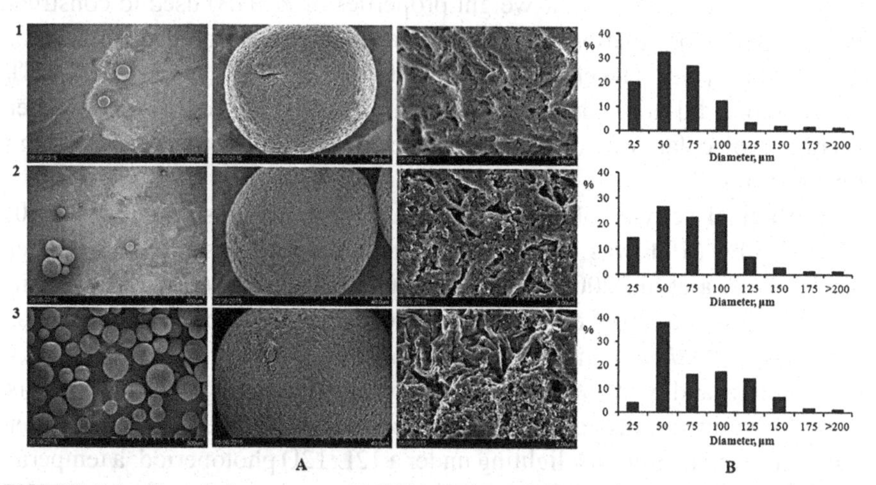

FIGURE 4.6 Cumulative release of MET into the soil from microparticles (a), films (b), microgranules (c) and pellets (d) with MET loadings of 10, 25 and 50% of the polymer weight in laboratory soil microecosystems with high plants. (Reprinted by permission from Springer Nature: Environmental Science and Pollution Research: Poly(3-hydroxybutyrate)/metribuzin formulations: characterization, controlled release, Volova, T., Zhila, N., Kiselev, E., Prudnikova, S., Vinogradova, O., Nikolaeva, E., et al. © 2016.)

The size of microparticles was influenced by MET loading: the average sizes of 10 and 25%-loaded microparticles were similar to each other – 54 μm; as the loading was increased to 50%, the average diameter of the particles increased to 70.7 μm. However, no relationship was found between the value of the ξ-potential, which is an important parameter of particles characterizing their stability in solutions, and the loading of microparticles with MET; ξ-potential of the particles with different MET loadings varied within a narrow range, between –26.2 and –33.2 mV. The yield of the particles from emulsions with different MET loadings was rather high, more than 60%, but MET encapsulation efficiency was low, no more than 33%.

Measurements of the initial substances (polymer and MET) and the experimental P(3HB)/MET formulations by DSC and X-Ray did not reveal any significant effect of MET on the physicochemical properties and, hence, the performance of the polymer. The results of measurements showed that the blending of the components did not cause their chemical binding, but produced a physical P(3HB)/MET mixture. X-ray structure analysis of P(3HB)/MET formulations showed an about 10% decrease in the degree of crystallinity (C_x) of the pellets, microgranules, and microparticles compared with the initial polymer (74%) and MET (90%); the decrease in the C_x of the films was more significant, reaching 51%. Thus, the embedding of MET into the polymer influenced crystallization of the polymer, making it somewhat more amorphous. Molecular-weight properties of P(3HB) used to construct P(3HB)/MET formulations of various shapes were compared with those of the initial polymer. The chromatograms of P(3HB), MET, and P(3HB)/MET formulations did not show any changes in the weight average and number average molecular weights and polydispersity caused by preparation of the formulations.

Herbicidal activity of the P(3HB)/MET formulations loaded with 10, 25, and 50% MET was studied in the laboratory ecosystems with higher plants (Volova et al., 2006b). Plastic containers were filled with field soil, which was used to grow two species of weeds: perennial creeping bentgrass (*Agrostis stolonifera*) and foxtail (*Setaria macrocheata*). P(3HB)/MET formulations and plant seeds were simultaneously buried in the soil. Plants were grown in a Conviron A1000 environmental chamber (Canada) under stable ambient conditions: lighting under a 12L:12D photoperiod; a temperature of 25–28°C, and humidity of 65%. Two groups were used as controls: in the positive control, the herbicide Sencor Ultra with MET concentrations corresponding to those of the experimental formulations was buried in the soil; in the negative control, no herbicide was added. A long-term experiment

(60 d) was conducted, with the condition and growth of the plants photo-monitored every week. Plant growth and productivity were evaluated by measuring fresh green biomass. The green biomass of the weeds was cut off near the ground every 7 days and weighed on the analytical balance of accuracy class 1 (Ohaus Discovery, Switzerland); the weighed portion of the plants (g) per area (m^2) was calculated. The density of the weeds was calculated based on the number of plants in a 54 cm^2 container, converted to m^2. P(3HB/MET) fungicidal activity was evaluated based on the time of plant death and the number of dead plants. The agrogenically-transformed soil (the village of Minino, the Krasnoyarsk Territory, Siberia, Russia) was placed into 250-mm^3 plastic containers (200 g soil per container). The soil was cryogenic-micellar agro-chernozem with high humus content in the 0–20-cm layer (7.9–9.6 %). The soil was weakly alkaline (pH 7.1–7.8), with high total exchangeable bases (40.0–45.2 mequiv/100 g). The soil contained nitrate nitrogen N-NO_3 – 6 mg/kg, and P_2O_5 – 6 and K_2O – 22 mg/100 g soil (according to Machigin). Results are presented in Figure 4.7.

All microparticles, irrespective of the amount of MET loading, were almost completely degraded after 30–40 days of incubation in soil (Figure 4.7a); the average degradation rates of the microparticles with the 10, 25, and 50% MET loadings were 0.15, 0.17, and 0.18 mg/d, respectively. Films showed the second highest degradation rate. After one month of incubation in soil, their residual mass was about 25, 15, and 10% of the initial mass of the films with 10, 25, and 50% MET loadings, respectively. The degradation rates of the films loaded with 10, 25, and 50% of MET were slower than those of the microparticles: 0.09 ± 0.004, 0.10 ± 0.005, and 0.18 ± 0.01 mg/day, respectively. Degradation rates of the granules were even lower: after 60 days of incubation in soil, their residual mass was 80, 60, and 40% of the initial mass, and the average degradation rates were 0.02 ± 0.003, 0.04 ± 0.002, and 0.06 ± 0.002 mg/day at MET loadings of 10, 25, and 50%, respectively. Similar mass loss dynamics was observed for the pellets, with the average degradation rates of 0.02 ± 0.001, 0.04 ± 0.002, and 0.08 ± 0.003 mg/day. The higher degradation rate of the films compared with microgranules made of P(3HB-co–3HV) and loaded with herbicide Zellek Super was determined in our previous study (Prudnikova et al., 2013). As the polymeric matrix was degraded, the molecular weight of the polymer decreased, while its polydispersity and degree of crystallinity increased, suggesting preferential disintegration of the amorphous phases of the polymer.

During the experiment, specimens were periodically taken out of the soil, and MET concentration in the soil was measured. MET was extracted

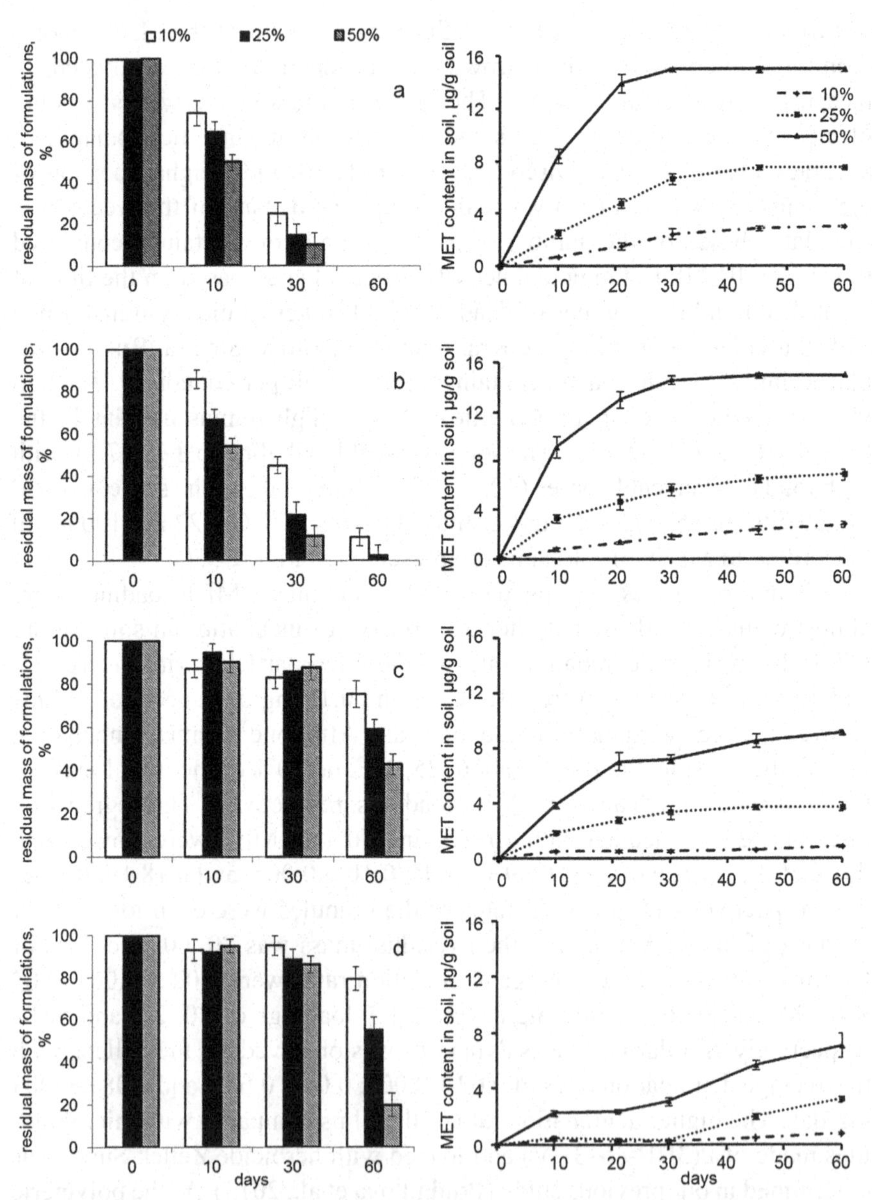

FIGURE 4.7 Degradation dynamics of P3HB/MET microparticles (a), films (b), granules (c), pellets (d) incubated in soil (A); cumulative release of MET into soil from microparticles (a), films (b), microgranules (c) and pellets (d) with MET loadings of 10, 25 and 50% of the polymer weight in laboratory soil microecosystems with high plants (B) . (Reprinted by permission from Springer Nature: Environmental Science and Pollution Research: Poly(3-hydroxybutyrate)/metribuzin formulations: characterization, controlled release, Volova, T., Zhila, N., Kiselev, E., Prudnikova, S., Vinogradova, O., Nikolaeva, E., et al. © 2016.)

with acetone from the total mass of the soil thrice. Before extraction, the soil was air dried for 48 h. The extracts were placed into the separating funnel, and distilled water, NaCl, a 10% aqueous solution of KOH, and dichloromethane were added to the funnel. The lower, dichloromethane, the layer was passed through anhydrous sodium sulfate. Extraction and filtration were performed two more times. The solvent was removed from the filtrate using a Rotavapor R–210 rotary evaporator (Switzerland), and the residue was dissolved in the known volume of acetone; MET was determined by gas chromatography. MET content in formulations determined as follows: a formulation was dissolved in chloroform. Polymer was precipitated with a double volume of hexane. Solvents containing MET were passed through anhydrous sodium sulfate. The solvents were removed using a Rotavapor R–210 rotary evaporator (Switzerland), and the residue was dissolved in the known volume of acetone; MET was determined by gas chromatography. The amount of MET released (RA) was determined as a percentage of the MET encapsulated in the polymer matrix, using the following formula (4.4):

$$RA=r/EA*100\%, \tag{4.4}$$

where EA is the encapsulated amount, mg; and r is the amount released, mg.

For describing herbicide release kinetics from different formulations, we used the Korsmeyer – Peppas model (formula 4.5):

$$M_t/M_\infty=Kt^n, \tag{4.5}$$

Here, M_t is the amount of the herbicide released at time t, M_∞ is the amount of the herbicide released over a very long time, which generally corresponds to the initial loading. K is a kinetic constant and n is the diffusional exponent. At $n=0.5$, the herbicide is released via diffusion, in accordance with the Fickian diffusion mechanism. At $n=1$, the release mechanism is described as the case-II transport, determined by relaxation processes and transitions in the carrier rather than by diffusion laws. This type of release occurs when the diffusion layer is dissolved, and the matrix is partly destroyed and degraded. Values of n between 0.5 and 1 indicate the superposition, or anomalous release. For the case of cylindrical pellets, $0.45\leq n$ corresponds to a Fickian diffusion mechanism, $0.45<n<0.89$ to non-Fickian transport, $n=0.89$ to Case II (relaxation) transport, and $n>0.89$ to super case II transport. To find the exponent n, the portion of the release curve where $M_t/M_\infty<0.6$ should only be used (Ritger and Peppas 1987; Peppas and Narasimhan 2014).

To measure residual (undegraded) polymer, the specimens (3 in a bag) were removed from the soil, thoroughly rinsed in distilled water, dried to constant weight, and weighed on the analytical balance of accuracy class 1 (Ohaus Discovery, Switzerland). Evaluation of polymer biodegradation was based on the mass loss of the specimen.

The dynamics of degradation of the polymer matrix, which determines MET release, influenced herbicide accumulation in soil (Figure 4.7a). The highest concentrations of MET were released from microparticles and films, which were comparable with MET concentration in soil from Sencor Ultra, and were measured after 20–30 days of incubation of the formulations loaded at 50, 25, and 10% MET. Concentrations reached about 15, 6.9–7.5, and 2.8–3 μg/g soil, respectively. For microgranules, MET concentration in the soil was somewhat lower – 7 ± 0.68, 3 ± 0.48 and 0.5 ± 0.03 μg/g soil after 20–30 days of incubation of the formulations loaded at 50, 25, and 10% MET; by the end of the experiment, the highest MET concentration in soil had reached 9.1 ± 0.19 μg/g soil in the experiment with the 50% MET loading. Similar MET release and concentrations in soil were obtained for pellets. Accumulation of the MET released from these forms was the slowest in the initial phase (about 30 days), reaching 0.2–4.2 μg/g soil by the end of the experiment. Thus, MET release to soil was determined by the loading degree and shape of the formulation. The 100% release of MET was observed from the microparticles, which were completely degraded during the experiment. Granules and pellets were degraded at slower rates, which affected MET release. MET release to soil occurred with the slowest rate (22–23% of the loaded amount) from 10% loaded granules and pellets, which were only 20–25% degraded. We showed the relationship between herbicide release rate and the level of loading and the type of the form in a previous study, in which we investigated P(3HB-co–3HV) microgranules and films loaded with the herbicide (Prudnikova et al., 2013).

Constant K and exponent n, characterizing kinetics of MET release from the experimental P(3HB)/MET formulations, which were obtained by using the Korsmeyer-Peppas model, are given in Table 4.1.

The time when MET is released with the highest rate is characterized by parameter t^{50} – the time needed for the herbicide content of the specimen to reach $M_t/M_\infty \leq 0.5$. MET release from microparticles was characterized by the anomalous case-II transport. The values of the exponent at different loadings varied between 0.91 and 0.99. Constant K, which contains diffusion coefficient and structural and geometric data on the formulations, varied between 0.0013 and 0.0024 h^{-1} as the loading was increased. MET embedded in microgranules was released via diffusion, in accordance with

TABLE 4.1 Constants Characterizing Metribuzin Release, According to Equation $M_t / M_\infty = Kt^n$, From the Experimental P(3HB)/MET Formulations of Different Geometries, Loaded at 10, 25, and 50% MET, Incubated in Laboratory Soil Ecosystems with Plants

Type of P(3HB)/MET formulation; MET loading, %	K(h^{-1})	n	R^2	t_{50}, d
Kinetics of metribuzin release from microparticles				
10	0.0013	0.98	0.99	21
25	0.0024	0.91	0.99	21
50	0.0021	0.99	0.99	10
Kinetics of metribuzin release from films				
10	0.103	0.27	0.92	30
25	0.005	0.77	0.96	21
50	0.0002	1.10	0.96	10
Kinetics of metribuzin release from microgranules				
10	0.020	0.24	0.82	60
25	0.019	0.47	0.99	45
50	0.021	0.46	0.91	45
Kinetics of metribuzin release from pressed pellets				
10	0.010	0.28	0.97	60<
25	0.0001	0.99	0.99	60<
50	0.0004	0.85	0.99	60<

Note: M_t is the amount of the herbicide released over time t; M_∞ is the amount of the herbicide corresponding to the initial loading; K is a kinetic constant, which contains structural and geometric data on the formulation; n is the parameter characterizing the mechanism of the release of the herbicide; t_{50} is the time of release of more than 50% herbicide. (Reprinted by permission from Springer Nature: Environmental Science and Pollution Research: Poly(3-hydroxybutyrate)/metribuzin formulations: characterization, controlled release, Volova, T., Zhila, N., Kiselev, E., Prudnikova, S., Vinogradova, O., Nikolaeva, E., et al. © 2016.)

the Fickian diffusion mechanism. The mode of MET release from the films and pellets differed depending on the loading. At the 10% loading, the values of the exponent were 0.27 for films and 0.28 for pellets. At higher loadings, MET release was characterized by the superposition of the case-II transport. Constant K decreased as exponent n increased. This relationship was reported by Akbuga (1993), Quadir et al. (2003), Sato et al. (1997). Changes in constant K for films and pellets suggested structural in homogeneity of the specimens that had the same geometry but different MET loadings. That was also supported by SEM images, which showed druses of larger areas, and by the increased degradation rates of the formulations with greater MET loadings. Degradation of films and pellets led to a change in the mechanism of MET release. The time when MET is released with the highest rate is characterized by parameter t_{50}. This parameter varies depending on the shape of the specimen and loading: as the loading is increased, the value of t_{50} is decreased (Table 4.1). Thus, variations in this parameter suggest the possibility of controlling the herbicide release rate by choosing the proper technique to produce the formulation and by varying its shape.

Microbial analysis of the soil in laboratory systems was conducted by using generally accepted methods. The number of ammonifying and copiotrophic bacteria (CFU/g soil) was determined on fish-peptone agar (FPA); the number of mineral nitrogen-assimilating prototrophic bacteria was determined on starch and ammonia agar (SAA); nitrogen-fixing bacteria were counted on Ashby's medium; oligotrophs were counted on soil extract agar (SA), and the number of micromycetes was determined on the wort agar (WA) (Netrusov et al. 2005). Mineralization coefficient was determined as a ratio between microorganisms assimilating mineral nitrogen and ammonifying bacteria. Oligotrophy coefficient was determined as a ratio of oligotrophic to ammonifying bacteria. Pure cultures of bacteria were isolated from soil samples and tested by conventional methods, based on their cultural and morphological properties and using standard biochemical tests mentioned in identification keys (Brenner et al. 2005; Dworkin et al. 2006). Dominant microorganisms were identified using MIKROLATEST® ID identification kits and 16S rRNA gene sequence analysis. Soil microscopic fungi were identified by their micro- and macro-morphological features (the structure and color of colonies, the structure of mycelium, the particularity of anamorph and teleomorph stages) (Sutton et al. 2001; Watanabe 2002). The soil had high mineralization and oligotrophy coefficients (1.52 and 11.74, respectively), indicating soil maturity and low contents of available nitrogen forms. The number of copiotrophic bacteria was 16.3 ± 5.1 million CFUs/g soil – 1.5 and 11.7 times lower than the number of prototrophic and oligotrophic bacteria, respectively, while the number of nitrogen-fixing bacteria was very high (26.1 ± 4.7 million CFUs/g soil).

As microorganisms play an essential role in pesticide cycling and transformation, we studied the effect of MET on the structure of microbial communities in laboratory systems. Studies of toxicity of MET showed that suppression of plant growth was due to the toxic effects on symbiotic microorganisms *Rhizobium* sp. MRL3, *Bradyrhizobium* sp. MRM6 and *Pseudomonas aeruginosa* PS1 isolated from the soil under lentil, mung bean, and mustard, respectively (Ahemad and Khan 2011a, b).

Free MET was added to the plant-free soil at concentrations of 3, 7.5, and 15 μg MET/g soil. In 7 days after application, no significant effect on the number of microorganisms was identified, but long-term effects of the herbicide (after 60 days) decreased the number of bacteria at all concentrations used. The number of ammonifying bacteria decreased by a factor of 1.8–3.6, prototrophic bacteria by a factor of 6.2–8.7, and nitrogen-fixing bacteria by a factor of 1.5–2 compared with the herbicide-free soil (Figure 4.8).

Incubation of P(3HB)/MET formulations with different concentrations of MET in plant-free soil increased the total number of bacteria by a factor of 2.7–3.4 compared with the initial soil. Thus, the growth of microorganism populations in the soil was influenced by the presence of the additional substrate, P(3HB), and slow release of MET from formulations, which reduced the time of exposure to high herbicide concentrations. Ultimately, that reduced the inhibitory effect of MET on soil microflora.

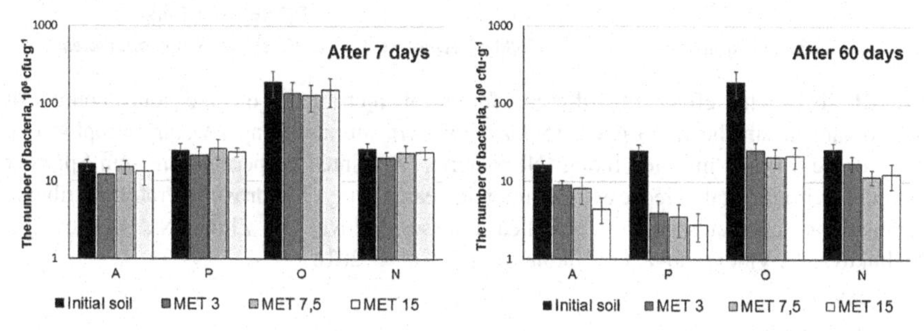

FIGURE 4.8 The effect of free metribuzin on the total counts of microorganisms in plant-free soil: ammonifying (A), prototrophic (P), oligotrophic (O) and nitrogen-fixing (N) bacteria. (Reprinted by permission from Springer Nature: Environmental Science and Pollution Research: Poly(3-hydroxybutyrate)/metribuzin formulations: characterization, controlled release, Volova, T., Zhila, N., Kiselev, E., Prudnikova, S., Vinogradova, O., Nikolaeva, E., et al. © 2016.)

Different results were obtained in experiments with MET added to the soil ecosystems with higher plants. The addition of MET as Sencor Ultra (3, 7.5, and 15 μg MET/g soil) did not significantly change the total counts of soil bacteria and fungi, i.e., MET did not either inhibit or stimulate the growth of microorganisms even at the highest concentration. The MET added as a component of P(3HB)/MET formulations stimulated the development of soil microflora: by the end of the experiment (60 days), the total counts of organotrophs and nitrogen-fixing bacteria had increased by a factor of 1.5–13.8 compared to their counts in the initial soil (Figure 4.9). Thus, soil oligotrophy coefficient decreased by a factor of 8–15.

The study of taxonomic diversity of soil microorganisms showed that the proportions of species in the microbial community had also changed with the introduction of MET. Morphological, physiological, and biochemical studies and molecular-genetic examination of the 16S and 28S rRNA gene

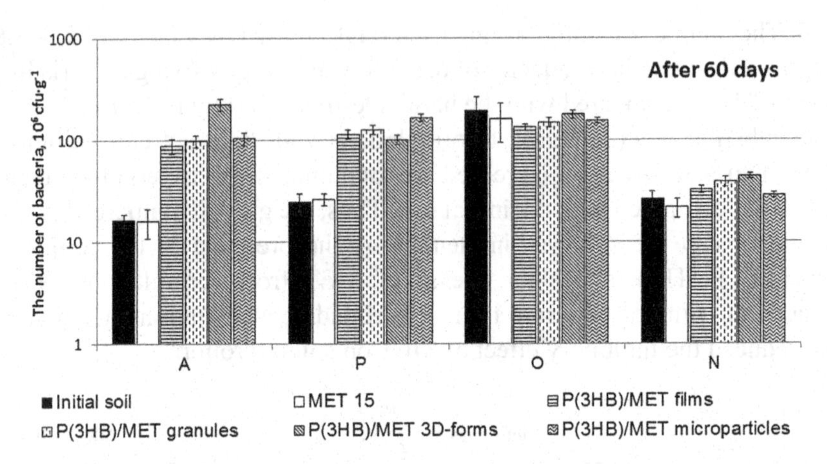

FIGURE 4.9 The effect of different forms of metribuzin on the total counts of microorganisms in the rhizosphere (*Setaria pumila*): ammonifying (A), prototrophic (P), oligotrophic (O) and nitrogen-fixing (N) bacteria. (Reprinted by permission from Springer Nature: Environmental Science and Pollution Research: Poly(3-hydroxybutyrate)/metribuzin formulations: characterization, controlled release, Volova, T., Zhila, N., Kiselev, E., Prudnikova, S., Vinogradova, O., Nikolaeva, E., et al. © 2016.)

fragments showed that the initial soil microbial community was dominated by actinobacteria (24%, including 19% of Streptomyces), *Arthrobacter* (18%) and *Corynebacterium* (12%) species; *Pseudoxanthomonas* were the major Gram-negative bacilli (12%) (Figure 4.10). By the end of the experiment (60 days), the composition of bacteria influenced by Sencor Ultra had changed due to an increase in the percentages of *Corynebacterium* (19–22%), *Arthrobacter* (20–24%), and *Bacillus* (15–17%). The percentage of *Pseudomonas* increased to 5–8%, but the total percent of Gram-negative bacilli decreased by 10–12% and actinobacteria decreased too (by 8–15%). The addition of P(3HB)/MET formulations increased the percentage of Gram-negative bacilli, including *Pseudomonas*, *Pseudoxanthomonas, Stenotrophomonas, and Variovorax*, to 31–33%. The proportion of spore-forming bacteria (*Bacillus* and *Paenibacillus*) also increased).

In all soil samples, the major microscopic fungi were *Penicillium* species, which constituted 52–65% (Figure 4.11). Fungi of the genera *Fusarium*, *Trichoderma*, and *Aspergillus* constituted 8–11% of the population of microscopic fungi in the initial soil and did not change significantly either. In the experiment with P(3HB)/MET formulations, an increase in the percentages of *Acremonium* and *Verticillium* was observed.

Thus, it was shown that the stimulating effect of P(3HB)/MET was caused by poly-3-hydroxybutyrate, which was a supplementary growth substrate

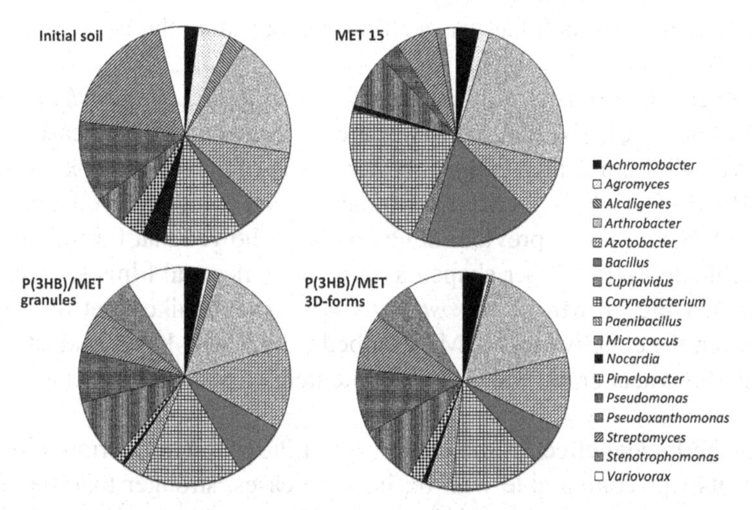

Figure 4.10 The effect of different forms of metribuzin on taxonomic composition of the bacterial community in the rhizosphere (*Setaria pumila*). (Reprinted by permission from Springer Nature: Environmental Science and Pollution Research: Poly(3-hydroxybutyrate)/metribuzin formulations: characterization, controlled release, Volova, T., Zhila, N., Kiselev, E., Prudnikova, S., Vinogradova, O., Nikolaeva, E., et al. © 2016.)

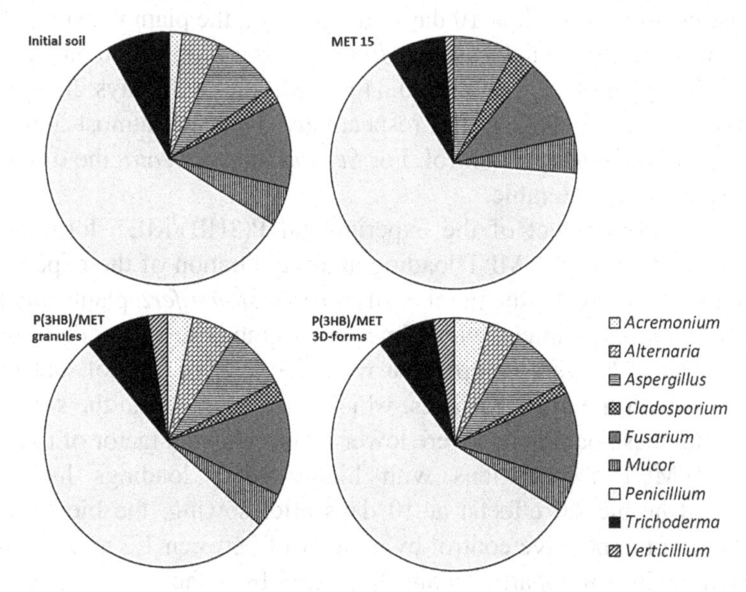

FIGURE 4.11 The effect of different forms of metribuzin on taxonomic composition of soil fungi in the rhizosphere (*Setaria pumila*). (Reprinted by permission from Springer Nature: Environmental Science and Pollution Research: Poly(3-hydroxybutyrate)/metribuzin formulations: characterization, controlled release, Volova, T., Zhila, N., Kiselev, E., Prudnikova, S., Vinogradova, O., Nikolaeva, E., et al. © 2016.)

for soil microflora, and that was a stronger factor than the inhibitory effect of MET (Volova et al., 2016b).

Further, the weeds *Agrostis stolonifera* and *Setaria macrocheata* were used to study the herbicidal activity of the experimental slow-release formulations of MET embedded in the polymer matrix of poly–3-hydroxybutyrate, P(3HB)/MET. All P(3HB)/MET formulations had comparable effects on the plants (Table 4.2). In a previous study, we also showed that formulations of the herbicide Zellek Super shaped as microgranules and films successfully suppressed the growth of *Agrostis stolonifera* (Prudnikova et al., 2013). Moreover, the effectiveness of MET embedded in CMC-KAO against weeds growing in wheat crops was shown in the field experiment by Kumar et al. (2010b).

The herbicidal effect of the experimental P(3HB)/MET formulations on the plants was comparable with or, in some cases, stronger than the effect achieved in the positive control. Analysis of the parameters of MET effect on the plant density and the weight of fresh green biomass (Table 4.2) showed that all experimental P(3HB)/MET formulations exhibited herbicidal activity. See photographs in Figure 4.12, which show *Agrostis stolonifera* and *Setaria macrocheata* plants at 10, 20, and 30 days post planting and application of the herbicide.

In the positive control, at 10 days after sowing, the plant density and the weight of the biomass of *Agrostis stolonifera* were 8333 ± 750 and 21.28 ± 1.26, at 20 days 6481 ± 713 and 10.64 ± 0.84, and at 30 days 2090 ± 187 plants/m^2 and 5.32 ± 0.32 g/ m^2, respectively. That was almost 5–6 times lower than in the negative control. For *Setaria macrocheata*, the difference was even more considerable.

The inhibitory effect of the experimental P(3HB)/MET formulations varied depending on the MET loading and the duration of the experiment. At 10 days after sowing, the number of *Agrostis stolonifera* plants and their biomass in the experiment with films, microgranules, and pellets loaded with MET at 10% were comparable with the positive control, but in the ecosystems with the microparticles, which were degraded in the soil at the highest rate, these parameters were lower by more than a factor of two.

P(3HB)/MET formulations with higher MET loadings had more pronounced herbicidal effects: at 10 days after sowing, the biomass was lower than in the positive control by a factor of between 1.7 and 4.1 in the ecosystems with microparticles and films and by a factor of between 1.3 and 1.8 in the experiments with microgranules and pellets. At 20 days, a considerable number of plants in all treatments were dead, and the green

TABLE 4.2 The Density of Plants and Weight of Fresh Green Biomass of the Weeds Grown in the Laboratory Microecosystems With Slow-Release P(3HB)/MET Formulations

P(3HB)/ MET formulation; MET loading, %	10 days	20 days	30 days	10 days	20 days	30 days
Biomass of *Agrostis stolonifera* (g/m^2)				Density of *Agrostis stolonifera* (number/m^2)		
Microparticles						
10	9.03 ± 0.47	4.08 ± 0.28	-	3536 ± 424	1597 ± 175	-
25	6.40 ± 0.37	3.26 ± 0.15	-	2506 ± 223	1276 ± 166	-
50	4.70 ± 0.20	1.49 ± 0.06	-	1840 ± 145	583 ± 52	-
Films						
10	20.07 ± 1.60	9.24 ± 0.55	4.36 ± 0.19	7859 ± 865	3618 ± 290	-
25	14.41 ± 1.01	7.94 ± 0.48	-	5642 ± 670	2209 ± 199	-
50	12.41 ± 0.75	-	-	4859 ± 435	-	-
Microgranules						
10	17.10 ± 1.33	10.42 ± 0.83	3.12 ± 0.22	6696 ± 870	4080 ± 326	1221 ±
25	15.41 ± 0.93	7.16 ± 0.58	-	6030 ± 540	2803 ± 252	-
50	11.07 ± 0.78	4.86 ± 0.23	-	4334 ± 475	1903 ± 133	-
Pellets						
10	17.10 ± 1.19	10.24 ± 0.72	5.36 ± 0.27	6696 ± 604	4009 ± 480	2098 ± 187
25	15.41 ± 0.93	9.94 ± 0.70	3.02 ± 0.13	6030 ± 840	3892 ± 500	1182 ± 154
50	12.41 ± 0.87	8.40 ± 0.59	1.12 ± 0.05	4859 ± 401	3289 ± 462	438 ± 31
Control (-)	72.81 ± 5.10	152.0 ± 10.63	198.0 ± 15.02	46296 ± 6020	46296 ± 694	60306 ± 844
Control (+)	21.28 ± 1.26	10.64 ± 0.84	5.32 ± 0.32	8333 ± 750	6481 ± 713	2090 ± 187
Biomass of *Setaria macrocheata* (g/m^2)				Density of *Setaria macrocheata* (number/m^2)		
Microparticles						
10	29.63 ± 1.76	-	-	741 ± 81	-	-
25	16.42 ± 1.15	-	-	370 ± 33	-	-
50	-	-	-	-	-	-
Films						
10	29.63 ± 1.58	7.61 ± 0.52	-	741 ± 67	185 ± 23	-
25	16.40 ± 1.15	-	-	555 ± 61	-	-
50	29.63 ± 1.47	-	-	741 ± 89	-	-
Microgranules						
10	22.81 ± 1.80	16.44 ± 1.15	7.61 ± 0.47	555 ± 50	370 ± 48	185 ± 17
25	16.40 ± 1.14	8.21 ± 0.56	-	370 ± 44	185 ± 16	-
50	16.40 ± 0.97	8.21 ± 0.50	-	370 ± 38	185 ± 17	-

TABLE 4.2 *(Continued)*

P(3HB)/ MET formulation; MET loading, %	10 days	20 days	30 days	10 days	20 days	30 days
Pellets						
10	29.63 ± 2.30	22.83 ± 1.37	16.44 ± 0.97	741 ± 89	555 ± 50	370 ± 30
25	28.41 ± 1.96	22.83 ± 1.29	16.44 ± 0.81	741 ± 70	555 ± 48	370 ± 36
50	29.63 ± 2.07	14.81 ± 1.08	-	741 ± 65	370 ± 40	-
Control (-)	326.48 ± 28.37	482.55 ± 37.61	557.4 ± 50.1	7962 ± 955	7962 ± 916	7962 ± 876
Control (+)	29.62 ± 1.67	22.7 ± 1.57	-	741 ± 59	555 ± 58	-

biomass was reduced much more dramatically than in the positive control. At 30 days, all plants were dead in the treatments and positive control. Similar results were obtained for *Setaria macrocheata* plants. The herbicidal activity of the P(3HB)/MET formulations also increased with the increase in the MET loading and with the duration of the experiment (Table 4.2). At 10 days after sowing, the plant density and the weight of fresh biomass were either comparable with or lower than the corresponding parameters in the positive control, depending on the MET loading and type of formulation. At 20 days, in the ecosystems with P(3HB)/MET microparticles and films, almost all plants were dead; in the ecosystems with microgranules and pellets, the herbicidal effects were less pronounced but stronger than in the positive control (by a factor of 1.5–2.8).

Thus, comparison of the herbicidal effect of the experimental MET formulations with that of the positive control (Sencor Ultra) showed that all experimental P(3HB)/MET formulations exhibited high herbicidal activity. The inhibitory effect of the experimental P(3HB)/MET formulations depended on the MET loading and duration of the experiment. In the early stages and at the lowest (10%) loadings, the effect of P(3HB)/MET was comparable to that of commercial Sencor Ultra. The effects of experimental formulations with higher MET loadings and in longer experiments were superior to the effect of Sencor Ultra. Thus, degradable poly–3-hydroxybuturate can be regarded as a promising material for designing slow-release formulations of the herbicide MET for soil applications (Volova et al., 2016b).

FIGURE 4.12 Photographs of *Agrostis stolonifera* (A) *Setaria macrocheata (B)* grown under laboratory conditions using P(3HB)/MET films with MET loadings of 10 (3), 25 (4) and 50% (5) of the polymer weight relative to negative control (1) and positive control (2). (Reprinted by permission from Springer Nature: Environmental Science and Pollution Research: Poly(3-hydroxybutyrate)/metribuzin formulations: characterization, controlled release, Volova, T., Zhila, N., Kiselev, E., Prudnikova, S., Vinogradova, O., Nikolaeva, E., et al. © 2016.)

The advancement of PHA biosynthesis processes has enabled the scale-up of the industrial production of these polymers (Kaur, Roy, 2015). It is impossible not to note that the increase of implementation of PHA polymers at the present time is still lowered by their high price. Searching for means of cost reduction of PHA based products is actual for PHA technical implementation. Not only pure PHA but also a mixture of these polymers with more obtainable materials can be a way of embedding agricultural preparations. Filling a polymer matrix with variable materials also gives access to the regulation of the polymeric carrier degradation and release of the agent into the environment.

Boyandin et al. (2016) constructed and researched sustained release formulations of herbicide MET with the application of mixtures of poly(3-hydroxybutyrate), the best-known polymer of PHA class, with synthetic and natural materials. Experimental formulations of herbicide MET embedded in matrices of degradable natural polymer poly(3-hydroxybutyrate) (P3HB) and its composites with PEG, polycaprolactone (PCL).

Poly(3-hydroxybutyrate) (P(3HB)) (the weight average molecular weight M_w 920 kDa; polydispersity 2.52) was synthesized according to previously described technology (Volova et al.,2013;2014). Poly-ε-caprolactone (PCL) with the number average molecular weight M_n 80 kDa in the form of granules and also poly(ethylene glycol) with M_w 300 kDa in the form of powder were produced by Aldrich (USA). Wood powder was obtained by birchwood disintegrating with the MD 250–85 wood-carving working bench (Stanko-Premier, Russia), followed by drying at 60°C for 120 h until it reached its constant weight and particle fraction selecting by means of screen sizing with 0.5 mm mesh. MET was purchased from "Ecolan" (Moscow, Russia). Empirical formula: $C_8H_{14}N_4OS$. Molecular mass: 214.3. IUPAC name: 4-amino–6-tert-butyl–4,5-dihydro–3-methylthio–1,2,4-triazin–5-one. CAS name: 4-amino–6-(1,1-dimethylethyl)–3-(methylthio)–1,2,4-triazin–5(4*H*)-one. Pure substance: colorless needles. Melting point of the active material: 126.2°C. Pure MET was used as positive control 1. A commercial MET formulation, Sencor Ultra, containing 600 g/kg of the active ingredient was purchased from Bayer CropScience AG (Monheim am Rhein, Germany), and used as positive control 2. For obtaining P(3HB)/filling material mixtures, polymers (P(3HB) and PCL) were ground using a ZM 200 mill (Retsch, Germany), and particle fractions under the size of 1 mm were selected. Particle size distribution of the obtained powder was determined with an analytical sieving machine AS 200 control (Retsch, Germany): particle fraction of size under 0.50 mm totaled 60%; of size from 0.50 to 1.00 mm – 40%. Pellets, 13 mm in diameter, containing 150 mg of pure P(3HB) or its mixtures with birch sawdust, powder of PCL or PEG in the ratio of 7:3 (105 mg of P(3HB) and 45 mg of the second component) were obtained by cold pressing powdered components using a Carver Auto Pellet 3887 press (Carver, U.S.); pressing force was 14 000 F. Pure P(3HB) powder and its mixtures with birch sawdust, PCL, and PEG powder at a source ratio of 7:3 (i.e., 70% of P(3HB) and 30% of the second component) were used for construction of pellets loaded with 25% MET. Specimens obtained by mixing of powders (150 mg of pure P(3HB) or 105 mg of P(3HB) and 45 mg of the second component) with 50 mg of MET were cold-pressed by the same way as it was described in the previous section.

Biodegradation of composite materials based on P3HB and release kinetics of MET in soil. MET was detected by gas chromatography using the gas chromatograph 7890/5975C (Agilent Technologies, U.S.) equipped with a mass spectrometer as it was described earlier. Kinetics of MET release from the polymeric samples was studied in laboratory systems with soil. For

MET concentration analysis in soil, – dry soil were an overflow of chloroform. Then the soil was glass-filtered using a vacuum pump, rinsing it with chloroform. United obtained extracts were concentrated on a rotary evaporator, then quantitatively transferred into smaller flasks using chloroform. Chloroform was removed in the rotary vacuum evaporator. The amount of MET released was determined as a percentage of the MET encapsulated in the polymer matrix. Mathematical analysis of release kinetics of MET was performed using the Korsmeyer – Peppas model.

Field soil (the village of Minino, the Krasnoyarsk Territory,) was cryogenic-mycelial agricultural black soil characterized by high humus level, weakly alkaline reaction of media (pH 7.1–7.8), and organic carbon 5.1%. Nitrate nitrogen (NO_3^-) concentration was 6 mg/kg, P_2O_5 – 60 mg/kg, K_2O – 220 mg/kg soil. Soil density had normal and mellow topsoil configuration (0.85–1.11g/cm^3). The total counts of organotrophic bacteria were (16.3 ± 5.1)×10^6 CFU/mL. Garden soil (the village of Subbotino, the Krasnoyarsk Territory) was agrogenically modified soil with very high humus level (17.4 %), pH 6.6, and organic carbon 9.9%. NO_3^- concentration was 122.0 mg/kg, P_2O_5 – 1512 mg/kg; K_2O – 800 mg/kg soil. The total counts of organotrophic bacteria were (66.3 ± 29.5)×10^6 CFU/mL. Experiments with MET release were conducted using the field and the garden soil. Other soil experiments were carried out with the field soil only.

Terms of exposition of pellets in soil totaled 35 days at the temperature of 21°C. A part of the samples were extracted from the soil at intervals of 7 days. For determining mass reduction, they were washed free from soil, placed in a thermostat by 40°C for 24 h for drying them to their constant weight and weighed on analytical scales (Ohaus Discovery, Switzerland). The experiment was carried in two alterations: with samples of P3HB/filling material (without MET) and samples of P3HB/filling material/MET. In the second alteration, the soil was analyzed for MET content after extraction of pellets.

As it is seen from Figure 4.13, among P(3HB)/PEG, P(3HB)/PCL, P(3HB)/wood powder, P3HB/PEG forms were the most actively degrading. In 35 days of observation, their mass reduction totaled 62.3 ± 7.0% from the original. It can most likely be explained by relatively fast erosion of water-soluble PEG from the form and then it was followed by fast polymer degradation. For P(3HB)/PCL sample, slow mass reduction is specific, which lowered by 49.4 ± 11.1% during the whole observation period. The least degradable were P(3HB)/wooden powder and pure P(3HB) forms, which lost their masses by 27.1 ± 1.2% and 30.3 ± 11.1%, respectively, during the experiment. These results show that the addition of filling material in

P3HB carriers can regulate the kinetics of the matrix degradation and thus influence the release of herbicide included in those forms.

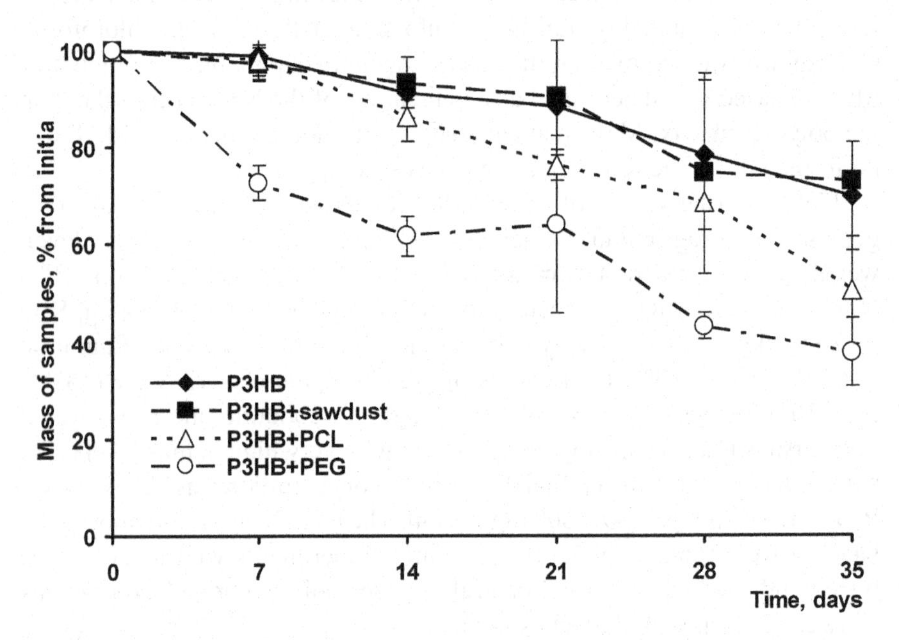

FIGURE 4.13 Mass decrease of pellets from P3HB and filling material at its exposing to soil in laboratory conditions (Boyandin et al., 2016).

For a comparative study of the release kinetics of MET from developed formulations, different laboratory systems were used. One of them was represented by sterile water; the others were soil ecosystems with microorganisms. During the matrices exposure process in water and MET structural release changes were noticed (appearance of small fissures, cavities), revealed by scanning electron microscopy.

Measurement of MET in sterile water (Figure 4.14) showed a quick increase of its concentrations in 14 days relevant for all samples; after that release curves gradually reach a plateau. Within 35 days, MET release arrived at 38.7 ± 5.7% from P3HB/PEG, 35.0 ± 5.8% from P3HB, 28.4 ± 4.5% from P3HB/sawdust and 16.8 ± 5.2% from P3HB/PCL formulations, respectively. The fastest preparation release was registered for the formulations from P3HB with water-soluble PEG: more than half of embedded MET in 14 days, and almost all substance by the end of the experiment were detected in the water. Thus, passive release of MET into water totals 70 days and more depending on the type of carrying material.

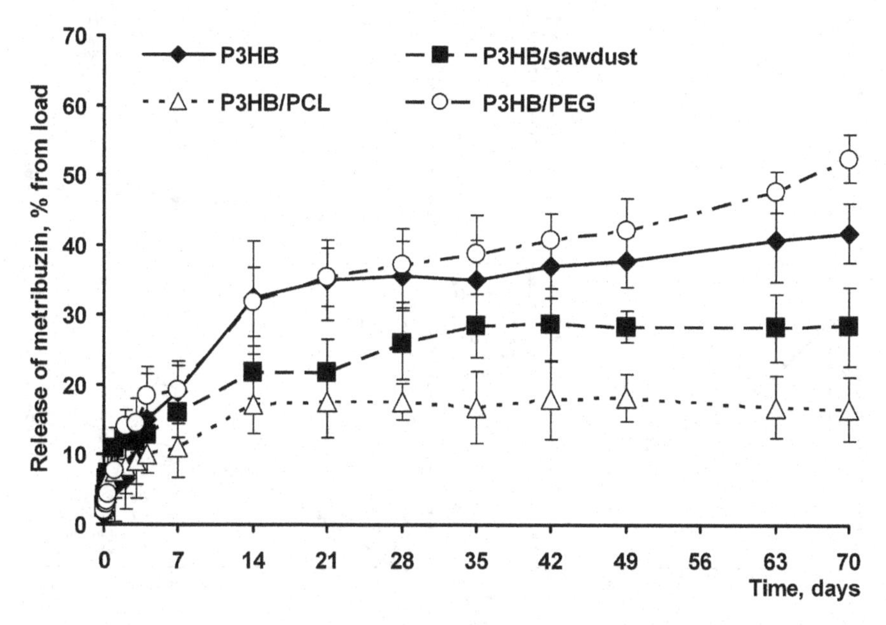

FIGURE 4.14 Metribuzin concentration in the course of exposing P3HB/filling material/metribuzin pellets to water (Boyandin et al., 2016).

A similar result was obtained by exposing developed MET forms to soil ecosystems in laboratory conditions during 35 days. As a result of P(3HB) degradation under the action of soil microflora, a mass decrease of samples was faster (Figure 4.15a, c). Samples with P(3HB)/PEG carrier degraded most actively. For these samples, reduction of mass by a half by the 7th day was registered; by the end of the experiment, remaining mass did not exceed 19% of the original. Kinetics of mass reduction of P(3HB), P(3HB)/PCL and P(3HB)/wood powder samples was comparable. After 14 days, a mass reduction of these samples was registered at the level of 20–25% for the field soil and 40% for the garden soil. By the end of the experiment, remaining mass of P(3HB)/PCL samples in the field soil totaled 41%, P(3HB) and P(3HB)/wood powder – 55% and 56% respectively.

Table 4.3 shows the kinetic parameters of herbicide release from studied formulations. Theoretical analysis of MET release into the field soil showed that the maximal K values were for P(3HB)/PEG pellets (0.072 h^{-1}) and the minimal for P(3HB) and P(3HB)/PCL pellets (0.003 h^{-1} and 0.0001 h^{-1}, respectively). The highest value of the exponent, 0.92, was for P(3HB)/PCL pellets indicating Case II transport mechanism for this type of the carrier. The lowest value was for P(3HB)/PEG matrices, 0.34, shows herbicide

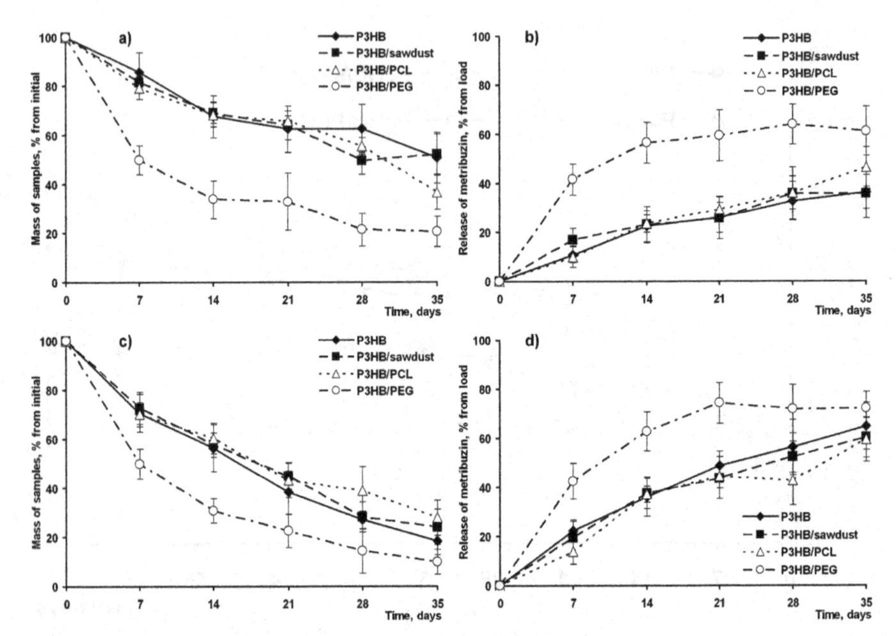

FIGURE 4.15 Remaining masses of P3HB/filling material/metribuzin formulations (a, c) and metribuzin concentration (b, d) in the course of exposing samples to field (a, b) and garden (c, d) soil (Boyandin et al., 2016).

diffusion according to Fick's law. Both other samples were characterized by an anomalous outflow of MET. For the garden soil, K varied from 0.007 for P(3HB) and P(3HB)/PEG pellets to 0.002 for P(3HB)/PCL pellets, respectively. For all samples, exponent values corresponded with anomalous outflow mechanisms that can be explained by the strong influence of microbial degradation of matrices.

TABLE 4.3 The Constants Derived From Fitting the Equation $M_t/M_\infty = Kt^n$ to Release of Metribuzin Into Water and Soil From P(3HB)/MET Formulations (Boyandin et al., 2016)

	Water			Field soil			Garden soil		
Formulations	**K(h^{-1})**	**n**	**R^2**	**K(h^{-1})**	**n**	**R^2**	**K(h^{-1})**	**n**	**R^2**
P(3HB)	0.011	0.49	0.95	0.003	0.74	0.97	0,007	0,67	0,99
P(3HB)/PEG	0.016	0.47	0.98	0.072	0.34	0.94	0,007	0,8	0,99
P(3HB)/PCL	0.031	0.25	0.94	0.0001	0.92	0.98	0,002	0,83	0,92
P(3HB)/Sawdust	0.031	0.33	0.98	0.013	0.49	0.95	0,005	0,71	0,96

*R^2 – Validity coefficient of approximation

Degradation of P(3HB)/filling material/MET pellets was followed by MET release from the polymeric carrier into soil (Figure 4.15b, d). The fastest MET release was recorded from P(3HB)/PEG samples, from which about 61% of the embedded herbicide was released into the field soil during the observation period. MET release was comparable with that from the other three forms; e.g., its concentration in the field soil gradually increased and reached 35–47% of the embedded amount. Specimens buried in the garden soil were degraded at a higher rate (Figure 4.15c) than those buried in the field soil (Figure 4.15a), thus increasing the rate of MET release (Figure 4.15b, d): 72% for the P(3HB)/PEG samples and 59–65% for other three forms. In both cases, however, the herbicide was also released with the highest rate from P(3HB)/PEG matrices; the rates of MET release from the other formulations were similar to each other. Thus, developed experimental forms of MET using degradable P(3HB) as a matrix are suitable for long-term (over 30 days) functioning after introduction into soil. Variation of filling material (PEG, PCL, wood powder) allows affecting P(3HB) degradability significantly and to control the release of MET from the form.

The efficiency of the experimental MET formulations was evaluated by growing plants in vitro. A perennial grass "creeping bent" (*Agrostis stolonifera*) was used as a test plant. Plastic containers of 250 cm^3 in volume and with the surface area of 50 cm^2 were filled with 200 g of field soil; pellets of MET were injected simultaneously with planting seeds of the test plant. In the control samples (positive control), 0.5 mg of free MET (0.5 mg of a pure substance or 0.83 mg of Sencor Ultra preparation) corresponding to the norm of 1 kg/ha were injected into each container; plants grown without herbicide in the soil served as negative control. In the test samples, a 2 mg fragment of a suitable matrix (P(3HB) or its composite with PCL, PEG or wood powder) containing 0.5 mg of embedded pesticide was buried in each pot. Seeds of *Agrostis stolonifera* L. (150 mg per pot) were sown in the soil. Plants were grown in a Conviron A1000 environmental chamber (Canada) under stable ambient conditions: lighting under a 12L: 12D photoperiod; a temperature of 20°C, and humidity of 65%. To evaluate the efficiency of the experimental MET forms, the appearance of the plants (e.g., wilt) was estimated, and the yield of the vegetative part of plants was evaluated by the weight of dry solids (drying at 105°C until the constant weight was reached in 24 h) at Days 7, 14, and 21 of the experiment.

Obtained results, showing the reliability of developed formulations of MET embedded into P3HB/filling material as a degradable matrix made it possible to assess their efficiency in an experiment with test weed higher plant – creeping bentgrass (*Agrostis stolonifera* L.).

Biomass assessment by dry substance weighing during the experiment served as the indicator of herbicidal effect (Table 4.4). Significant herbicidal effect of developed forms was registered. In the negative control sample, grown biomass totaled 72 mg in 20 days; in both positive control samples, with pure substance and commercial preparation Sencor Ultra, moderate growth of plants not exceeding 6% from the negative control sample was reported. In all cases of embedded herbicide introduction, no growth appeared.

TABLE 4.4 Influence of Metribuzin Carrying Form on Biomass Increase *Agrostis Stolonifera* L. (Dry Biomass, mg) (Boyandin et al., 2016)

Age of plants, days	Initial soil (negative control)	Free metribuzin (positive control 1)	Sencor® Ultra (positive control 2)	Formulations P3HB/filling material/metribuzin			
				P(3HB)	P(3HB)/ PEG	P(3HB)/ PCL	P(3HB)/ wood powder
7	3.3 ± 0.1	0	0	0	0	0	0
14	20.6 ± 2.6	1.40 ± 0.91	1.23 ± 0.84	0	0	0.09 ± 0.09	0.23 ± 0.21
20	72.6 ± 6.8	4.13 ± 0.34	2.31 ± 0.96	0	0	0	0

Thus, the constructed forms of herbicide embedded into a polymeric carrier had a significant negative effect on the growth of the test weed, and it was somewhat stronger than the effect of free forms of MET.

So, in the result of the experiment, the strongly pronounced herbicidal effect of developed formulations in comparison with negative control (plant growth without MET implementation) and the more significant effect in comparison with free-form preparation use was registered. The use of biodegradable poly(3-hydroxybutyrate) and filling materials as a platform for embedding MET made it possible to develop sustained-release forms of herbicide preparations for pre-emergence application through insertion into soil. The efficiency of the herbicidal effect of the developed formulations on a model pest plant, possibility of polymeric carrier degradation control, as well as control over MET release with the application of different types of filling materials have been revealed.

4.4 HERBICIDAL ACTIVITY OF SLOW-RELEASE HERBICIDE FORMULATIONS IN WEED-CONTAMINATED WHEAT STANDS

Herbicides constitute one of the largest group of pesticides. Herbicides are widely used to suppress the growth of unwanted weeds which are reported to compete with the desired crops. Apart from its beneficial characteristics, the abundant usage of herbicides is often associated with various hazardous implications. It has been reported that herbicidal usage can lead to biomagnification and add its deleterious effects on the normal flora and fauna causing risk and imbalance to the ecosystems. In recent years, inappropriate usage of herbicides has led to the development of a resistant mechanism in weeds. Hence, in order to cope with these limitations, scientific communities are engaged to create new formulations which can act efficiently on the undesirable weeds and promote crop productivity with minimal toxicity to the ecosystem. In recent years, encapsulation of herbicides in the polymer matrix is gaining importance owing to the fact that control release of herbicides leads to profound activity in comparison with the sole herbicides.

Although the increasing number of studies is focused on new-generation slow-release targeted herbicide formulations, the published studies mainly describe the methods of embedding herbicides and materials used as matrices. Very few data can be found on the herbicidal efficacy of new formulations and outcomes of studies conducted with weed-infested crops. In the work (Zhila et al., 2017), the present study was designed and investigated the herbicidal activity of MET, and tribenuron-methyl formulations in wheat stands *Triticum aestivum*, cv. Altaiskaya 70 infested by weeds – white sweet clover *Melilotus albus* and lamb's quarters *Chenopodium album* – under laboratory conditions.

Two herbicides were used: MET and tribenuron-methyl. MET – (MET) ($C_8H_{14}N_4OS$) (State Standard Sample 7713–99 – the state standard accepted in Russia (Blok–1, Moscow) – has a systemic effect against a wide range of weeds infesting vegetable and grain crop stands. Tribenuron-methyl **(TBM)** ($C_{15}H_{17}N_5O_6S$) (State Standard Sample 8628–2004 – the state standard accepted in Russia (Blok–1, Moscow) – is a systemic herbicide that has a selective effect on a wide range of weeds; it is highly effective against weeds at different stages of their development (from the tillering phase to the formation of the second stem internode); it both has a foliar action and can penetrate into plants through their roots. TBM has no aftereffects; it can be used in all types of crop successions. It is used to control broad-leaved weeds (poppy, chamomile, perennial Canada thistle, cruciferous plants, black bindweed, common chickweed, etc.) in cereals. TBM acts by inhibiting acetolactate kinase – the enzyme that catalyzes the synthesis of branched-chain amino acids (isoleucine and valine). Specific density (g/mL): 1.46. The lifespan in soil (in the field) is between 5 and 20 days, according to the data obtained in laboratory research in the European Union.

P(3HB) was used as a matrix for preparing films and microgranules. Each polymer matrix was loaded with 25% (w/w) MET or TBM. Films were prepared as follows: a chloroform solution containing 2% (w/v) of P(3HB) was mixed with MET or tribenuton-methyl solutions (the polymer/herbicide mass ratios in the film were 75:25). The polymer/MET or polymer/TBM solution systems were placed onto the MR Hei-Standart magnetic stirrer (Heidolph, Germany) operated at a speed of 300 rpm for 2–3 h (until completely dissolved). The homogeneous polymer/herbicide solution was filtered and poured into the degreased mold under a bell-glass (to protect it from draught and dust). The films stayed under the bell-glass for 24 h at room temperature, and then they were placed into a vacuum drying cabinet (Labconco, U.S.) for 3–4 days, until complete solvent evaporation took place. The films were then weighed on an analytical balance. The film thickness was measured with an EDM–25–0.001 digital micrometer (LEGIONER, Germany). The films were 25 ± 0.3 μm thick. Squares of 25 mm^2 in area (5 mm × 5 mm) were then cut from the films.

Polymer microgranules loaded with MET or TBM were prepared from the solution of the herbicide and P(3HB) in chloroform. The system was mixed to achieve homogeneity, by using a Silent Crusher high-speed homogenizer (Heidolph, Germany). A Pumpdrive 5001 peristaltic pump (Heidolph, Germany) was used to drop the polymer/herbicide solutions into a sedimentation tank that contained hexane, where the polymer was crystallized and granules formed. The polymer concentration in the solution = 10% (w/v), needle size = 20G, and thickness of the precipitate layer (h) = 200 mm. The polymer/herbicide mass ratio was 75:25. The average diameter of the granules with the MET encapsulation efficiency close to 100% was 2–3 mm.

The herbicidal activity of the developed experimental formulations was studied using laboratory ecosystems. An agrogenically-transformed soil (collected at the village of Minino, Krasnoyarsk Territory, Siberia, Russia) was placed into 500 cm^3 plastic containers (500 g soil per container). In the present investigation, soft spring wheat (*Triticum aestivum*, cv. Altaiskaya 70) was used as subject crop, and white sweet clover *Melilotus albus* and lamb's quarters *Chenopodium album* were used as weeds.

P(3HB)/MET or P(3HB)/TBM films and microgranules were weighed on an analytical balance of accuracy class 1, placed in close-meshed gauze bags (3 specimens of the same formulation per bag), and the seeds of wheat (100 g per 1 m^2), white sweet clover (24 g per 1 m^2), and lamb's quarters (20 g per 1 m^2) were simultaneously buried in the soil. In the positive control, we treated soil with commercial formulations of MET (Sencor Ultra) (or TBM (Stalker) at the beginning of the experiment; the amounts of the herbicides applied corresponded to the recommended application rates – 0.180–0.96 kg/ha or 0.015–0.025 kg/ha, respectively (the same concentrations as those of the active ingredients in the experimental formulations). Wheat stands infested by weeds but not treated with herbicides were used as negative control.

The experimental set up was incubated in a Conviron A1000 environmental chamber (Canada) for 50 days. A six-step diurnal cycle of the temperature, lighting, and humidity was maintained: night – early morning – late morning – early afternoon – late afternoon – evening. The temperature was varied with 10°C at night and 18°C during the daytime for the first seven weeks of the experiment and 14°C at night and 22°C during the day for the following five weeks. The illumination was varied between 0 and 300 μmol/m^2/s, in 100 μmol/m^2/s increments. The moisture content of the soil was maintained to be 50%. The experimental set up was recorded and photographed every week regularly to monitor its growth. The herbicidal activity was measured as the rate of suppression of weed growth above ground level with plant density per unit area. The effect of herbicidal activity on the growth of wheat plants was estimated and measured with productive growth of the fresh green biomass.

The herbicidal activity of the developed P(3HB/MET) films and microgranules was studied in wheat stands infested with the weed – white sweet clover. The complete 50 days experimental outcomes are illustrated in Figure 4.16 and 4.17. The study was compared with both positive and negative control. On the 10th day of the experiment, the density of the white sweet clover plants and their biomass in control (without application of the herbicide) reached 6537 plants/m^2 and 9.8 g/m^2, respectively. Those values were significantly higher than the corresponding values in the positive control (5185 plants/m^2 and 5.1 g/m^2) and treatments (4352–5556 plants/m^2 and 4.3–6.1 g/m^2), where the weed development was evidently inhibited. The effect was more significant on the 20th day which showed a decline in the number of clover weeds to 1481 and 1790 plants per m^2 with the treatment of P(3HB)/MET films and microgranules, respectively; in the positive control, the weed density was somewhat higher (2037 plants/m^2). At the end of the experiment, at Day 50, complete suppression of clover plants was observed in the herbicide-treated ecosystems. It is essential that the clover plant density and the amount of the weed aboveground biomass were significantly lower in the ecosystems with the two experimental formulations – P(3HB)/MET films and microgranules – than in the ecosystem with the commercial formulation.

FIGURE 4.16 Photographs of wheat (*Triticum aestivum*) stands infested with white sweet clover (*Melilotus albus*) and treated with different P(3HB)/MET formulations: *a* – microgranules, *b* – films, *c* – positive control, *d* – negative control (untreated). (Reprinted with permission from Zhila, N., Murueva, A., Shershneva, A., Shishatskaya, E., & Volova, T., (2017). Herbicidal activity of slow-release herbicide formulations in wheat stands infested by weeds. J. Env. Sci. Health, P. B., 52, 729–735. © 2017 Taylor & Francis.)

Effective weed control caused an increase in wheat productivity. The vegetative organs of wheat reached 198.6 and 182.1 g/m^2 with the treatments with P(3HB)/MET films and microgranules, respectively (Figure 4.17), whereas in the positive control, biomass was recorded to be 168.6 g/m^2 followed by the negative control which showed the least biomass with 135.5 g/m^2. These results clearly indicated the effectiveness of the developed experimental formulations of MET in combating weeds and also influencing the growth of the desired plants.

A similar experiment with wheat stands infested by the weed lamb's quarters (*Chenopodium album*) was performed to study the herbicidal activity of the systemic herbicide **TBM** embedded in films and microgranules. The results of weed suppression and the state of the wheat plants during the 30-day experiment are shown in Figures 4.18 and 4.19.

Seven days after sowing, in the negative control (with no herbicide application), the biomass and density of the weed reached 4815 plants/m^2 and 3.9 g/m^2, respectively. Those values were considerably higher than the corresponding values in the treatment groups (585–613 plants/m^2 and 2.5–2.6 g/m^2) and positive control (926 plants/m^2 and 3.2 g/m^2). The effect of the free herbicide was somewhat weaker than that of the experimental formulations. At Day 20, most of the weeds were dead, and in the treatment groups, their density dropped to 185–194 plants/m^2 and biomass to 1.1 g/m^2. Further, at the end of the 30th day, complete suppression of lamb's quarters weeds was observed. The drastic decline in the density of weeds favored the growth and development of wheat plants. Interestingly, on the 30th day, the wheat stands treated with P(3HB)/TBM films, microgranules, and free **TBM**, had almost similar amounts of the biomass with 116, 117 and 119 g/m^2 respectively, whereas in the negative control group, the biomass could merely reach 50 g/m^2 as shown in Figure 4.19

Thus, the results obtained in the present investigation revealed that experimental formulations of two herbicides (MET and **TBM**) embedded in the degradable matrix of poly-3-hydroxybutyrate showed significant herbicidal activity and also influenced the growth of wheat. Overall, the study is aimed at developing experimental formulations for enhanced herbicidal activity. The results with experimental formulations of two herbicides (MET and **TBM**) embedded in the degradable matrix of poly-3-hydroxybutyrate showed significant activity against weed infestation. The activity was significant in comparison with commercial formulations. The study concludes with the fact that degradable poly-3-hydroxybutyrate can be a promising material for constructing pre-emergence slow-release herbicide formulations.

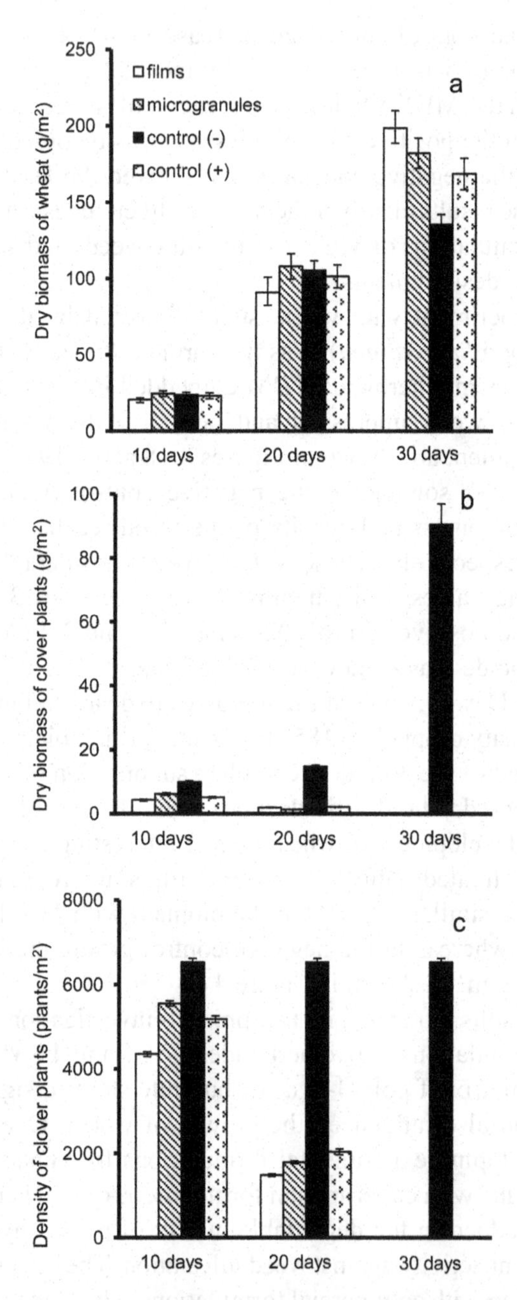

FIGURE 4.17 Dry biomass weight of wheat (*Triticum aestivum*) (a) and dry biomass weight (b) and density (c) of white sweet clover (*Melilotus albus*) plants in laboratory soil ecosystems treated with different formulations of the herbicide metribuzin (Zhila's data).

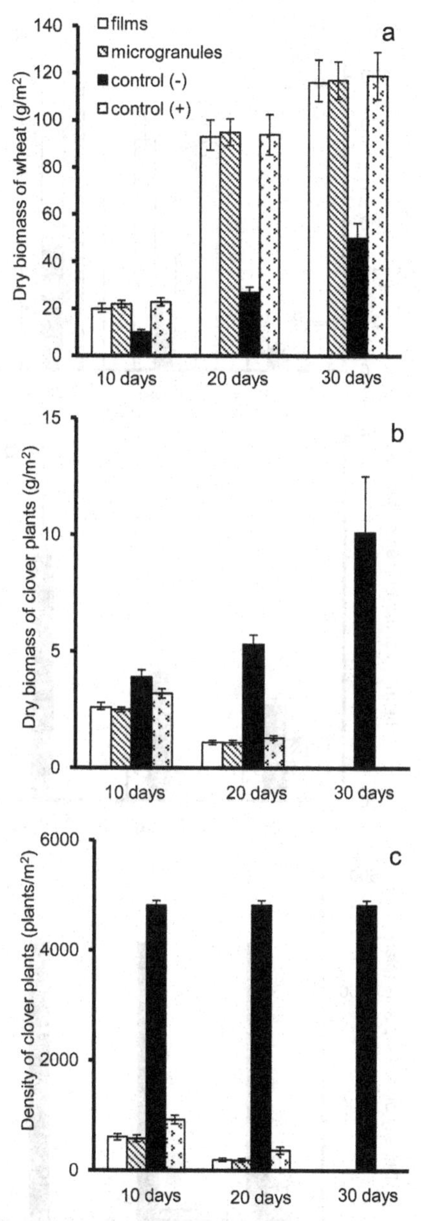

FIGURE 4.18 Photographs of wheat (*Triticum aestivum*) stands infested with lamb's quarters and treated with different P(3HB)/TBM formulations: *a* – microgranules, *b* – films, *c* – positive control, *d* – negative control (untreated). (Reprinted with permission from Zhila, N., Murueva, A., Shershneva, A., Shishatskaya, E., & Volova, T., (2017). Herbicidal activity of slow-release herbicide formulations in wheat stands infested by weeds. J. Env. Sci. Health, P. B., 52, 729–735. © 2017 Taylor & Francis.)

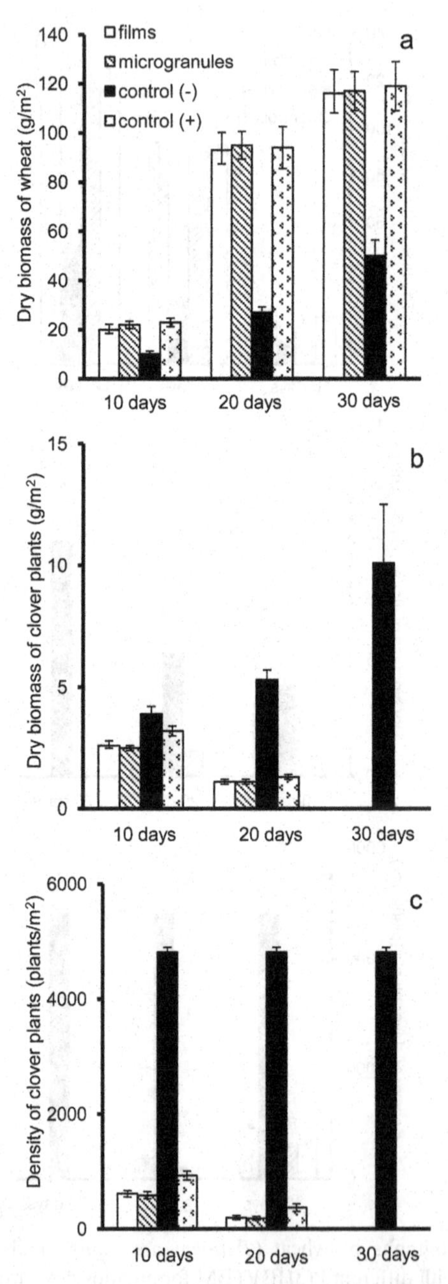

FIGURE 4.19 Dry biomass weight of wheat (*Triticum aestivum*) (a) and dry biomass weight (b) and density (c) of lamb's quarters (*Chenopodium album*) plants in laboratory soil ecosystems treated with different formulations of the herbicide tribenuron-methyl (Zhila's data).

KEYWORDS

- **carboxymethyl cellulose-kaolinite composite**
- **ethylene glycol**
- **MET**
- **methacrylic acid**
- **polyhydroxyalkanoates**
- **TBM**

REFERENCES

Ahemad, M., & Khan, M. S., (2011a). Effect of pesticides on plant growth promoting traits of greengram-symbiont, *Bradyrhizobium* sp. strain MRM6. *Bull. Environ. Contam. Toxicol.*, *86*, 384–388.

Ahemad, M., & Khan, M. S., (2011b). Toxicological assessment of selective pesticides towards plant growth promoting activities of phosphate solubilizing *Pseudomonas aeruginosa*. *Acta Microbiol. Immunol. Hung.*, *58*, 169–187.

Akbuga, J., (1993). Use of chitosonium malate as a matrix in sustained-release tablets. *Int. J. Pharm.*, *89*, 19–24.

Boyandin, A. N., Zhila, N. O., Kiselev, E. G., & Volova, T. G., (2016). Constructing slow-release formulations of metribuzin based on degradable poly(3-hydroxybutyrate). *J. Agric. Food Chem.*, *64*, 5625–5632.

Brenner, D. J., Krieg, N. R., Staley, J. T., & Garrity, G. M., (2005). *Bergey's Manual® of Systematic Bacteriology. Vol. 2: The Proteobacteria, (Part B, C)* (2459 p.). Springer Science & Business Media. ISBN 978-0-387-28022-6 (Part B); ISBN 978-0-387-29298-4 (Part C); DOI: 10.1007/0-387-28022-7 (Part B); DOI: 10.1007/0-387-29298-5 (Part C).

Chowdhury, M. A., (2014). The controlled release of bioactive compounds from lignin and lignin-based biopolymer matrices. *Int. J. Biol. Macromol.*, *65*, 136–147.

Dailey, O. D., Dowler, C. C., & Glaze, N. C., (1990). Evolution of cyclodextrin complex of pesticides for use in minimization of groundwater contamination. In: Bode, L. E., Hazen, J. L., & Chasin, D. G., (eds.), *Pesticide Formulations and Application Systems* (Vol. 10, pp. 26–36). ASTM STP 1078, American Society for Testing and Materials: Philadelphia.

Dworkin, M., Falkow, S., Rosenberg, E., Schleifer, K. H., & Stackebrandt, E., (2006). *The Prokaryotes, Vol. 5: Proteobacteria: Alpha and Beta Subclasses* (910 p.). Springer-Verlag: New York. ISBN: 0-387-30745-1 DOI: 10.1007/0-387-30745-1

Fedtke, C., (1981). Nitrogen metabolism in photosynthetically inhibited plants. In: Borthe, H., & Trebst, A., (eds.), *Biology of Inorganic Nitrogen and Sulphur* (pp. 260–265). Springer-Verlag: Berlin.

Fernández-Pérez, M., Villafranca-Sánchez, M., & Flores-Céspedes, F., (2010). Prevention of chloridazon and metribuzin pollution using lignin-based formulations. *Environ. Pollut.*, *158*, 1412–1419.

Fernández-Pérez, M., Villafranca-Sánchez, M., & Flores-Céspedes, F., (2015). Lignin-polyethylene glycol matrices and ethylcellulose to encapsulate highly soluble herbicides. *J. Appl. Polym. Sci.*, *132*, 41422.

Fernández-Pérez, M., Villafranca-Sánchez, M., Flores-Céspedes, F., & Daza-Fernández, I., (2011). Prevention of herbicides pollution using sorbents in controlled release formulations. In: Kortekamp, A., (ed.), *Herbicides and Environment* (pp. 157–172). In Tech.

Fernandez-Urrusuno, R., Gines, J. M., & Morillo, E., (2000). Development of controlled release formulations of alachlor in ethylcellulose. *J. Microencapsul.*, *17*, 331–342.

Fisyunov, A. V., (1984). *Guide to Weeds* (p. 237). Kolos: Moscow (in Russian).

Flores-Céspedes, F., Pérez-García, S., Villafranca-Sánchez, M., & Fernández-Pérez, M., (2013). Bentonite and anthracite in alginate-based controlled release formulations to reduce leaching of chloridazon and metribuzin in a calcareous soil. *Chemosphere*, *92*, 918–924.

Grillo, R., Melo, N. F. S., Lima, R., Lourenco, R. W., Rosa, A. H., & Fraceto, L. F., (2010). Characterization of atrazine-loaded biodegradable poly(hydroxybutyrate-co-hydroxyvalerate) microspheres. *J. Polym. Environ.*, *18*, 26–32.

Grillo, R., Pereira, A. E. S., Melo, N. F. S., Porto, R. M., Feitosa, L. O., Tonello, P. S., et al., (2011). Controlled release system or ametryn using polymer microspheres: Preparation, characterization and release kinetics in water. *J. Hazard. Mater.*, *186*, 1645–1651.

Ikladious, N., & Messiha, N., (1984). Tributyltin esters of malic anhydride copolymers as antifouling coatings. *J. Control. Rel.*, *1*, 119–126.

Kumar, J., Nisar, K., Shakil, N. A., & Sharma, R., (2010). Residue and bio-efficacy evaluation of controlled release formulations of metribuzin against weeds in wheat. *Bull. Environ. Contam. Toxicol.*, *85*, 357–361.

Lobo, F. A., Aguirre, C. L., Silva, M. S., Grillo, R., Melo, N. F. S., Oliveira, L. K., et al., (2011). Poly(hydroxybutyrate-co-hydroxyvalerate) microspheres loaded with atrazine herbicide: Screening of conditions or preparation, physic-chemical characterization, and *in vitro* release studies. *Polym. Bull.*, *67*, 479–495.

Maqueda, C., Villaverde, J., Sopeña, F., Undabeytia, T., & Morillo, E., (2008). Novel system or reducing leaching of the herbicide metribuzin using clay-gel-based formulations. *J. Agric. Food. Chem.*, *56*, 11941–11946.

McCormick, C. L., & Anderson, K. V., (1984). Synthesis and characterization of chitin pendently substituted with the herbicide metribuzin. In: Zikakis, J., (ed.), *Chitin, Chitosan and Related Enzyme* (pp. 41–53). Academic Press: New York.

McCormick, C. L., (1985a). Modifications of naturally occurring polymers with herbicides. *Sci. Lett.*, *446*, 76–81.

McCormick, C. L., (1985b). Controlled activity polymers with pendent metribuzin. Effect of structure on hydrolytic release. *Ann. N. Y. Acad. Sci.*, *446*, 76–92.

Netrusov, A. I., Egorov, M. A., Zakharchuk, L. M., & Kolotilova, N. N., (2005). *Workshop on Microbiology: A Textbook for Students of Higher Educational Institutions* (608 p.). Academy Publ.: Moscow (in Russian). ISBN 5-7695-1809-X.

Peppas, N. A., & Narasimhan, B., (2014). Mathematical models in drug delivery: How modeling has shaped the way we design new drug delivery systems. *J. Control. Release*, *190*, 75–81.

Prudnikova, S. V., Boyandin, A. N., Kalacheva, G. S., & Sinskey, A. J., (2013). Degradable polyhydroxyalkanoates as herbicide carriers. *J. Polym. Environ.*, *21*, 675–682.

Quadir, M. A., Rahman, M. S., Karim, M. Z., Akter, S., Awkat, M. T., & Reza, M. S., (2003). Evaluation of hydrophobic materials as matrices for controlled-release drug delivery. *Pak. J. Pharm. Sci.*, *16*, 17–28.

Rehab, A., Akelah, A., & El-Gamal, M. M., (2002). Controlled-release systems based on the intercalation of polymeric metribuzin onto montmorillonite. *J. Polym. Sci. P. A: Polym. Chem.*, *40*, 2513–2525.

Ritger, P. L., & Peppas, N. A., (1987). A simple equation for description of solute release. 1. Fickian and non-Fickian release from non-swellable devices in the form of slabs, spheres, cylinders or discs. *J. Control. Release.*, *5*, 23–36.

Sahoo, S., Manjaiah, K. M., Datta, S. C., Ahmed, S. T. P., & Kumar, J., (2014). Kinetics of metribuzin release from bentonite-polymer composites in water. *J. Environ. Sci. Health P. B.*, *49*, 591–600.

Sato, H., Miyagawa, Y., Okabe, T., Miyajima, M., & Sunada, H., (1997). Dissolution mechanism of diclofenac sodium from wax matrix granules. *J. Pharm. Sci.*, *86*, 929–934.

Savenkova, L., Gercberga, Z., Muter, O., Nikolaeva, V., Dzene, A., & Tupureina, V., (2002). PHB-based films as matrices for pesticides. *Process Biochem.*, *37*, 719–722.

Shabaev, A. I., (2000). Features of soil cultivation in various zones and agro landscapes of the Volga region. *Agriculture*, *5*, 13–15 (in Russian).

Shcheglov, Y. V., (1961). *Development of a Chemical Way to Control Weeds in Crops of Millet and Miliary Cultures* (pp. 64-69). In: *Application of Herbicides and Plant Growth Stimulants.* Publishing House of Academy of Sciences of Byelorussian SSR (in Russian).

Sopeña, F., Cabrera, A., Maqueda, C., & Morillo, E., (2005). Controlled release of the herbicide norflurason into water from ethylcellulose. *J. Agricult. Food Chem.*, *53*, 3540–3547.

Sopeña, F., Maqueda, C., & Morillo, E., (2008). Microencapsulation of alachlor for reducing its pollution in soil-water system. *Abstracts of XVI International Conference on Bioencapsulation*, p. 58.

Suave, J., DallAgnol, E. C., Pezzin, A. P. T., Meier, M. M., & Silva, D. A. K., (2010). Biodegradable microspheres of poly(3-hydroxybutyrate)/poly(ε-caprolactone) loaded with malathion pesticide: Preparation, characterization, and *in vitro* controlled release testing. *J. Appl. Polym. Sci.*, *117*, 3419–3427.

Sutton, D., Fothergill, A., & Rinaldi, M., (2001). *Key Pathogenic and Opportunistic Fungi* (468 p.). Mir: Moscow (in Russian). ISBN 5-03-003308-4.

Undabeytia, T., Recio, E., Maqueda, C., Morillo, E., Gómez-Pantoja, E., & Sánchez-Verdejo, T., (2011). Reduced metribuzin pollution with phosphatidylcholine-clay formulations. *Pest Manag. Sci.*, *67*, 271–278.

Voinova, O. N., Kalacheva, G. S., Grodnitskaya, I. D., & Volova, T. G., (2009). Microbial polymers as a degradable carrier for pesticide delivery. *Appl. Biochem. Microbiol.*, *45*, 384–388.

Volova, T. G., Voinova, O. N., Kalacheva, G. S., & Grodnitskaya, I. D., (2008). The prospects of the use of restorable polyesters for designing safe pesticides. *Dokladi Biologicheskikh Nauk.*, *419*, 100–103 (in Russian).

Volova, T. G., Zhila, N. O., Vinogradova, O. N., Nikolaeva, E. D., Kiselev, E. G., Shumilova, A. A., et al., (2016a). Constructing herbicide metribuzin sustained-release formulations based on the natural polymer poly–3-hydroxybutyrate as a degradable matrix. *J. Environ. Sci. Health, P. B.*, *51*, 113–125.

Volova, T., Kiselev, E., Shishatskaya, E., Zhila, N., Boyandin, A., Syrvacheva, D., et al., (2013). Cell growth and accumulation of polyhydroxyalkanoates from CO_2 and H_2 of a hydrogen-oxidizing bacterium, *Cupriavidus eutrophus* B–10646. *Bioresour. Technol.*, *146*, 215–222.

Volova, T., Kiselev, E., Vinogradova, O., Nikolaeva, E., Chistyakov, A., Sukovatiy, A., & Shishatskaya, E., (2014). A glucose-utilizing strain, *Cupriavidus euthrophus* B–10646: Growth kinetics, characterization and synthesis of multicomponent PHAs. *PLoS One.*, *9*, e87551.

Volova, T., Zhila, N., Kiselev, E., Prudnikova, S., Vinogradova, O., Nikolaeva, E., et al., (2016b). Poly(3-hydroxybutyrate)/metribuzin formulations: Characterization, controlled release properties, herbicidal activity, and effect on soil microorganisms. *Environ. Sci. Pollut. Res.*, *23*, 23936–23950.

Watanabe, T., (2002). *Pictorial Atlas of Soil Fungi: Morphologies of Fungi and Key Species* (2nd edn.). CRC Press LLC, Boca Raton.

Zhang, S., Yang, G., Zheng, Z., & Chen, Y., (2009). On-line preconcentration and analysis of metribuzin residues in corn fields by use of a molecularly imprinted polymer, *Chromatographia*, *69*, 615–619.

Zhila, N., Murueva, A., Shershneva, A., Shishatskaya, E., & Volova, T., (2017). Herbicidal activity of slow-release herbicide formulations in wheat stands infested by weeds. *J. Env. Sci. Health, P. B.*, *52*, 729–735.

CHAPTER 5

CHARACTERIZATION OF EXPERIMENTAL FORMULATIONS OF THE FUNGICIDE TEBUCONAZOLE AND THEIR EFFICACY

Modern high-performance crop farming is impossible without fungicides. A most promising way to enhance agricultural production and improve its quality is to protect crops against diseases and pests. Although chemicals and technological measures are widely used to protect plants, global losses caused by pests constitute about 35% of the potential crop yield; in developing countries, they are estimated at 48%. About one-third of these losses are caused by diseases of plants. Numerous harmful fungi, bacteria, and viruses impair the quality of agricultural products and cause poisoning of animals and people. Mycotoxins, which are produced by some disease agents, pose a serious danger (Peresypkin et al., 1991; Kravchenko and Tutelyan, 2005; Zakharova et al., 2006).

Chemical substances used to control fungal diseases of plants are called fungicides (*fungus* + *caedo* "killing; killer"). Fungicides hold the third place in sales volume and usage because phytopathogenic fungi are responsible for about 10% of the crop loss. Such important sources of plant protein as maize, wheat, and rice, whose yields constitute over 55% of the total yield of cereal crops, are particularly susceptible to the adverse effects of pathogens. Annual production of grain as the major food source has exceeded 2250 million tons; cereal crops occupy areas totaling more than 670 million ha (Cereals, 2008).

Fusarium infection is among the most common diseases of cereal crops, which is caused by soil pathogenic fungi of the genus *Fusarium*. Application of fungicides decreases the incidence of Fusarium infection and reduces the levels of mycotoxins in commercial grain (Schmale, 2003). Yield losses caused by Fusarium infection may reach between 5 and 30%. Fusarium root rot may affect the ears and grain of cereal crops, contaminating the grain with mycotoxins and making it unsuitable and even unsafe food for humans and animals. *Fusarium* fungi are producers of very potent mycotoxins, the most common of which are fusarium toxins such as deoxynivalenol (vomitoxin), zearalenone, and T–2 mycotoxin (Kravchenko, Tutelyan, 2005; Binder et al., 2007).

To protect crops against fungal infections, farmers employ cultural control methods as well as seed treatment and soil application of chemicals inhibiting the growth of plant pathogens (Pavlyushin et al., 2009).

The spread of Fusarium infection is mainly associated with the extensive development of agroecosystems and an increase in the area occupied by cereals. This upsets the ecological balance in the soil-plant system, as crops alone influence the soil environment. The composition of the soil microbial community changes: the percentage of harmful microflora increases with the growing abundance of microscopic fungi, which produce toxins hazardous to plants, animals, and even humans. Uncontrolled use of fungicides may cause irreversible qualitative shifts in the human environment in the nearest future. It is important to develop a system of measures to minimize or even eliminate one of the unwanted effects of pesticides – contamination of the soil cover. Residual amounts of pesticides may be present in food, aquatic environments, soil, and other systems (Davletov et al., 2010).

Fungicides used in agriculture must meet a number of requirements:

- they must be effective against causative agents of diseases and kill them at the lowest possible concentration, being harmless to crops;
- they must be inexpensive and simple to use;
- they must be stable and remain unaffected by chemical and physical factors of the surrounding medium; and
- they must be able to suppress various infectious agents.

Modern fungicides do not meet all of these requirements. Classifications of modern fungicides and other antifungals are based on their effects on the agents of diseases, their chemical origins, and techniques of applying them, or, sometimes, on their affinity for water, which is determined by the physicochemical properties of the compound. According to their role in protection, fungicides are classified as protective and curative (eradicating) ones. Protective fungicides prevent a fungus from attacking the plant but do not kill it. Contact (protective) fungicides, which are localized on the plant surface, are able to affect the fungus only when contacting it directly, and they mainly inhibit reproduction of the fungi. Systemic fungicides are absorbed into the plant at concentrations safe for it and protect the parts of the plant far away from the application site. In accordance with their chemistry, fungicides are divided into antibiotics (biofungicides), organic fungicides, and inorganic ones. Fungicides can be intended for treating plants during their growing season or during dormancy, for treating seeds and seedlings, and for soil application. Based on their mode of action, fungicides are divided into two groups: fungicides affecting pathogenesis in host plants and fungicides directly affecting vital biochemical processes in cells of the causative agent. Fungicides of the latter group selectively inhibit the corresponding enzymes, which serve as biological catalysts in living cells of fungi. The mode of action of most fungicides has not been sufficiently understood. Fungicides act on plant pathogens by affecting the course of biochemical reactions, controlling or blocking them. Organophosphorus fungicides inhibit synthesis of lipids – components of membranes, such as phosphatidylcholine. Such fungicides as triazoles, morpholines, pyrimidines, imidazoles, and piperazines inhibit biosynthesis of ergosterol – one of the main components of cell membranes. It is assumed that hydroxypyrimidines and alanine derivatives inhibit synthesis of nucleic acids and antibiotics inhibit protein synthesis; oxathiin fungicides affect tissue respiration; derivatives of benzimidazole and thiophanate affect cell division.

Modern fungicides are divided into the following groups:

- copper fungicides, which are used to treat and prevent various diseases in fruit and vegetable crops: copper sulfate, Bordeaux mixture, HOM, "Oxikhom," "Curzate," "Ordan," and "Abiga-Peak";
- sulfur fungicides, which are used to treat tree bark and dust berry-producing shrubs: "Thiovit Jet," "Cumulus," and colloidal sulfur;
- strobilurins, which are used to control diseases of apple and pear trees and prevent storage diseases of fruits: "Strobi," "Profit Gold," "Acrobat MZ," "Ridomil," and "Tattoo";
- carboxins, which are used to treat seeds: "Vitaros" and "Previcur";
- benzimidazoles, which are used to treat seeds, to process fruits before storage, and to cure berry-producing shrubs: "Fundazol," "Fundazim," and "Benorad;" and

A special group comprises plant-derived fungicides such as "Phytosporin," "Albit," "Phytolavin," "Trichodermin," "Gliocladin," and "Alirin-B."

Among the great diversity of modern fungicides, triazole chemicals occupy a special place, as they have a broad spectrum of fungicidal activity and regulate plant development (Fletcher et al., 1999; Child et al., 1993). Triazole fungicides now constitute 30% of the fungicide market. These fungicides are potential inhibitors of fungal sterol–14-α-demethylase, affecting sterol metabolism and, finally, causing function deficit of membrane systems of the mycopathogen (Lamb et al., 2001; Hartwig et al., 2012). On the other hand, triazole fungicides are also plant toxicants (Ahemad and Khan, 2011a, b). No specific effector targets for triazole fungicides have been found in plant cells; their systemic toxic effects are exhibited as hormonal imbalance (Yang et al., 2014), impairment of nitrogen metabolism, lower germination rate, impairment of root growth and development (Serra et al., 2013). Among triazole fungicides are the commonly used fungicides containing tebuconazole (TEB) as the active ingredient. TEB-based formulations are widely used to protect crops, including socially and economically important cereal crops (wheat, maize, etc.). In addition to being a fungicide, TEB also regulates plant growth. Commercial TEB formulations include "Raxil Ultra" – a high-concentration systemic fungicide for seed treatment, which is effective against diseases of cereal and industrial crops – and "VIAL TRUST" – a bi-component systemic fungicide for treating seeds of cereal crops and soybean, which contains two active ingredients: TEB and thiabendazole.

TEB is an effective multifunctional systemic fungicide used to protect a number of agricultural crops (wheat, barley, rice, rape, maize, vineyards, etc.) against powdery mildew, rust, rots, and leaf blotches. TEB rapidly penetrates into the plants through both their vegetative organs and roots. It inhibits ergosterol synthesis, preventing the formation of cell membranes, and disrupts metabolic processes, causing the death of pathogens.

However, this fungicide is potentially toxic to plants and inhibits plant growth if it is used in large quantities to spray leaves or treat seeds. The adverse effects of the fungicide are expressed as low germinating capacity of seeds, inhibition of the growth of seedlings, and, especially, inhibition of root elongation. The unfavorable regulatory effect of TEB occurs via the following mechanism: triazoles, TEB, in particular, shift the balance of phytohormones in plant tissues and inhibit biosynthesis of gibberellins, causing a temporary increase in the content of abscisic acid in plants (Yang et al., 2014).

Another way of using TEB formulations is to treat seeds prior to sowing, but this may decrease seed germination ability, suppress the growth of seedlings, and inhibit root growth (Yang et al., 2014). Moreover, TEB used as seed dressing is rather quickly depleted, and the vegetative organs of plants have to be sprayed with the fungicide. However, when TEB is used as suspensions or emulsions to spray the vegetative organs of plants, the active ingredient is released too quickly to be sufficiently effective, and the fungicide has to be applied again, in greater quantities (Tyuterev et al.,

2005). As TEB is potentially phytotoxic, this method of application of the fungicide inhibits plant growth, causing economic losses and posing a threat to the health of people and the environment (Zhang et al., 2015).

Commercial formulations of TEB ("Raxil Ultra," "Bunker," "Tebu") are able to control 95–100% of stinking smut and 90–100% of loose smut of wheat, 95–100% of smut diseases of barley, 80–100% of smut of oat, 75–95% of smut of maize and millet, and 70–100% of smut of sorghum. However, the efficacy of these formulations in controlling pathogens causing root rots varies between 25 and 80%, and they kill between 2 and 40% of the fungi causing seed rot. These formulations are not effective against pathogens causing black mold of ears and germs (Abelentsev et al., 2006).

Thus, new fungicide formulations need to be developed to increase the efficacy of fungicides and minimize their harmful effects on the environment.

5.1 NEW-GENERATION FORMULATIONS TO CONTROL FUSARIUM INFECTION OF CEREALS

Fusarium infection, causing the development of root and foot rot of cereal crops, results in considerable yield decrease and impairment of grain quality (Danilenkova et al., 2004; Zakharenko, 2003).

Modern means of protection of cereal crops against root rots should be based on the ecosystem approach to reduce the adverse effects of human-made pollutants and to produce health foods. The most common way to control Fusarium root rot is to treat seeds with fungicides prior to sowing, but this treatment cannot protect plants throughout their growing season; moreover, it may decrease the germinating capacity of seeds and inhibit the development of the roots (Yang et al., 2014). Therefore, it is important to develop and use ecofriendly targeted slow-release formulations capable of inhibiting the development of plant pathogens without posing significant risks to beneficial biota and the entire environment.

One of the modern approaches to plant protection is the use of composite formulations or tank-mixes. The tank-mixes usually contain both biological and chemical pesticides and, thus, exert strong protective and curative effects. Moreover, they do not significantly inhibit plant growth and are less toxic to plants. Consequently, fewer treatments are needed, and control measures become less expensive.

Construction of formulations in which agrochemicals are embedded in the degradable matrix seems to be a propitious approach. The gradual degradation of the matrix in soil should enable steady and targeted release of the active ingredient. This is the way to reduce the amounts of the chemicals added to the soil and decrease their uncontrolled distribution and accumulation in agroecosystems. It is very important to find proper material to be used as a matrix for embedding agrochemicals.

Research aimed at designing slow-release TEB formulations was started quite recently. Asrar et al. (2004) described microparticles prepared from poly(methyl methacrylate) and poly(styrene-co-maleic anhydride) and loaded with TEB. Yang et al. (2014) described encapsulation of TEB in ethyl cellulose microcapsules to be used to suppress the pathogen causing loose smut. Qian et al. (2013) reported encapsulation of TEB in silica nanospheres. The study by Khalikov et al. (2013) described the preparation of nanoparticulate powders with enhanced water solubility by mixing TEB with water-soluble polymers in a planetary-type mill. The authors used arabinogalactan (AG), pectin (P), cyclodextrin (CD), polyvinylpyrrolidone (PVP), and hydroxyethyl starch (HES). TEB/AGandTEB/HES samples showed higher fungicidal activity against root rot pathogens (*Biopolaris* and *Fusarium*) than the commercial formulation "Raxil Ultra." Those studies described TEB encapsulation efficiency and its release duration and mechanism in laboratory systems such as sterile water and soil.

Several studies compared the efficacy of slow-release TEB formulations with that of free TEB against such plant pathogens as wheat rust *Puccinia recondita* (Asrar et al., 2004), root rot pathogens *Biopolaris* and *Fusarium* (Khalikov et al., 2013), maize head smut (Yang et al., 2014), and wheat powdery mildew (Zhang et al., 2015). TEB encapsulated in microcapsules of ethyl cellulose did not affect the germinating capacity of maize seeds adversely, in contrast to "Raxil Ultra." Phytohormonal analysis showed that the microencapsulated and continuously released TEB had a beneficial effect on the balance of phytohormones during maize seed germination. Encapsulated TEB provided better protection against maize head smut than the conventionally used "Raxil Ultra" (Yang et al., 2014). Zhang et al. (2015) reported a study in which cells of cyanobacterium *Synechocystis* sp. strain *PCC* 6803 used to encapsulate TEB were coated with the rubber-like urea/formaldehyde material. That formulation remained 80% effective against powdery mildew for 12 days, as TEB was slowly and steadily released from microgranules, while the efficacy of "Raxil Ultra" was 52.25%. In another study (Asrar et al., 2014), TEB was encapsulated into polymeric microcapsules and used to spray plants. The microparticles provided better protection against wheat rust *Puccinia recondite* than commercial foliar-applied TEB, "Raxil Ultra."

The efficacy of pesticides is determined not only by the type of the active ingredient and its activity but also by the type of formulation (Tropin, 2007), which must preserve the useful properties of the active ingredient, prolong its effectiveness, and minimize its adverse effects on beneficial biota and the entire environment.

The newest trend in research is the development of environmentally safe new-generation pesticides with targeted and controlled release of active ingredients embedded in biodegradable matrices and/or covered with biodegradable coatings. The main condition for constructing such formulations is the availability of appropriate materials with the following properties: safety for the environment, chemical compatibility with agrochemicals, and controlled degradation in soil followed by formation of non-toxic products.

5.2 POLYHYDROXYALKANOATES AS DEGRADABLE MATRIX FOR CONSTRUCTING SLOW-RELEASE TEBUCONAZOLE FORMULATIONS

Research aimed at constructing such PHA-based formulations is still in its initial stages. A search of the literature revealed few studies that reported the use of PHAs to construct slow-release fungicide formulations. Savenkova et al. (2002) described P(3HB) films loaded with fungicides. Development of slow-release fungicide formulations is a relatively new branch of research, and data about the potential and effectiveness of such formulations are limited. There are published data on the use of fungicides encapsulated in polystyrene/maleic anhydride (Asrar et al., 2004), calcium carbonate (Qian et al., 2011), chitosan (Brunel et al., 2013), ethyl cellulose (Yang et al., 2014), and urea–formaldehyde (Zhang et al., 2015).

The purpose of study work (Volova et al., 2016) was to construct and investigate slow-release formulations of the fungicide TEB embedded in films and pellets made from P(3HB) – a degradable polymer.

Raxil Ultra (Bayer CropScience, Russia) is a systemic fungicide based on the active ingredient TEB, which has a broad spectrum of activity. It provides effective protection against various diseases in cereals and helianthus. This fungicide disinfects seeds and partially decontaminates soil and plant residues around the seeds. Its chemical formula is $C_{16}H_{22}ClN_3O$, and its structural formula is shown in Figure 5.1. IUPAC name: (RS)- 1-(4-Chlorophenyl)- 4,4-dimethyl–3-(1H, 1,2,4-triazol–1-ylmethyl)pentan- 3-ol. CAS name: α-(2-(4-chlorophenyl)ethyl)-alpha-(1,1-dimethylethyl)–1H–1,2,4-triazol–1-ethanol. Molar mass (g/mol): 307.82. Solubility in water: 36 mg/L at 20°C. Melting point is 105°C. The time of degradation in soil is 177 days.

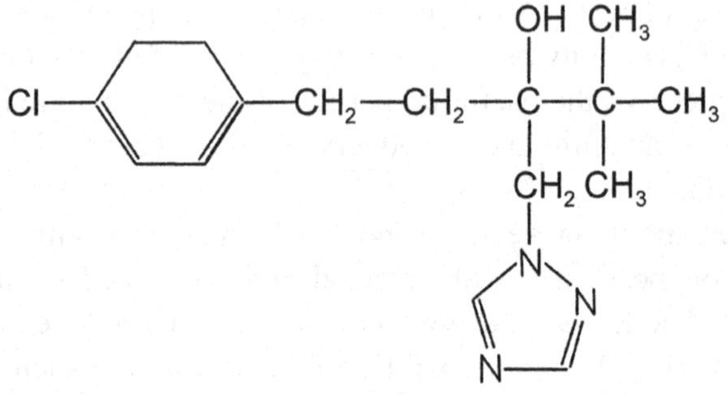

FIGURE 5.1 The structural formula of tebuconazole.

Polymer/TEB mixtures were prepared by using purified P(3HB) and chemically pure TEB (Russian Federal Standard GSO 7669–99, purity 99.1%). A polymer sample was dissolved in chloroform, and a solution of TEB in chloroform was added to the polymer solution. We used 2% (w/v) polymer solutions in chloroform. The viscosity of the solution at 25°C was 20.02 cP. The viscosity of the solutions was measured by using a HAAKE Hoppler Falling Ball Viscometer C (Thermo Scientific, Germany). The P(3HB)/TEB solution was mixed on an MR Hei-Standart magnetic stirrer (Heidolph, Germany) for 3–4 h (until completely dissolved) and heated to 35–40°C under reflux condenser. The powder system was prepared as follows: the polymer was ground in a ZM 200 ultracentrifugal mill (Retsch, Germany). The fractional composition of the polymeric powder was determined by using an AS 200 control analytical sieve shaker (Retsch, Germany): the fraction of the particles under 0.50 mm comprised 65%, and the fraction of the particles between 0.80 and 1.00 mm constituted 45%. The powdered polymer and TEB were mixed mechanically. Samples of the two powders of different fraction compositions (0.10 mm to 1.0 mm) were weighed on the analytical balance, homogenized with a laboratory stirrer for 2 min, and mixed with a TEB sample.

The polymeric systems (solutions and powders) were used to construct TEB-loaded films and pellets. The P(3HB) matrix was loaded with 25% (w/w) TEB. Films loaded with the fungicide were prepared as follows: the P3HB/TEB solution was cast in Teflon-coated metal molds, and then solvent evaporation occurred. The films were left to stay at room temperature in a laminar flow cabinet for 24 h, and then they were placed into a vacuum drying cabinet (Labconco, U.S.) for 3–4 days, until complete solvent evaporation took place. Films were cut into disks of 10 cm and 20 cm in diameter and 5×5 mm squares, which were weighed on the analytical balance of accuracy class 1 Discovery (Ohaus, Switzerland). The film thickness (0.080±0.005 mm) was measured with a digital micrometer (LEGIONER EDM–25–0.001, Germany). The pellets were prepared from the mixture of P3HB and TEB powders by cold pressing, using a Carver Auto Pellet 3887 press (Carver, U.S.) under pressing force of 8 000. Pellets were 5 and 13 mm in diameter and were loaded with 25% (w/w) active ingredient.

TEB is a volatile and thermostable compound; it was detected with gas chromatography. Plotting of calibration curves was based on Russian Federal Standard GSO 7669–99 (purity 99.1%). Measurements were done on a chromatograph mass spectrometer (7890/5975C, Agilent Technologies, U.S.). The ion chromatogram and mass spectra of the components are shown in Figure 5.2. Calibration curves were prepared by using two modes ("split" and "splitless") and samples with TEB concentrations ranging between 1.0 ng/μL and 1.4 μg/μL. One μL of each TEB sample was introduced into the chromatograph. At least three parallel measurements were performed, and the average area of the chromatographic peak was determined for each concentration. We plotted the calibration curves of the relationship between the area of the chromatographic peak (calculated automatically, in relative units) and TEB concentration in the sample, μg/μl. The detection limit of the MS-detector for TEB was 0.1 ng/μL; the standard error of the method was 1.0%.

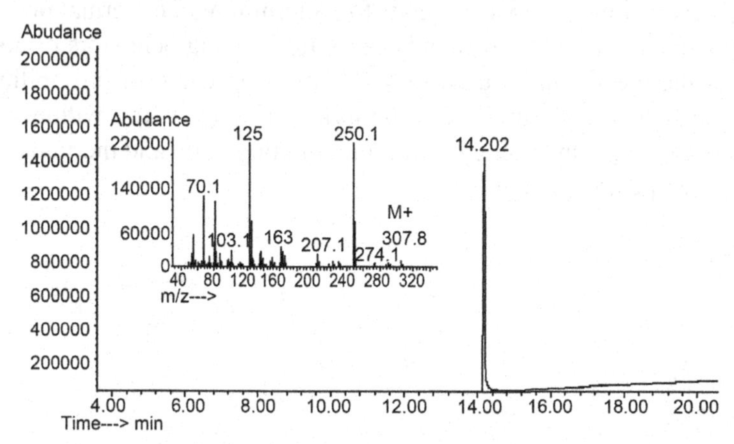

FIGURE 5.2 Ion chromatography and mass spectrum of tebuconazole. RT=14.202 – tebuconazole. (Reprinted by permission from Springer Nature: Environmental Science and Pollution Research: Characterization of biodegradable poly–3-hydroxybutyrate films and pellets loaded with the fungicide tebuconazole, Volova, T., Zhila, N., Vinogradova, O., Shumilova, A., Prudnikova, S., & Shishatskaya, E. © © 2015.)

P(3HB)/TEB solutions and powders were used to produce 0.080±0.005 mm thick 200 mg films and 200 mg pellets, which are shown in Figure 5.3.

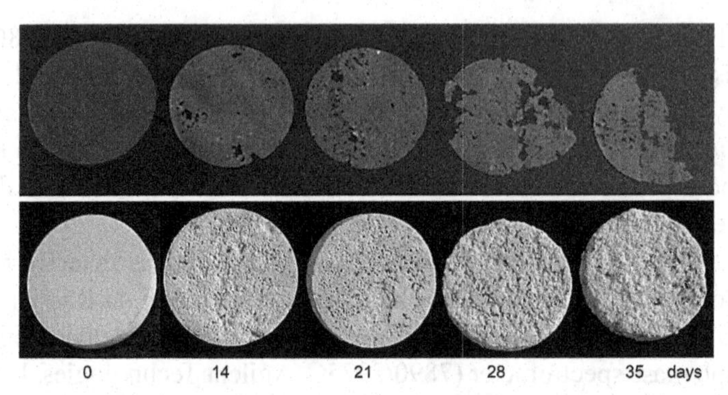

FIGURE 5.3 Photographs of films and pellets: before the loading experiment, loaded with TEB, and during degradation in the laboratory soil system. (Reprinted by permission from Springer Nature: Environmental Science and Pollution Research: ., Characterization of biodegradable poly–3-hydroxybutyrate films and pellets loaded with the fungicide tebuconazole, Volova, T., Zhila, N., Vinogradova, O., Shumilova, A., Prudnikova, S., & Shishatskaya, E. © © 2015.)

Physicochemical properties of initial components and P3HB/TEB listed in Table 5.1 and shown in Figures 5.4 and 5.5 suggest that loading of the polymer matrices with the fungicide influenced the degree of crystallinity (C_x) and temperature properties of the polymer systems. DSC was used as one of the most informative methods for determining the thermal properties of polymers and polymer blends because the melting behaviors of components of the blend can be used to determine the degree of miscibility and interactions of the components. Thermograms were taken within a wide temperature range, including the polymer melting point and thermal decomposition temperature (Figure 5.4).

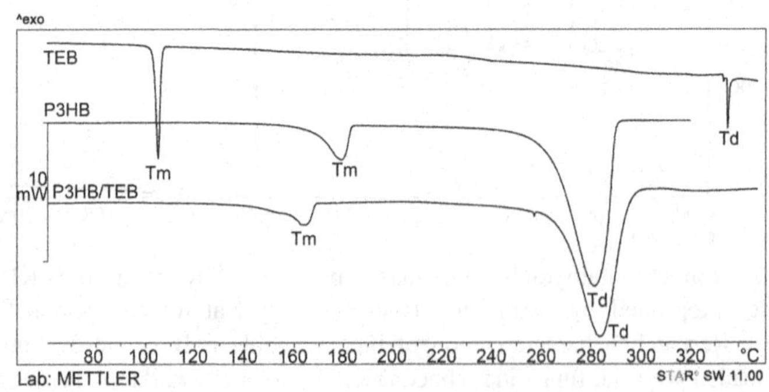

FIGURE 5.4 DSC curves of TEB, P3HB and P3HB/TEB (75:25%): Peaks of melting and thermal decomposition . (Reprinted by permission from Springer Nature: Environmental Science and Pollution Research: ., Characterization of biodegradable poly–3-hydroxybutyrate films and pellets loaded with the fungicide tebuconazole, Volova, T., Zhila, N., Vinogradova, O., Shumilova, A., Prudnikova, S., & Shishatskaya, E. © © 2015.)

Analysis of the number and shapes of endothermic peaks on the thermograms showed the formation of a stable mixture of the polymer and TEB, which was not separated under heating, as the thermogram had only one peak of melting and one peak of decomposition. The thermogram of the mixture had neither a melting peak nor a thermal decomposition peak of TEB: 104.7 and 296–304°C, respectively. Embedding of TEB in P(3HB) decreased its melting temperature from 180°C to 164°C, i.e., by 16°C. The T_m decrease suggests an increase in the viscosity of the melts and, thus, inhibition of polymer crystallization. In consequence, small crystals may form in the polymer, and these crystals may begin to melt at a lower temperature than the initial polymer, thus decreasing the melting temperature of the mixture and making the structure of the polymer more amorphous. This assumption was also supported by a decrease in the enthalpy of melting (Table 5.1): the peaks were somewhat smeared (Figure 5.4), which is typical of melting of amorphous regions. X-ray structure analysis showed that the loading of P(3HB) with TEB increased the degree of crystallinity of the polymer from the initial 74% to 80%.

TABLE 5.1 Physicochemical Properties of P(3HB), TEB, P(3HB)/TEB

Sample	M_w, kDa	Đ	C_x,%	T_{melt}, °C	T_{degr} °C	Enthalpy of melting, (J/g)
P3HB	526	3.46	74	180	282	89.5
TEB	0.308	1.00	96	105	335	96.1
P3HB*/TEB**	711* 0.492**	4.39* 1.43**	80	164	283	70.1

*and ** show the presence of two peaks in molecular weight measurements.

(Reprinted by permission from Springer Nature: Environmental Science and Pollution Research: ., Characterization of biodegradable poly–3-hydroxybutyrate films and pellets loaded with the fungicide tebuconazole, Volova, T., Zhila, N., Vinogradova, O., Shumilova, A., Prudnikova, S., & Shishatskaya, E. © © 2015.)

FTIR analyses show that the most informative range of the wavenumbers was that between 1450 and 1700 1/cm. Pure P3HB has no absorption peaks within this range. The only absorption peaks (bands) observed in the P3HB/TEB system were those associated with the specific structural groups of TEB (Figure 5.5). No groups that would form due to a chemical interaction between P3HB and the fungicide were detected. Peaks of the existing groups became higher, indicating that the polymer/TEB system was a mechanical mixture of P(3HB) and TEB.

Thus, results of DSC, X-Ray, and FTIR suggest that there were no chemical bonds between TEB and the polymer and that the system was a mechanical mixture of components. The decrease in the temperature and

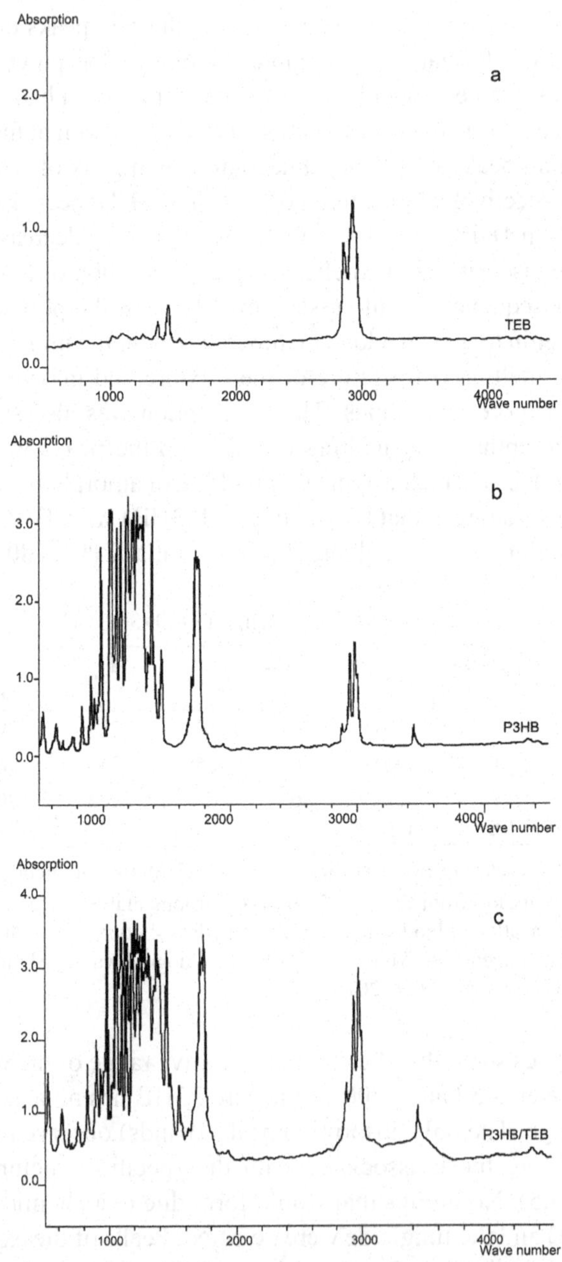

FIGURE 5.5 Infrared-spectra of TEB (a), P3HB (b) and P3HB/TEB (c); fragments of spectra with higher wave number and absorption resolution. (Reprinted by permission from Springer Nature: Environmental Science and Pollution Research: ., Characterization of biodegradable poly–3-hydroxybutyrate films and pellets loaded with the fungicide tebuconazole, Volova, T., Zhila, N., Vinogradova, O., Shumilova, A., Prudnikova, S., & Shishatskaya, E. © © 2015.)

enthalpy of melting suggests that the active ingredients of the chemicals behave as fillers of the polymer matrix.

Release kinetics of TEB from the polymeric matrices was studied *in vitro* in the aqueous and soil laboratory systems. In the first case, the films and pellets were sterilized and placed into 500-ml sterile conical flasks filled with sterile distilled water (100 ml). The number of films or pellets in a flask was determined in such a way that the samples in each flask contained equal total amounts of the active ingredient (50 mg). The flasks were incubated at 25°C in an Innova 44 New Brunswick temperature controlled incubator shaker at 150 rpm. Samples for analysis were collected periodically, under aseptic conditions, and an aliquot of water was added to the flask to maintain a constant volume of liquid in it. TEB was extracted with chloroform three times to determine its concentration. The chloroform extracts were passed through sodium sulfate. Chloroform was removed in a rotary vacuum evaporator. The amount of TEB released (RT) was determined as a percentage of the TEB encapsulated in the polymer matrix, using the following formula:

$$RT=(r/EA) \times 100\% \quad (5.1)$$

where EA is the encapsulated amount, mg, and r is the amount released, mg.

In the soil system, vegetable garden soil (200 g) was placed in 250-mm^3 containers (three replicates for each experimental point), and samples of formulations, which had been weighed in the same way as in the experiment with the aqueous system, were introduced into the soil. We used agrogenically-transformed soil (the Krasnoyarsk Territory, the village of Subbotino). The containers were incubated in a temperature-controlled cabinet at a constant temperature of 21 ± 0.1°C and soil moisture content of 50%. To monitor degradation of the polymeric matrix, the samples were regularly removed from the soil, thoroughly rinsed in water, and dried to constant weight (samples were weighed on the analytical balance of accuracy class 1 (Discovery (Ohaus, Switzerland). The samples were methanolized, and their polymer content was determined on the gas chromatograph equipped with a GCD plus mass spectrometric detector ("Hewlett-Packard," U.S.). TEB was extracted from the soil with chloroform and purified, and, then, its concentration in the soil was determined by gas chromatography.

TEB release from polymer films and pellets was studied in aqueous and soil laboratory systems. Figure 5.6 shows profiles of fungicide release to water from the films and pellets under aseptic conditions, with no P3HB degrading microorganisms present in the medium. The water-soluble TEB was released at a higher rate from the films than from the pellets. After 60 days of incubation of the samples in the water, 46% of the fungicide was released from the films, while only 20% leaked from the pellets (Figure 5.6).

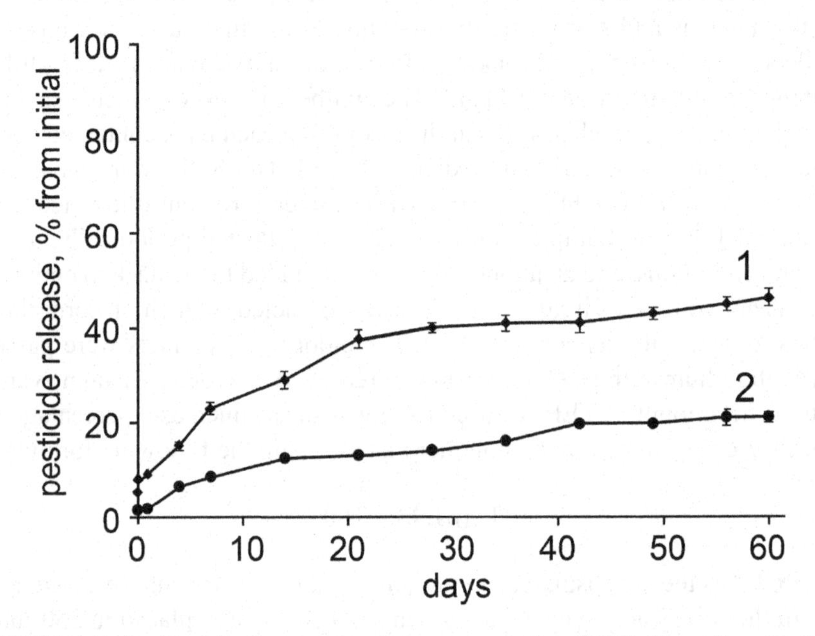

FIGURE 5.6 Profiles of tebuconazole release to water from P3HB films (1) and pellets (2) . (Reprinted by permission from Springer Nature: Environmental Science and Pollution Research: Characterization of biodegradable poly–3-hydroxybutyrate films and pellets loaded with the fungicide tebuconazole, Volova, T., Zhila, N., Vinogradova, O., Shumilova, A., Prudnikova, S., & Shishatskaya, E. © © 2015.)

In the soil, the TEB release rate was higher (Figure 5.7). As the polymer incubated in the soil was degraded by microorganisms and this process influenced the fungicide release kinetics, we needed to study the structure of soil microbial community in the initial soil and monitor it during the experiment. The soil that we used in our experiments was rich in humus (17.4%) and contained high concentrations of nitrate-nitrogen N-NO_3 – 122.0 mg/kg, available phosphorus P_2O_5 – 151.2 mg/100 g, and potassium K_2O – 80 mg/100 g soil. The pH level was close to neutral (6.6). The total number of CFU was 96.5 million in 1 g of soil. The percentage of copiotrophic bacteria was very high (over 60%). Together with the high concentrations of biogenic elements (N, P, K), this is indicative of intense transformations of organic matter in the soil. Low mineralization and oligotrophy coefficients (0.07 and 0.46, respectively) confirm this conclusion.

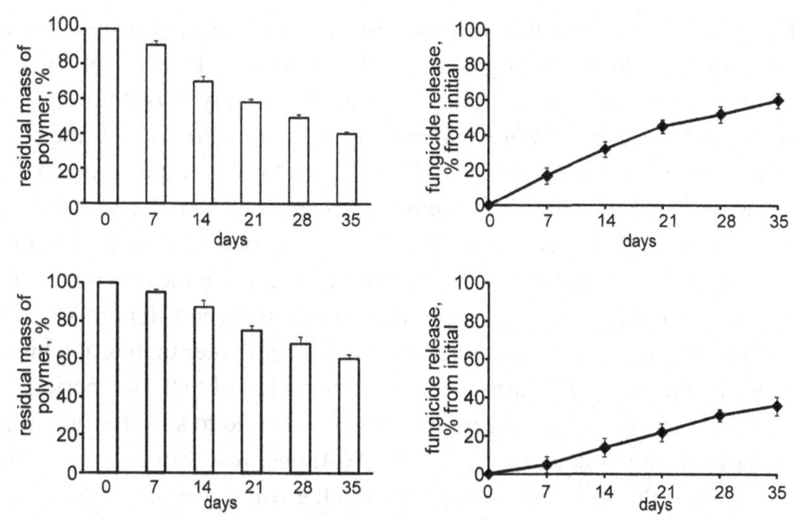

FIGURE 5.7 Dynamics of P3HB matrix degradation (a, c) and tebuconazole release (b, d) from films (a, b) and pellets (c, d) incubated in the laboratory soil ecosystem. (Reprinted by permission from Springer Nature: Environmental Science and Pollution Research: Characterization of biodegradable poly–3-hydroxybutyrate films and pellets loaded with the fungicide tebuconazole, Volova, T., Zhila, N., Vinogradova, O., Shumilova, A., Prudnikova, S., & Shishatskaya, E. © © 2015.)

The initial soil microbial communities and the microbial communities after 35 days of incubation of TEB-loaded films and pellets in soil were examined by plating soil samples on solid medium. The number of ammonifying copiotrophic bacteria (CFU/g soil) was determined on fish-peptone agar (FPA); the number of mineral nitrogen-assimilating prototrophs was determined on starch and ammonia agar (SAA); nitrogen-fixing bacteria were counted on Ashby's medium; and oligotrophs were counted on soil extract agar (SA) (Netrusov et al. 2005). Mineralization coefficient was determined as a ratio between microorganisms assimilating mineral nitrogen and ammonifying bacteria. Oligotrophy coefficient was determined as a ratio of oligotrophic to ammonifying bacteria. Concentrations of microorganisms (CFU/g soil) were calculated for the control soil and the soil layer on the surface of the specimens. All platings were performed in triplicate from 10^7 dilutions of soil suspension. The plates were incubated for 3–7 days at 30°C. The number of microorganisms was determined taking into account the dilutions. Dominant microorganisms were isolated and identified by conventional methods, based on their cultural and morphological properties and using standard biochemical tests mentioned in identification keys (Holt et al., 1997; Boone et al., 2005).

The presence of available nitrogen in the soil was responsible for the low numbers of nitrogen-fixing bacteria (3.5 million CFU in 1 g of soil). As shown in Figure 5.8A, initial soil microbial community was dominated by four genera of bacteria – *Bacillus, Micrococcus, Corynebacterium,* and *Pseudomonas* – which comprised 43%, 16%, 12%, and 8%, respectively. Minor genera were *Arthrobacter, Cellulomonas, Curtobacterium, Mycobacterium,* and *Streptomyces* (between 2 and 7%). Thus, soil conditions were favorable for the intense transformation of organic matter, including the polymer, which is a substrate for many soil microorganisms (Jendrossek and Handrick, 2002).

In the soil system, TEB was released from the polymer matrix at a higher rate than in the water (Figure 5.7). For a considerably shorter period (35 days), about 60% of the loaded fungicide was released to soil from the films and 36% from the pellets, or 1.23 and 1.80 times, respectively, more TEB was released to the soil over 35 days than to the water over 60 days.

TEB release occurred simultaneously with the destruction of the polymer matrix. For the first 14 days of incubation, the mass of the films and pellets had decreased by 13 and 30%, respectively. Then, P3HB degradation rate increased, and after 35 days, P(3HB) films were almost 60% degraded, while pellet degradation reached only 40%. The changes in the state of the samples are shown in Figure 5.7. As films and pellets were degraded and their mass decreased, their surface became uneven; we observed the formation of small pores and cracks (Figure 5.3). The number and size of defects grew during the experiment. The number or perforations increased and, finally, the films broke up. The pellets retained their mass and shape for longer periods of time, i.e., were less destructible.

After the P(3HB)/TEB samples had been incubated in soil for 35 days, the composition of the soil microbial community changed as follows. The total number of CFU in 1 g soil did not decrease dramatically, reaching 88.7 million. However, the proportions of the trophic groups of microorganisms and their composition changed relative to those of the initial microbial community. The proportion of copiotrophic bacteria decreased, while the proportions of prototrophic and nitrogen-fixing bacteria increased. These results are indicative of high rates of organic matter mineralization and polymer biodegradation; degradation products were used by soil microflora as supplementary substrate. Rapid degradation of organics and assimilation of this substrate by bacteria in the soil containing high levels of nitrogen-containing organic matter caused stabilization of the numbers of organotrophic bacteria and an increase in the proportions of prototrophic and nitrogen-fixing bacteria. The composition of the major microbial species changed, too (Figure 5.8B). The counts of spore-forming bacilli dropped dramatically, by 20%, and the counts of Gram-positive cocci decreased by 12%. The number of *Pseudomonas* and *Corynebacterium* increased by 11 and 12%, respectively. The total counts of Gram-negative bacilli and actinobacteria increased; among them were representatives of *Pseudomonas, Stenotrophomonas, Variovorax*, and *Streptomyces*, which, as we have found before, degrade PHAs very effectively (Volova et al., 2007, 2010; Boyandin et al., 2012, 2013).

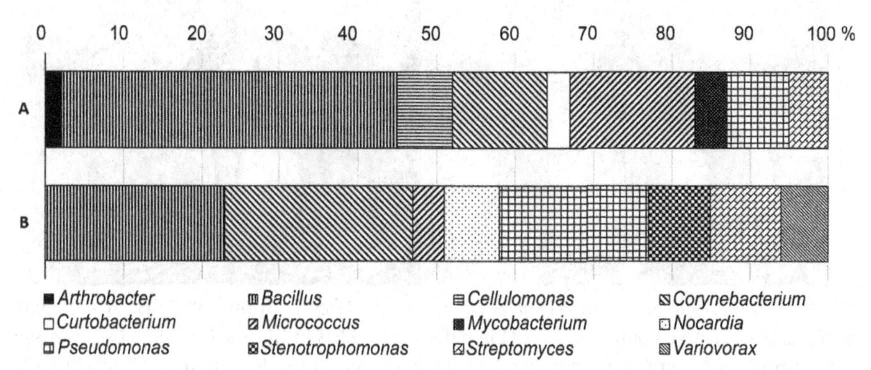

FIGURE 5.8 Major bacteria in the soil before (A) and after incubation of P(3HB)/TEB (B). (Reprinted by permission from Springer Nature: Environmental Science and Pollution Research: Characterization of biodegradable poly–3-hydroxybutyrate films and pellets loaded with the fungicide tebuconazole, Volova, T., Zhila, N., Vinogradova, O., Shumilova, A., Prudnikova, S., & Shishatskaya, E. © © 2015.)

Fungi of the genus *Fusarium* (*F. moniliforme* and *F. solani*) were extracted from the field soil and grown on the malt extract agar (MEA), (Sigma-Aldrich, U.S.) in Petri dishes at a temperature of 25°C for 5–7 days. Then, 5-mm diameter slabs of agar were aseptically drilled from the culture regions with actively growing colonies. A slab with the fungal culture and a film or pellet with encapsulated fungicide were placed at opposite sides of the Petri dish containing sterile MEA. The mass of the film or pellet in a Petri dish was 100 mg, and the TEB/P(3HB) ratio was 1:3. The dishes were incubated in the temperature-controlled cabinet at 25°C for 72 h; then, we measured the radius of the fungal mycelium and determined the degree of fungus growth inhibition relative to the control. As a negative control, we measured the radius of the fungal mycelium in the dish without the fungicide. As a positive control, we used 2 ml of the commercial TEB formulation (Raxil Ultra) placed in 10-mm diameter wells; this amount was equivalent to the amounts of TEB in films and pellets. All procedures were done in triplicate.

Figure 5.9 shows photographs of test organisms – phytopathogenic fungi of the genus *Fusarium* – *F. moniliforme* and *F. solani*, growing on solid medium in the presence of TEB-loaded films and pellets.

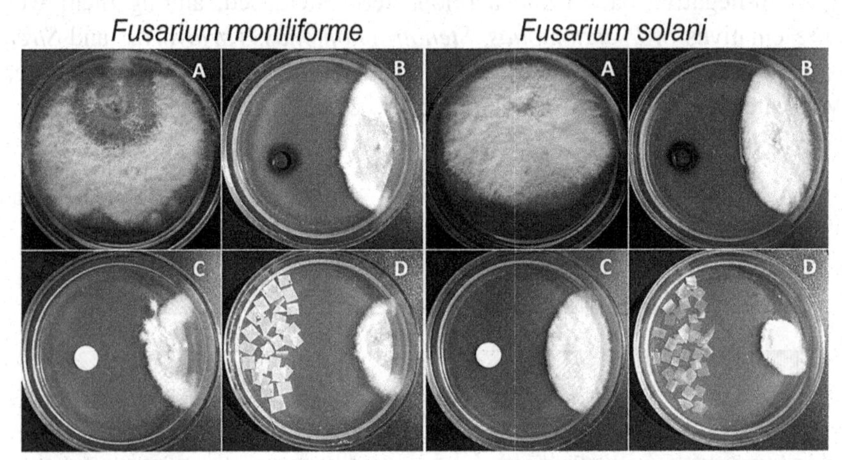

FIGURE 5.9 Sensitivity of *Fusarium* species to different forms of TEB; A – negative control, B – positive control, C – TEB in pellets; D – TEB in films. (Reprinted by permission from Springer Nature: Environmental Science and Pollution Research: ., Characterization of biodegradable poly–3-hydroxybutyrate films and pellets loaded with the fungicide tebuconazole, Volova, T., Zhila, N., Vinogradova, O., Shumilova, A., Prudnikova, S., & Shishatskaya, E. © © 2015.)

Fungus *F. moniliforme* showed higher sensitivity to TEB. The antifungal activity of the TEB formulations investigated in this study was comparable with that of the fungicide-loaded pellets against *F. solani* in the positive control. Moreover, films showed higher antifungal activity than pellets against both species used in the assay, because of their larger surface and, hence, contact with the greater area of the nutrient medium surface and better diffusion of the substance. Thus, the inhibitory effect of the TEB formulations tested in our study was comparable with the effect of free TEB.

Experimental slow-release formulations of the fungicide TEB were constructed by using poly(3-hydroxybutyrate) as a degradable polymeric matrix. The P(3HB)/TEB films and pellets were prepared from polymer solutions and powder and examined by DSC, X-ray structure analysis, and Fourier transform infrared spectroscopy, which showed that the P3HB/TEB system was a mechanical mixture of the components. TEB release was studied in the aqueous and soil systems. The soil microbial community was characterized, and polymer matrix degradation was monitored in experiments with TEB formulations; the TEB release kinetics and polymer degradation rate were found to be influenced by the geometry of the P(3HB)/TEB system. Experiments with the cultures of fungi *F. moniliforme* and *F. solani* showed that the polymer-embedded TEB had antifungal activity comparable with that of free TEB. The formulations of TEB embedded in the slowly degrading P(3HB) matrix constructed in this study hold promise for the development of slow-release fungicide formulations.

5.3 FUNGICIDAL ACTIVITY OF EXPERIMENTAL TEBUKONAZOLE FORMULATIONS IN SOIL MICROECOSYSTEMS

The most common commercial TEB formulations are suspensions or emulsions, which are used to spray the vegetative organs of plants. Standard formulations are present at higher concentrations than required for activity at the time of application in order to guarantee the duration of effect at later time intervals. As TEB is potentially phytotoxic, this method of application of the fungicide inhibits plant growth, causing economic losses and posing a threat to the health of people and the environment. Another way of using TEB chemicals is to treat seeds prior to planting, but this may decrease seed germination ability, suppress the growth of seedlings, and inhibit root growth. The TEB used as seed dressing is quickly depleted, and the vegetative organs of plants have to be sprayed with the fungicide. Thus, new TEB formulations need to be developed to increase the efficacy of TEB and minimize its harmful effects on the environment.

The purpose of study Volova et al. (2017) was to investigate release profile of TEB embedded in films, microgranules, and pellets of degradable poly–3-hydroxybutyrate incubated in soil microecosystems and the fungicidal activity of the formulations against the plant pathogen of the genus *Fusarium*.

The chemicals used in this study were the systemic fungicide Raxil Ultra (Bayer Crop Science, Russia), with TEB as the active ingredient, and chemically pure TEB (Russian Federal Standard GSO 7669–99, purity 99.1%). Films, pellets, and granules with TEB loadings of 10% and 50% were prepared as experimental formulations, as described in Chapter 4. The effect of the experimental TEB formulations on plant pathogenic fungi was studied in laboratory soil microecosystems. Soil microecosystems were prepared as follows. The agrogenically-transformed soil (the village of Minino, the Krasnoyarsk Territory, Russia) was placed into 250-cm^3 plastic containers (200 g soil per container). The soil used in the experiment is a typical agricultural soil occurring in

Siberian agroecosystems. The soil had high mineralization and oligotrophy coefficients (1.52 and 11.74, respectively), indicating soil maturity and low contents of available nitrogen forms. The number of copiotrophic bacteria was 16.3±5.1 million CFUs g^{-1} – 1.5 and 11.7 times lower than the number of prototrophic and oligotrophic bacteria, respectively, while the number of nitrogen-fixing bacteria was very high (26.1 ± 4.7 million CFUs g^{-1}).

TEB formulations (films, granules, pellets – 6 mg per container) were buried in the soil. The positive control was TEB in the form of commercial formulation Raxil Ultra, and the negative control was soil without TEB. The analysis of the chemical composition of the soil included measuring pH of the aqueous extract (following Russian Federal Standard 26423–85) and concentrations of nitrate-nitrogen (by the method developed at the Central Research Institute for Agrochemical Support of Agriculture, CRIASA, following Russian Federal Standard 26488–85), mobile phosphorus and exchangeable potassium (by the method developed by Machigin and modified at CRIASA, following Russian Federal Standard 26204–91).

Figure 5.10 shows photographs of the experimental slow-release formulations of TEB embedded in the polymer matrix of degradable poly–3-hydroxybutyrate (P3HB). The films prepared by solvent evaporation from the 2% polymer solution were 25 ± 0.3-μm thick; 2–3 mm-diameter granules were prepared from the 10% P3HB solution by the microdrop technique; pellets prepared from polymer powder and TEB were 3 mm in diameter and 1 mm thick. The formulations were loaded with TEB at 10 or 50% of the polymer matrix weight. The surface properties of the TEB-loaded pellets were not significantly different from the surface properties of the pellets without TEB (Table 5.2). Some differences were observed between surfaces of P(3HB) and P(3HB)/TEB films: the water contact angle of the P(3HB) films was 68.4° while the water contact angles of P(3HB)/TEB films with 10% and 50% TEB loading were 75.9 and 72.1°, respectively, i.e., the surfaces of the TEB-loaded films were somewhat more hydrophobic.

Their surface energy and its dispersive and polar components were also different. Analysis of microstructure showed that the surfaces of all formulations with the higher TEB loading were more uneven, with 1–2-μm pores, bumps, and erosions (Figure 5.11). These changes were more pronounced on films.

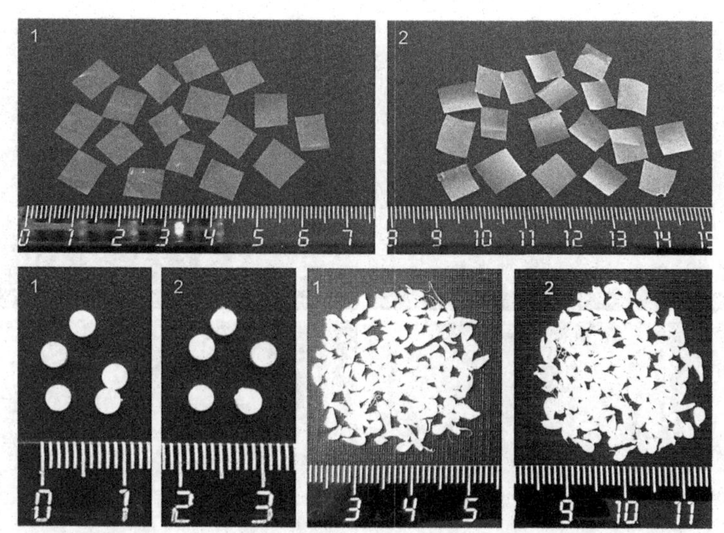

FIGURE 5.10 Photographs of P3HB/TEB films, pellets, and granules with TEB loadings of 10 and 50% of the polymer weight: 1 and 2, respectively. (Reprinted with permission from Volova, T. G., Prudnikova, S. V., Zhila, N. O., Vinogradova, O. N. Shumilova, A. A., Nikolaeva, E. D., Kiselev, E. G., & Shishatskaya, I. I., Efficacy of tebuconazole embedded in biodegradable poly–3-hydroxybutyrate to inhibit the development of Fusarium moniliforme in soil microecosystems. Pest Manag. Sci., 73, 925–935. © 2016 John Wiley and Sons.)

TABLE 5.2 Surface Properties of Experimental P3HB/TEB Films and Pellets with Different TEB Loadings

Sample, TEB loading	Water contact angle, degrees	Surface energy, mN m^{-1}	Dispersive component, mN m^{-1}	Polar component, mN m^{-1}
Initial P(3HB):				
film	68.4±4.3	59.2±1.9	43.8±0.9	15.4±0.9
pellet	77.0±2.7	58.4±3.2	45.5±0.9	5.7±0.9
P(3HB)/TEB films				
TEB loading (%):				
10	75.9±3.1	44.2±0.3	38.5±0.2	5.6±0.2
50	72.1±0.5	47.9±0.6	41.4±0.5	6.5±0.1
P(3HB)/TEB pellets				
TEB loading (%):				
10	77.0±3.4	58.2±2.9	44.8±0.93	5.6±0.97
50	75.0±5.4	56.7±3.7	43.4±0.70	5.1±0.89

(Reprinted with permission from Volova, T. G., Prudnikova, S. V., Zhila, N. O., Vinogradova, O. N. Shumilova, A. A., Nikolaeva, E. D., Kiselev, E. G., & Shishatskaya, I. I., (2017). Efficacy of tebuconazole embedded in biodegradable poly–3-hydroxybutyrate to inhibit the development of Fusarium moniliforme in soil microecosystems. Pest Manag. Sci., 73, 925–935. © 2016 John Wiley and Sons.)

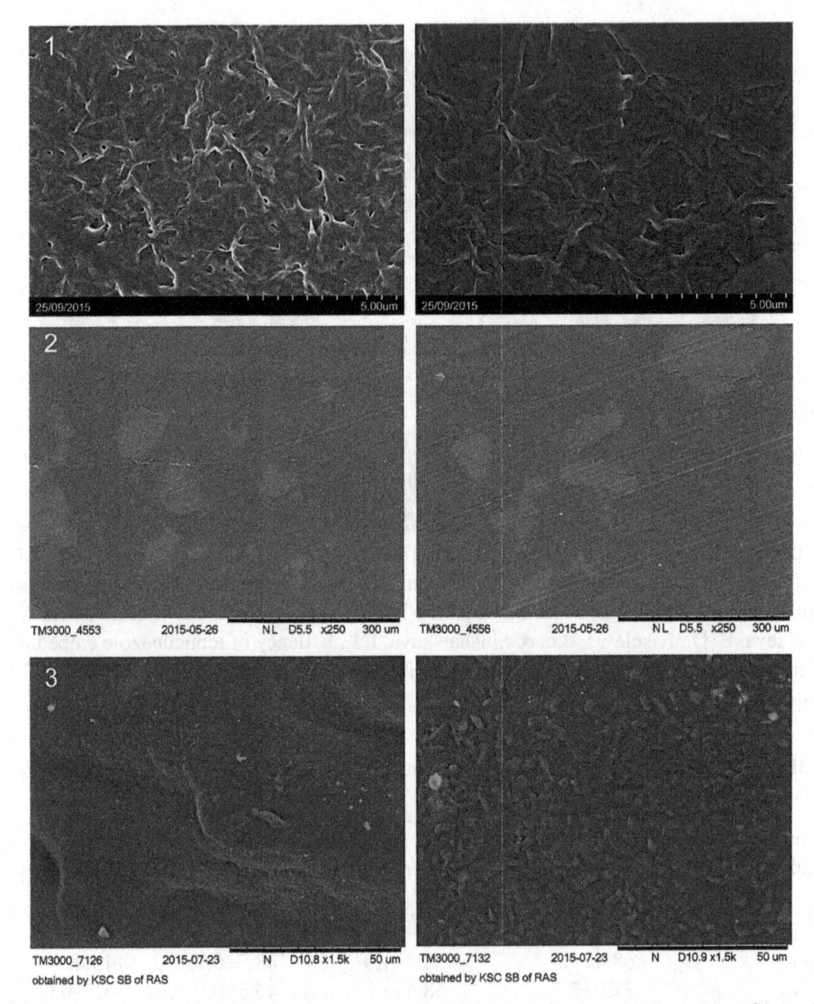

FIGURE 5.11 SEM images of the experimental P3HB/TEB formulations: 1 – films, 2 – pellets, 3 – granules, with TEB loadings of 10 and 50% of the polymer weight. (Reprinted with permission from Volova, T. G., Prudnikova, S. V., Zhila, N. O., Vinogradova, O. N. Shumilova, A. A., Nikolaeva, E. D., Kiselev, E. G., & Shishatskaya, I. I., (2017). Efficacy of tebuconazole embedded in biodegradable poly–3-hydroxybutyrate to inhibit the development of Fusarium moniliforme in soil microecosystems. Pest Manag. Sci., 73, 925–935. © 2016 John Wiley and Sons.)

Based on morphological, physiological, and biochemical studies and molecular-genetic examination of the 16S and 28S rRNA gene fragments, we determined the major microorganisms in the soil samples (Figure 5.12). The microbial community was dominated by actinobacteria such as *Streptomyces* (24%), *Arthrobacter* species (18%), and *Corynebacterium* species

(12%); *Pseudoxanthomonas* were the major Gram-negative bacilli (12%) (Figure 5.12 A). Microscopic fungi were dominated by *Penicillium* species (58–65%); fungi of the genera *Fusarium, Trichoderma*, and *Aspergillus* constituted 8–11% of the population of microscopic fungi in soil samples (Figure 5.12 B). *Fusarium* species isolated from the initial soil samples were represented by *F. solani* and *F. lateritium*. No *F. moniliforme* was detected in the initial microbial community.

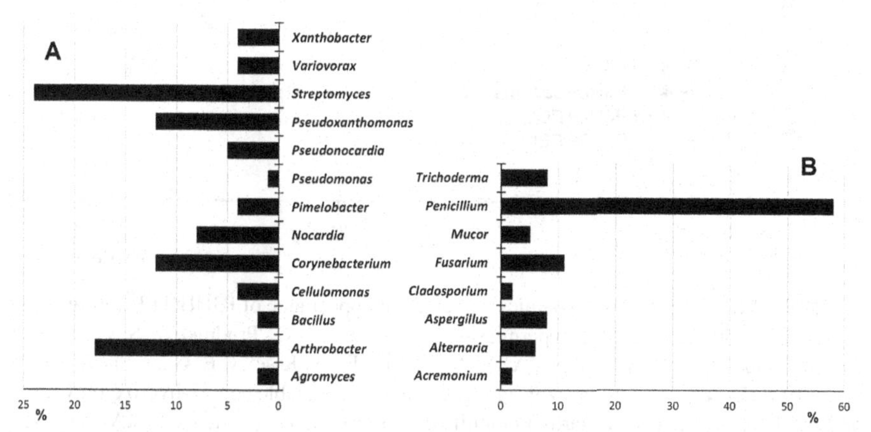

FIGURE 5.12 Dominant bacteria (A) and fungi (B) in the initial soil samples. (Reprinted with permission from Volova, T. G., Prudnikova, S. V., Zhila, N. O., Vinogradova, O. N. Shumilova, A. A., Nikolaeva, E. D., Kiselev, E. G., & Shishatskaya, I. I., (2017). Efficacy of tebuconazole embedded in biodegradable poly–3-hydroxybutyrate to inhibit the development of Fusarium moniliforme in soil microecosystems. Pest Manag. Sci., 73, 925–935. © 2016 John Wiley and Sons.)

Results of studying P(3HB)/TEB degradation dynamics and TEB release kinetics are shown in Figures 5.13 and 5.14. Dynamics of degradation of P(3HB)/TEB formulations was influenced by their geometry, and the percentage of TEB loaded into them (Figure 5.13).

The highest degradation rate was recorded for P(3HB)/TEB films: after 7 days of incubation in the soil, their weight loss reached 40% of their initial weight, irrespective of the TEB loading. Then, the degradation process was influenced by the percentage of loaded TEB. The weight of the specimens with 50% TEB loading had dropped to 2–5% of their initial weight by Day 28 while the specimens with 10% TEB loading showed a comparable weight loss at Day 51. The average degradation rate of the films was 0.106 mg d^{-1}. The second fastest degradation rates were recorded for microgranules. P3HB/TEB microgranules with different TEB percentages showed similar average degradation rates: 0.081–0.083 mg d^{-1}. At the end of the experiment,

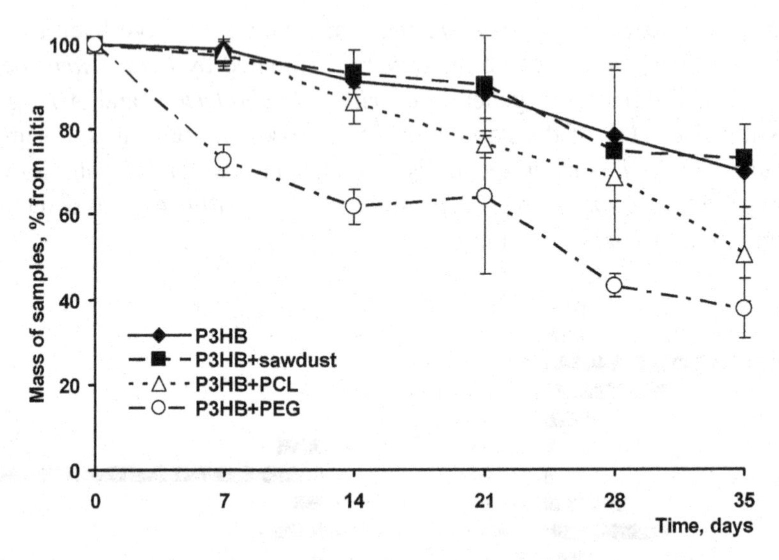

FIGURE 5.13 Degradation dynamics and external appearance of P3HB/TEB formulations incubated in soil. (Reprinted with permission from Volova, T. G., Prudnikova, S. V., Zhila, N. O., Vinogradova, O. N. Shumilova, A. A., Nikolaeva, E. D., Kiselev, E. G., & Shishatskaya, I. I., (2017). Efficacy of tebuconazole embedded in biodegradable poly–3-hydroxybutyrate to inhibit the development of Fusarium moniliforme in soil microecosystems. Pest Manag. Sci., 73, 925–935. © 2016 John Wiley and Sons.)

their residual weight was no higher than 15% of the initial weight. The pressed pellets were degraded at the slowest rate. Formulations with 10 and 50% TEB loadings showed the average degradation rates of 0.030 and 0.055 mg d^{-1}, respectively, and their residual weights were about 70 and 40% of the initial weights.

There were lag phases in the degradation of all polymer formulations: 2–3 d for films, 4–6 d for microgranules, and over 2 weeks for pellets. The reason for the occurrence of the lag phase is that it takes some time for microorganisms to get attached to the surface of the polymer samples and get adapted to the synthesis of the depolymerizing enzyme systems corresponding to the substrate (polymer). We monitored changes in the structure and physicochemical properties of the polymer specimens during degradation and observed changes in the surface structure of all specimens in the early degradation phase. In the later phases, films broke into fragments, while pellets underwent surface erosion, with the polymer weight dropping and crystallinity rising. The increase in the degree of crystallinity was caused by the quicker degradation of the amorphous region of the polymer, which is a more readily available growth substrate to the microorganisms than the crystalline region.

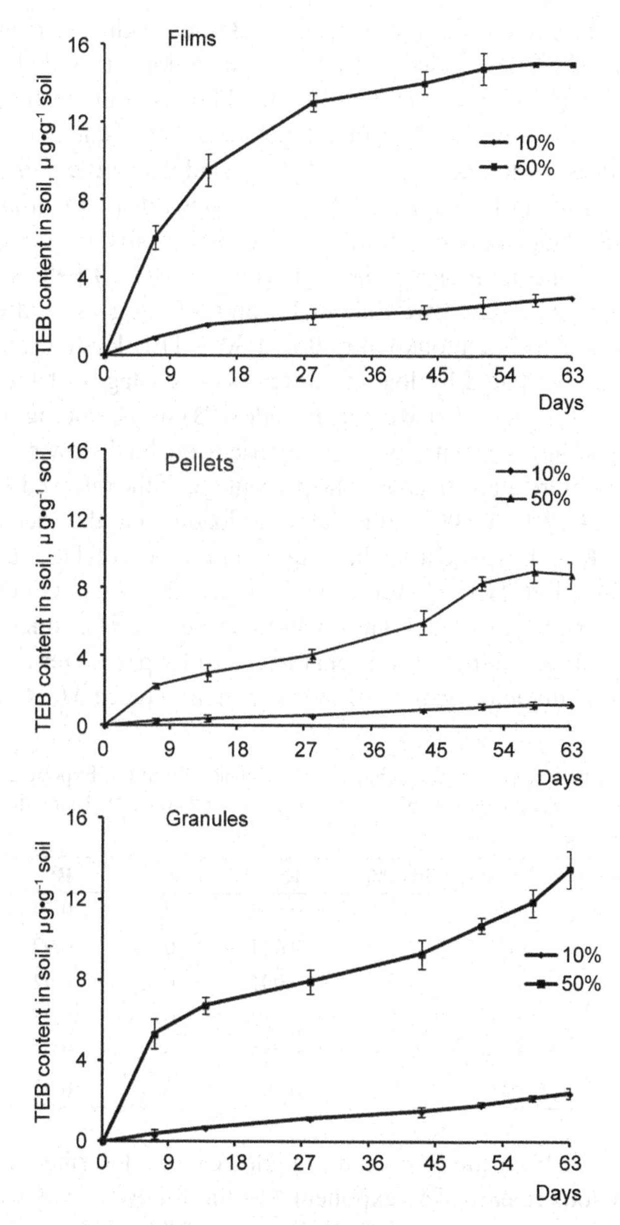

FIGURE 5.14 Cumulative release of TEB into the soil from different P(3HB)/TEB forms (films, granules, pellets) with 10 and 50% TEB. (Reprinted with permission Volova, T. G., Prudnikova, S. V., Zhila, N. O., Vinogradova, O. N. Shumilova, A. A., Nikolaeva, E. D., Kiselev, E. G., & Shishatskaya, I. I., (2017). Efficacy of tebuconazole embedded in biodegradable poly–3-hydroxybutyrate to inhibit the development of Fusarium moniliforme in soil microecosystems. Pest Manag. Sci., 73, 925–935. © 2016 John Wiley and Sons.)

The geometry of the form influenced TEB release kinetics (Figure 5.14). The amount of TEB released from the most rapidly degraded specimens (films) reached its maximum in the soil after 43 days of incubation: 15 and 3 μg g^{-1} soil in the samples with formulations loaded at 10 and 50% TEB. That was comparable to the concentration of TEB added to the soil in the free form. The degrading P(3HB)/TEB microgranules released similar amounts of TEB, increasing its concentration in the soil to about 2.4 and 13.5 μg g^{-1} soil. TEB concentrations in the soil with P3HB/TEB pellets (degraded at the slowest rate) were lower: about 1.1 and 8.3 μg g^{-1} soil after 51 days of incubation of formulations with 10 and 50% TEB loadings. Thus, TEB release can be regulated by loading different percentages of the fungicide (10 and 50% of the polymer weight) into P(3HB) films, microgranules, and pellets. Embedding TEB into polymer matrices enabled slow release of the fungicide (for more than 60 days). The percentage of the released TEB varied between 40–50 and 90–100% of the amount loaded into the specimens.

Constant K and exponent n, characterizing kinetics of TEB release from the experimental P(3HB)/TEB formulations, which were obtained by using the Korsmeyer-Peppas model, are given in Table 5.3. The time when TEB is released with the highest rate is characterized by parameter t^{50} – the time needed for the fungicide content of the specimen to reach $M_t/M_\infty \leq 0.5$.

TABLE 5.3 Kinetic Constants of Tebuconazole Release From the Experimental P(3HB)/TEB Formulations in Laboratory Soil Microecosystems, Obtained by Using the $M_t/M_\infty = Kt^n$ Equation

Formulation type	TEB loading, %	K(h^{-n})	n	R^2	t^{50}, d
Films	10	0.005	0.81	0.99	14
	50	0.014	0.66	0.99	12
Granules	10	0.001	0.85	0.92	43
	50	0.032	0.49	0.93	28
Pellets	10	0.045	0.30	0.98	63
	50	0.044	0.36	0.99	43

TEB release from the films and granules can be described as the zero-order anomalous release. The exponent had the following values: for films loaded at 10%, n = 0.81 and for films loaded at 50%, n – 0.66; for granules loaded at 10%, n = 0.85 and for granules loaded at 50%, n = 0.49. In both cases, as the loading was increased, the exponent decreased and constant K, which contains a diffusion coefficient and the geometric and structural characteristics of the specimens, increased. Changes in constant K are indicative of the heterogenic structure of the specimens of the same geometry but with different loadings. SEM data lead to the same conclusion. Pellets have more stable structural and geometric parameters, as suggested by the values of K, and the value of exponent n characterizes the release mechanism as Fickian diffusion.

The release time of the greater percentage of the fungicide, kinetic constant, and exponent of the specimens correlate with polymer degradation kinetics. For the films with 10 and 50% TEB loadings, t^{50} was 14 and 12 d, respectively, and the polymer weight loss was 59% and 69%, respectively. For the granules with 10 and 50% TEB loadings, t^{50} was 43 and 28 d, respectively, and the polymer weight loss was 49% and 56%, respectively. Pellets were more resistant to degradation, and TEB release from pellets loaded at 10% was 40% for 63 d. For 50% loaded pellets, t^{50} was 43 d. The polymer weight loss for the time when TEB release occurred at the highest rate was 32 and 29% for the 10%- and 50%-loaded pellets, respectively.

Thus, TEB release was influenced by the geometry of the specimens and their structure, which was related to the percentage of the loaded fungicide and the process employed to fabricate the formulation.

Results characterizing the fungicidal activity of the experimental P(3HB)/TEB formulations are shown in Figures 5.15 and 5.16. The parameters evaluated in this study were the total number of the CFUs and production of the purple or pink pigment, which identifies *F. moniliforme* and indicates its physiological activity. Microbiological analysis showed that after 7 days, the number of *F. moniliforme* CFUs decreased in all treatments, but it remained the dominant species, and the number of *F. moniliforme* colonies on the dishes with the nutrient medium was higher than the number of colonies of saprotrophic microscopic fungi. Application of the commercial formulation Raxil Ultra decreased the number of *F. moniliforme* CFUs by a factor of 400–500 compared with the initial number of the CFUs of this plant pathogen – to 1.8–3.2 thousand CFUs g^{-1} soil – and kept it at that level for the following 7 weeks, to the end of the experiment. The polymer-embedded TEB was less effective in the early phase of incubation, but the number of *Fusarium* CFUs decreased steadily throughout the experiment, as TEB was gradually released from the polymer matrix.

P(3HB)/TEB granules and films were more effective than pellets. In the soil with microgranules, the *F. moniliforme* population decreased by a factor of 70–90 compared with the initial population after 7 days of incubation. During the first 2 weeks, P3HB/TEB films were less effective than granules, but after 14 days, the effects of the two types of formulations did not differ

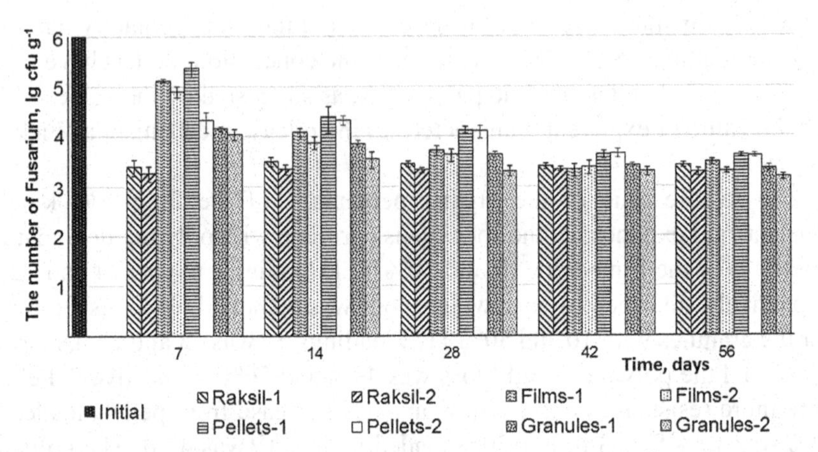

FIGURE 5.15 The fungicidal effect of various forms of TEB on the plant pathogen *Fusarium moniliforme*; tebuconazole concentration in the formulation is 600 mg (1) or 3000 mg (2). (Reprinted with permission from Volova, T. G., Prudnikova, S. V., Zhila, N. O., Vinogradova, O. N. Shumilova, A. A., Nikolaeva, E. D., Kiselev, E. G., & Shishatskaya, I. I., (2017). Efficacy of tebuconazole embedded in biodegradable poly–3-hydroxybutyrate to inhibit the development of Fusarium moniliforme in soil microecosystems. Pest Manag. Sci., 73, 925–935. © 2016 John Wiley and Sons.)

significantly; after 4–6 weeks, the effectiveness of the P(3HB)/TEB granules and films became comparable with that of the commercial formulation Raxil Ultra (Figure 5.16). P(3HB)/TEB pellets had a weaker fungicidal effect than films or granules, as the release of the fungicide from the pressed pellets was slower during the first month of incubation in the soil. After 8 weeks, the fungicidal effect of the TEB embedded in polymer matrices was not significantly different from the effect of Raxil Ultra. The higher TEB loading and the increase in the TEB concentration in control with Raxil Ultra did not significantly influence its fungicidal effect.

TEB exhibited a fungistatic effect not only towards plant pathogens but also towards all saprotrophic fungi. The microbiological examination of the structure of microbial communities during the experiments showed that the use of Raxil Ultra inhibited the growth of such aboriginal fungi as *Penicillium*, *Alternaria*, and *Aspergillus*. In 3 days after the commercial formulation was added to the soil, no growth of colonies was observed on Petri dishes. After 7 days of incubation, the fungi resumed growth, but the total number of microscopic fungi in the positive control was lower than in the treatments (with P3HB/TEB formulations) (Figure 5.16).

A possible explanation for this is that the concentration of TEB added to the soil as the active ingredient of the commercial formulation was initially

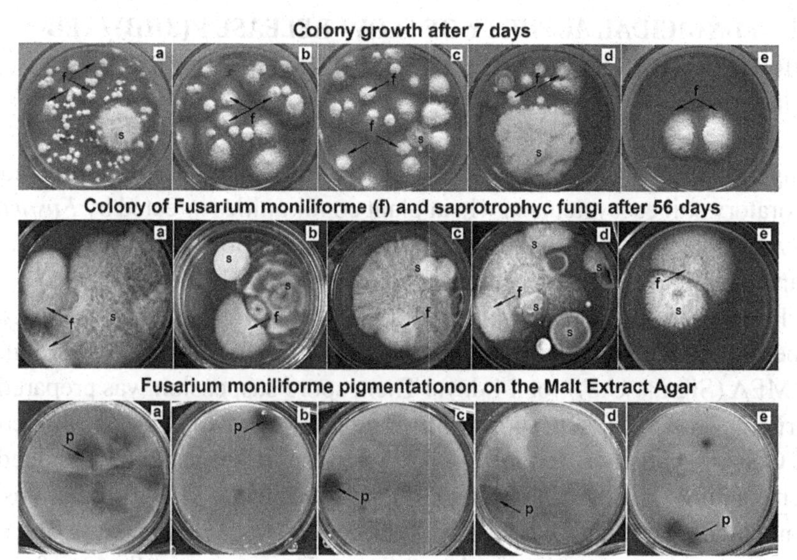

FIGURE 5.16 The growth of *Fusarium moniliforme* colonies; *a* – negative control, *b* – P3HB/TEB films, *c* – P(3HB)/TEB pellets, *d* – P(3HB)/TEB granules, *e* – Raxil Ultra. Arrows indicate colonies of *Fusarium* (*f*) and purple-pink pigment formation (*p*); *S* – colonies of saprotrophic fungi. (Reprinted with permission from Volova, T. G., Prudnikova, S. V., Zhila, N. O., Vinogradova, O. N. Shumilova, A. A., Nikolaeva, E. D., Kiselev, E. G., & Shishatskaya, I. I., (2017). Efficacy of tebuconazole embedded in biodegradable poly–3-hydroxybutyrate to inhibit the development of Fusarium moniliforme in soil microecosystems. Pest Manag. Sci., 73, 925–935. © 2016 John Wiley and Sons.)

much higher than the concentration of TEB released from the experimental formulations. TEB release from the P(3HB)/TEB formulations occurred as the polymer was degraded, and the amount of the TEB in soil increased gradually. The initially low TEB concentrations could not inhibit the growth of aboriginal fungi, mainly *Penicillium*, which dominated aboriginal saprophytic fungi.

The total counts of bacteria varied very slightly between treatments and did not differ significantly from the bacterial counts in the initial soil and controls. Hence, none of the TEB formulations inhibited the growth of saprotrophic bacteria in soil.

Thus, fungicide TEB embedded in the polymer matrix of P(3HB) in the form of films, microgranules, and pellets was effective against plant pathogens of the genus *Fusarium*, and that was a long-lasting effect (for 8 weeks) (Volova et al., 2017). After 2–4 weeks of incubation in soil, P3HB/TEB formulations were as effective as the commercial formulation Raxil Ultra. The experimental forms of TEB embedded in the slowly degraded P(3HB) can be used as a basis for developing slow-release fungicide formulations.

5.4 FUNGICIDAL ACTIVITY OF SLOW-RELEASE P(3HB)/TEB FORMULATIONS IN WHEAT STANDS INFECTED BY *FUSARIUM MONILIFORME*

Fungicidal activity of experimental TEB formulations was investigated in laboratory soil ecosystems in wheat plant communities infected by *Fusarium moniliforme*. TEB was embedded in the matrix of poly–3-hydroxybutyrate, shaped as films and microgranules (Volova et al., 2018).

Fungi of the genus *Fusarium* Link (*F. moniliforme* J. Sheld) were used in experiments. The inoculum was prepared by growing fungi in culture tubes on MEA (Sigma, U.S.) for 14 days. Then, spore suspension was prepared in sterile tap water, 5.2×10^7 spores ml^{-1}. The number of spores was counted in the Goryaev chamber. Ten ml *F. moniliforme* spore suspension was added to each container with soil. The effect of the experimental TEB formulations on plant pathogenic fungi was studied in laboratory soil microecosystems. Soil microecosystems were prepared as follows. The agrogenically-transformed soil (from the village of Minino, the Krasnoyarsk Territory, Siberia, Russia) was placed into 500-cm^3 plastic containers (500 g soil per container). Wheat seeds were sown into the soil, 100.45 g seeds per 1 m^2. The plants were grown in the Conviron A1000 growth chamber (Canada) for 30 d under stable conditions: illumination 100–300 μmoles/m^2/s under the 12L:12D photoperiod, the temperature of 18–25°C, and humidity of 65%.

Two experiments were carried out, each lasting 30 days. In Experiment 1, wheat seeds were infected with plant pathogens. There were five experimental groups. In Group 1 (negative control), seeds sown into the soil had not been treated with the fungicide, and no TEB was added to the soil. In Group 2 (positive control), at the time of planting, the soil was treated with commercial formulation Raxil at a concentration comparable with the TEB concentrations in the treatments: 3 μg TEB/g soil. In two treatment groups, the seeds had not been treated with the fungicide prior to sowing, but P(3HB)/TEB films (Group 3) and granules (Group 4) were buried in the soil. In Group 5, the seeds were soaked in the Raxil solution for 10 min prior to sowing; no TEB was added to the soil. In Experiment 2, the efficacy of the P(3HB)/TEB formulations were tested under harsher conditions. In addition to using the infected wheat seeds, we also added spores of the plant pathogen *F. moniliforme* into the soil. There were four experimental groups. In Group 1, the wheat seeds had not been treated with the fungicide before sowing; no TEB was added to the soil. In the treatment groups, the seeds sown into the soil had not been treated with the fungicide, but P(3HB)/TEB films

(Group 2) and granules (Group 3) were buried in the soil. In Group 4 (control), at the time of planting, the soil was treated with commercial formulation Raxil at a concentration comparable with the TEB concentrations in the treatments: 3 μg TEB/g soil.

The fungicidal activity of the experimental P(3HB)/TEB formulations was compared with the activity of commercial formulation Raxil by evaluating changes in the following parameters: the counts of *F. moniliforme* in soil; the percentage of plant roots affected by the rot disease; the amount of aboveground biomass of the plants (determined by weighing the biomass preliminarily dried to constant weight). These parameters were measured at Days 10, 20, and 30 of the experiment.

Counting of the total microscopic fungi, including *F. moniliforme*, was performed by plating soil suspension onto Petri dishes with MEA, which was supplemented with chloramphenicol (100 μg L^{-1} of the medium) to suppress cell growth. All platings were performed in triplicate from 10^2–10^5 dilutions of soil suspension. The dishes were incubated at a temperature of 25°C for 7–10 days. Microscopic analysis of the colonies was done using an AxioStar microscope (Carl Zeiss). Microscopic fungi were identified by their cultural and morphological properties, with identification guides (Sutton et al., 2001; Watanabe, 2002). The degree to which the roots were damaged by Fusarium infection was evaluated as follows: the plant roots were carefully removed from the soil, rinsed first in running water and then three times in sterile tap water, and placed onto paper filters wetted to the maximum water holding capacity in Petri dishes. The Petri dishes were incubated in the thermostat at 25°C. The degrees to which the roots were infected by the fungi *Fusarium*, *Alternaria*, and *Bipolaris* were determined at Day 5–10, by microscopic examination of mycelium and spore formation of the fungi on wheat roots. The initial degree of infection of wheat seeds by plant pathogens (internal infection) was determined by germinating the seeds in Petri dishes on sterile nutrient medium MEA (Russian Federal Standard 12044–93). The state of the plants and their growth were evaluated by photographing the plant communities and the roots.

Phytosanitary analysis of wheat seeds germinated on the nutrient medium showed the presence of infections caused by the fungi of the genera *Fusarium* Link, *Alternaria* Nees, and *Bipolaris* Shoem. Wheat seeds infected by plant pathogens constituted 9.5 ± 1.2%, 5.6 ± 0.2% of which were infected by *Fusarium* species. Thus, natural infections of the seeds were caused not only by *Fusarium* species, which were detected in the initial soil, but also by the phytopathogenic microscopic fungi that developed when the seeds containing internal infection were germinated.

Analysis of microscopic fungi in soil samples showed that in the initial soil, their total counts reached $(28.3 \pm 9.4)10^3$ CFU·g^{-1}, while at the end of the experiments, this number had dropped by a factor of 2.3, to $(12.3 \pm 2.5)\times10^3$ CFU·g^{-1}. That was caused by changes in the structure of the microbial community that occurred due to the selective effects of wheat on the rhizosphere microflora: Wheat roots exuded organic substances readily available to bacteria, leading to an increase in their abundance. The total bacterial counts increased considerably: from $(16.3\pm5.2)\times10^6$ CFU·g^{-1} in the initial soil to $(164.7\pm8.5)\times10^6$ CFU·g^{-1} in the soil samples analyzed after 30 days of the experiment. Thus, as wheat plants were growing, the structure of the microbial community was changing. The decrease in the counts of microscopic fungi seemed to be caused by their competition with rhizosphere bacteria.

The degrees of infection caused by phytopathogenic fungi in the initial soil differed considerably between the two experiments, and that resulted in different efficacy of TEB. In Experiment 1, wheat was grown in the soil with natural infection mainly caused by *Fusarium* fungi, whose counts reached 3.1×10^3 CFU·g^{-1}. In Experiment 2, when the soil was additionally inoculated with *F. moniliforme* spores, the counts of plant pathogens in the initial soil were higher by three orders of magnitude – 1×10^6 spores g^{-1} soil.

The evaluation of the fungicidal effect of experimental TEB formulations in Experiment 1, with the soil containing relatively low concentrations of plant pathogens, did not reveal any significant differences between the effects of the experimental formulations and commercial formulation Raxil. Both P(3HB)/TEB and Raxil decreased not only the counts of such plant pathogens as *Fusarium*, *Alternaria*, and *Bipolaris*, but also the total counts of microscopic fungi – by a factor of 1.7–2.3 compared to the TEB-free soil (negative control). A different result was achieved in Experiment 2, with soil additionally inoculated with *F. moniliforme* spores. At the time of sowing, the total counts of saprotrophic fungi and the abundance of plant pathogens of the genus *Fusarium* (including *F. moniliforme* and minor species) reached 25.2×10^3 CFU g^{-1} and 1×10^6 CFU g^{-1}, respectively, but at Day 30, in the negative control, the total counts of the introduced *Fusarium* fungi dropped by three orders of magnitude – to 21.2×10^3 CFU g^{-1}, and the counts of indigenous saprotrophic microflora decreased by a factor of 5.8–4.9×10^3 CFU g^{-1}. That was most probably caused by trophic and competitive interactions between the introduced species and the indigenous microflora (Simberloff and Stiling, 1996; Ricciardi et al., 2013). In the soil with Raxil, the counts of phytopathogenic and saprotrophic fungi were 8.4×10^3 and 9.2×10^3 CFU g^{-1}, respectively. Neither of the experimental P(3HB)/TEB formulations inhibited the growth of saprotrophic fungi, and both were 3.0–3.6 times more effective against *F. moniliforme* than Raxil. Thus, in the soil with a high level of *F. moniliforme* infection, the fungicidal activity of the experimental P(3HB)/TEB formulations was higher than that of the commercial Raxil.

The roots of the initially infected wheat plants grown in the soil infected by plant pathogens were damaged by rot. In Experiment 1, with naturally infected soil, Fusarium infection was detected in all plant groups, including the groups with TEB added to the soil, in the first 10 days (Figure 5.17). The reason for that was the internal infection of the seeds, which developed in the early, seedling, stage. Then, in the negative control group, the infection of the roots caused by phytopathogenic microscopic fungi increased. Between Days 10 and 30, the percentage of roots damaged by rot increased from 17 to 30% (of the total root mass). The major contribution to the etiology of root rot was made by Fusarium infection – 50–80% of all infections. In the soil with commercial formulation Raxil, the degree of root rot infection was significantly lower than in the negative control, but the infection of the roots increased, reaching 25% by the end of the experiment. The pre-sowing seed treatment with Raxil restrained the development of the overall root infection in the first 10 days, but then rot infection damaged more roots and persisted at a high level – 21–27%. The experimental P(3HB)/TEB films were effective against all root rots, including fusarium infection, restraining their development. Between Days 10 and 20, the efficacy of this formulation was comparable to that of Raxil. Moreover, TEB embedded in the polymer matrix showed extended fungicidal effect, and between Days 20 and 30, root infection did not increase, in contrast to the groups with Raxil and pretreated seeds. P(3HB)/TEB granules did not show any fungicidal effect in the first 10 days. TEB release from the granules occurred at a slower rate than from the films, and TEB concentration in the soil was too low: 1.1 μg/g soil versus 2.7 μg/g soil in the soil with P(3HB)/TEB films. At Day 30, however, inhibition of root rot development in this group was comparable to the effect of commercial formulation Raxil.

In the soil with commercial formulation Raxil, the degree of root rot infection was significantly lower than in the negative control, but the infection of the roots increased, reaching 25% by the end of the experiment. The pre-sowing seed treatment with Raxil restrained the development of the overall root infection in the first 10 days, but then rot infection damaged more roots and persisted at a high level – 21–27% (Figure 5.17). The experimental P(3HB)/TEB films were effective against all root rots, including fusarium infection, restraining their development. Between Days 10 and 20, the efficacy of this formulation was comparable to that of Raxil. Moreover,

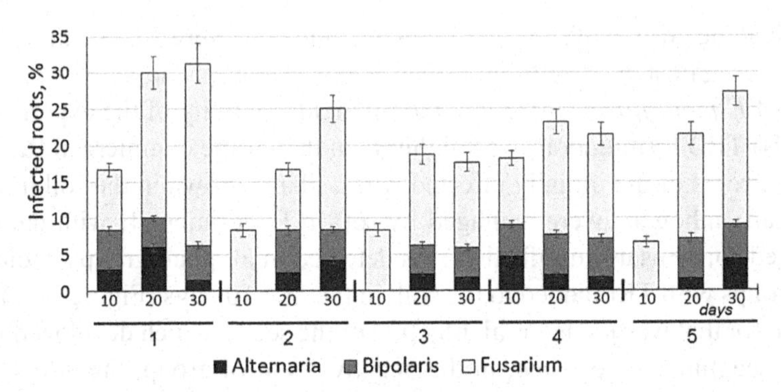

FIGURE 5.17 The effect of TEB delivery mode on the percentage of wheat roots damaged by rot: 1 – negative control, 2 – positive control (Raxil applied to soil), 3 – P(3HB)/TEB films, 4 – P(3HB)/TEB granules, 5 – pre-sowing treatment of seeds with Raxil. (Reprinted by permission from Springer Nature: Environmental Science and Pollution Research: Fungicidal activity of slow-release P(3HB)/TEB formulations in wheat plant communities infected by Fusarium moniliforme. Volova, T. G., Prudnikova, S. V., & Zhila, N. O. © 2017.)

TEB embedded in the polymer matrix showed extended fungicidal effect, and between Days 20 and 30, root infection did not increase, in contrast to the groups with Raxil and pretreated seeds. P(3HB)/TEB granules did not show any fungicidal effect in the first 10 days. TEB release from the granules occurred at a slower rate than from the films, and TEB concentration in the soil was too low: 1.1 μg/g soil versus 2.7 μg/g soil in the soil with P(3HB)/TEB films. At Day 30, however, inhibition of root rot development in this group was comparable to the effect of commercial formulation Raxil.

In Experiment 2, with *F. moniliforme* spores added to the soil, the percent of the damaged roots was considerably higher (Figure 5.18). In the group with no TEB added to the soil (negative control), after 30 days, the percentage of infected roots reached 61.5%, the roots damaged by Fusarium infection caused by *F. moniliforme* constituting 53.8%. TEB in the form of commercial Raxil (positive control) was effective against root rots in the early stage. Between Days 10 and 20, the total percentage of infected roots was 1.8–1.9 times lower and the percentage of fusarium infection-damaged ones 3.3–2.2 times lower than in the negative control group. Later, however, the fungicidal effect of Raxil became weaker, and the percentage of infected roots increased. Similarly to Experiment 1 (with the naturally infected soil), the fungicidal effects of the two experimental P(3HB)/TEB formulations was comparable to that of commercial formulation Raxil in the first 20 days, but it lasted longer and restrained the development of root rots, including fusarium infection, during the final stage (Days 20–30). At Day 30, the total percentage of the infected roots was 1.6 times lower and the percentage of Fusarium infection 1.4 times lower than in the group with Raxil.

FIGURE 5.18 The effect of TEB delivery mode on the percentage of wheat roots damaged by rot in the experiment with *Fusarium moniliforme* spores added to the soil: 1 – negative control, 2 – positive control (Raxil applied to soil), 3 – P(3HB)/TEB films, 4 – P(3HB)/TEB granules. (Reprinted by permission from Springer Nature: Environmental Science and Pollution Research: Fungicidal activity of slow-release P(3HB)/TEB formulations in wheat plant communities infected by Fusarium moniliforme. Volova, T. G., Prudnikova, S. V., & Zhila, N. O. © 2017.)

Figure 5.19 shows photographs of wheat roots at Day 30 of the experiment in the groups with different TEB delivery methods, illustrating the beneficial effect of the experimental P(3HB)/TEB formulations.

Results of evaluating the productivity of wheat communities growing on soils infected by *F. moniliforme* to various degrees, with different percentages of roots damaged by root rot, are shown in Figure 5.20. In Experiment 1, with milder damage to roots, measurements of aboveground biomass in the early stage (10 days) in the negative and positive control groups and in the group with pretreated seeds gave comparable values. In the treatment groups, the biomass was somewhat (15–20%) lower (Figure 5.20A). At Day 20, however, all groups showed comparable values. In the later stage (at Day 30), the biomass of the plants grown without TEB amounted to 180 g/m^2, which was 40% lower than in the group with Raxil and pretreated seeds and 60% lower than in the treatment groups. The difference in biomass between the groups with pretreated seeds and Raxil, on the one hand, and the groups with the experimental P(3HB)/TEB formulations, on the other, reached about 15–17%.

The effectiveness of the experimental P(3HB)/TEB formulations was more noticeable in Experiment 2, with higher degrees of soil infection and root damage caused by rot (Figure 5.20 and 5.21). At Day 30, in the group with Raxil, the aboveground biomass reached 190 g/m^2, while in the treatment groups, it was 26% higher – 233–240 g/m^2.

Differences in the fungicidal activity of TEB could be seen not only in the dissimilar levels of soil infection caused by the plant pathogen and percentages of rot-damaged roots but also in different plant growth, evaluated by the increase in aboveground biomass. In experiments with different TEB formulations and, hence, different fungicidal activities, the increase in plant biomass was 15–17 to 40–60% higher than in the groups where TEB was applied by using conventional techniques. As fusarium infection causes root rot in the plants of any age, we examined the state and degree of infection of the wheat roots during the experiments with different modes of TEB delivery. The commonly used seed treatment or soil treatment with Raxil solution showed a significant decrease in the percentage of rot-damaged roots, which, though, increased in later stages of the experiment. In the early stage (between Days 10 and 20), the percentage of rot-damaged roots in the soil with TEB embedded in the slowly degraded P(3HB) matrix was similar to that in the soil with Raxil. However, the efficacy of P(3HB)/TEB formulations lasted longer, and in later stages (between Days 20 and 30), the percentage of rot-damaged roots in that group did not grow, in

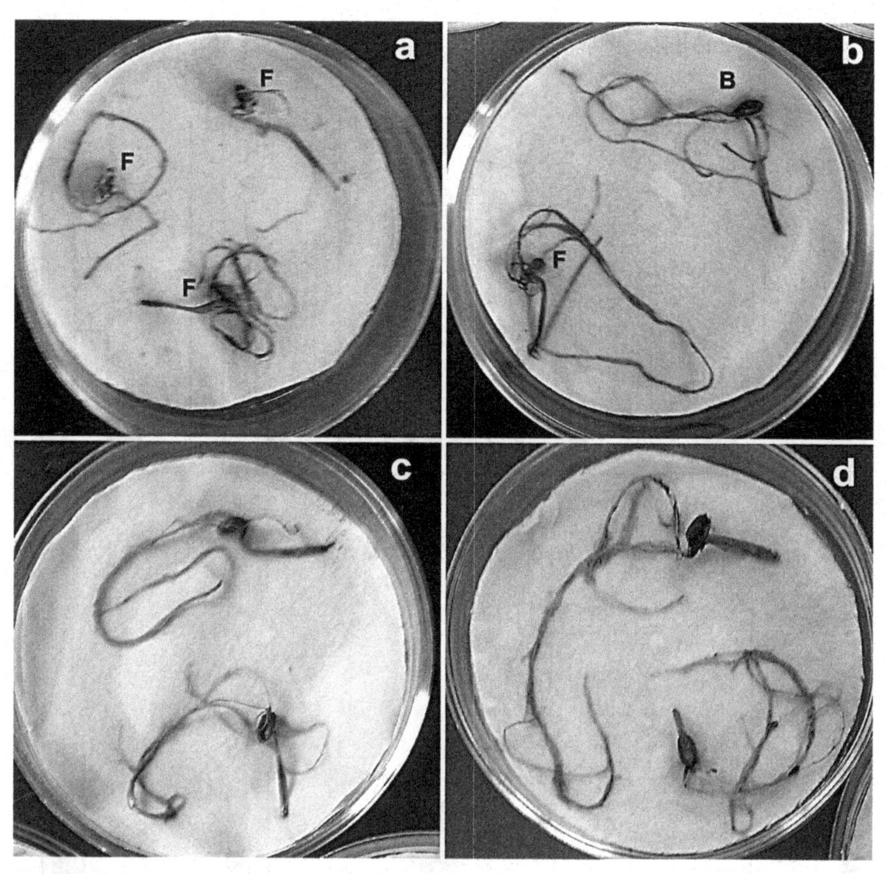

FIGURE 5.19 Wheat root rot under different modes of TEB delivery: a – negative control, b – Raxil Ultra, c – P(3HB)/TEB films, d – P(3HB)/TEB granules; F – *Fusarium* infection, B – *Bipolaris* infection. (Reprinted by permission from Springer Nature: Environmental Science and Pollution Research: Fungicidal activity of slow-release P(3HB)/TEB formulations in wheat plant communities infected by Fusarium moniliforme. Volova, T. G., Prudnikova, S. V., & Zhila, N. O. © 2017.)

contrast to the group with the soil treated with Raxil and in the group with the pre-treated seeds.

Thus, the fungicidal activity of the experimental slow-release formulations of TEB embedded in the matrix of degradable poly–3-hydroxybutyrate against fusarium infection of wheat was comparable to that of TEB in commercial formulation Raxil in early stages. In the later stages, P(3HB)/TEB formulations more effectively suppressed the development of *Fusarium* in soil and inhibited the growth of plant root rot.

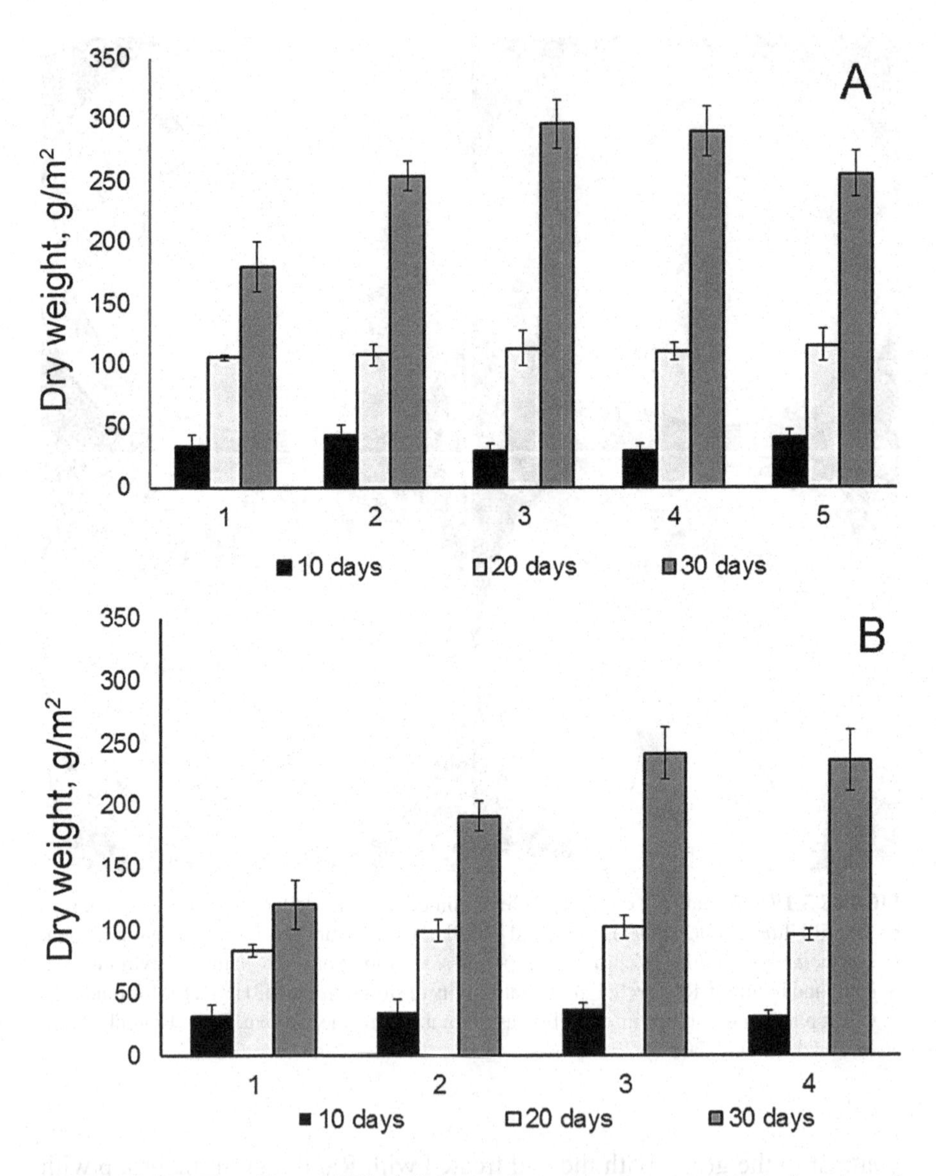

FIGURE 5.20 The effect of TEB delivery mode on the increase in wheat aboveground biomass on the naturally infected soil (a) and on the soil to which *Fusarium* was added (b): 1 – negative control, 2 – positive control (Raxil applied to soil), 3 – P(3HB)/TEB films, 4 – P(3HB)/TEB granules, 5 – pre-sowing treatment of seeds with Raxil. (Reprinted by permission from Springer Nature: Environmental Science and Pollution Research: Fungicidal activity of slow-release P(3HB)/TEB formulations in wheat plant communities infected by Fusarium moniliforme. Volova, T. G., Prudnikova, S. V., & Zhila, N. O. © 2017.)

FIGURE 5.21 The effect of TEB delivery mode on the productivity of wheat crops on the naturally infected soil (a): 1 – P(3HB)/TEB films, 2 – P(3HB)/TEB granules, 3 – positive control (Raxil applied to soil), 4 – negative control and on the soil to which *Fusarium* was added (b): 1 – P(3HB)/TEB films, 2 – P(3HB)/TEB granules, 3 – pre-sowing treatment of seeds with Raxil, 4 – positive control (Raxil applied to soil), 5 – negative control (Volova's photos).

KEYWORDS

- **fusarium infection**
- **fusarium moniliforme**
- **Raxil solution**

REFERENCES

Abelentsev, V. I., (2006). Assortment of disinfectants—spectrum of fungicidal action, biological effectiveness, problems, choice of preparation. *Achievements of Science and Technology*, *9*, 44–48 (in Russian).

Ahemad, M., & Khan, M. S., (2011a). Effect of pesticides on plant growth promoting traits of greengram-symbiont, *Bradyrhizobium* sp. strain MRM6. *Bull. Environ. Contam. Toxicol.*, *86*, 384–388.

Ahemad, M., & Khan, M. S., (2011b). Toxicological assessment of selective pesticides towards plant growth promoting activities of phosphate solubilizing *Pseudomonas aeruginosa*. *Acta Microbiol. Immunol. Hung.*, *58*, 169–187.

Asrar, J., Ding, Y., La Monica, R. E., & Ness, L. C., (2004). Controlled release of tebuconazole from a polymer matrix microparticle: Release kinetics and length of efficacy. *J. Agric. Food Chem.*, *52*, 4814–4820.

Binder, E. M., Tan, L. M., Chin, L. J., Handl, J., & Richard, J., (2007). Worldwide occurrence of mycotoxins in commodities, feeds and feed ingredients. *Anim. Feed Sci. Technol.*, *137*, 265–282.

Boone, D., Castenholz, R., Garrity, G., Brenner, D., Krieg, N., & Staley, J., (2005). *Bergey's Manual® of Systematic Bacteriology* (722 p.). Springer Science & Business Media. ISBN 978-0-387-21609-6 DOI: 10.1007/978-0-387-21609-6.

Boyandin, A., Prudnikova, S., Filipenko, M., Khrapov, E. A., Vasilev, A. D., & Volova, T. G., (2012). Biodegradation of polyhydroxyalkanoates by soil microbial communities of different structures and detection of PHA degrading microorganisms. *Appl. Biochem. Microbiol.*, *48*, 28–36.

Boyandin, A., Prudnikova, S., Karpov, V., Ivonin, V. N., Đỗ, N. L., Nguyễn, T. H., et al., (2013). Microbial degradation of polyhydroxyalkanoates in tropical soils. *Int. Biodeterior. Biodegrad.*, *83*, 77–84.

Brunel, F., El Gueddari, N. E., & Moerschbacher, B., (2013). Complexation of copper(II) with chitosan nanogels: Toward control of microbial growth. *Carbohyd. Polym.*, *92*, 1348–1356.

Child, R., Evans, D., Allen, J., & Arnold, J. M., (1993). Growth response in oilseed rape (*Brassica napus* L.) to combined applications of the triazole chemicals triapenthenol and tebuconazole and interactions with gibberelli. *J. Plant Growth Regul.*, *13*, 203–212.

Danilenkova, G. N., (2004). All-Russian forum of plant protection. *Plant Protection and Quarantine*, *1*, 4–8 (in Russian).

Davletov, H. D., Chikisheva, G. E., & Galiakhmetov, R. N., (2010). Search for environmentally less harmful fungicides in a series of benzimidazole derivatives. *Chemical Journal of Bashkiriya.*, *17*, 28–32 (in Russian).

Fletcher, R. A., Gilley, A., Sankhla, N., & Davis, T. D., (1999). Triazoles as plant growth regulators and stress protectants. In: Janick, J., (ed.), *Horticultural Reviews* (pp. 55–138). John Wiley and Sons Inc.: Oxford.

Hartwig, T., Corvalan, C., Best, N. B., Budka, J. S., Zhu, J. Y., Choe, S., & Schulz, B., (2012). Propiconazole is a specific and accessible brassinosteroid (BR) biosynthesis inhibitor for *Arabidopsis* and maize. *PLoS One.*, *7*, e16625.

Holt, J., Krieg, N., & Snit, P., (1997). *Bergey's Manual® of Systematic Bacteriology. In 2* Volumes (1232 p.). Mir: Moscow (in Russian). ISBN: 5-03-003112-X.

Jendrossek, D., & Handrick, R., (2002). Microbial degradation of polyhydroxyalkanoates. *Annu. Rev. Microbiol.*, *56*, 403–432.

Khalikov, S. S., Dushkin, A. V., Dvaleti, R. D., & Evseenko, V. I., (2013). Innovative fungicides based on tebuconazole: Creation and mechanochemical processes. *Fundamental Research*, *10*, 2695–2700 (in Russian).

Kravchenko, L. V., & Tutelyan, V. A., (2005). Biological safety. Mycotoxins – natural impurities of wheat. *Nutrition Issues*, *3*, 3–13 (in Russian).

Lamb, D. C., Cannieux, M., Warrilow, A. G., Bak, S., Kahn, R. A., Manning, N. J., et al., (2001). Plant sterol 14α-demethylase affinity for azole fungicides. *Biochem. Biophys. Res. Commun.*, *284*, 845–849.

Netrusov, A., Egorov, M., Zakharchuk, L., & Kolotilova, N., (2005). *Workshop on Microbiology: A Textbook for Students of Higher Educational Institutions.* Moscow, (in Russian).

Pavlyushin, V. A., (2010). Scientific support of plant protection and food security of Russia. *Protection and Quarantine of Plants*, *2*, 11–15 (in Russian).

Peresypkin, V. F., Tyuterev, S. L., & Batalova, T. S., (1991). *Diseases of Cereal Crops Using Intensive Technologies for Their Cultivation* (p. 272). Agropromizdat: Moscow, (in Russian).

Qian, K., Shi, T., He, S., Luo, L., Liu, X., & Gao, Y., (2013). Release kinetics of tebuconazole from porous hollow silica nanospheres prepared by miniemulsion method. *Micropor. Mesopor. Mater.*, *169*, 1–6.

Qian, K., Shi, T., Tang, T., Zhang, S., Liu, X., & Cao, Y., (2011). Preparation and characterization of nano-sized calcium carbonate as controlled release pesticide carrier for validamycin against *Rhizoctonia solani. Microchim. Acta.*, *173*, 51–57.

Ricciardi, A., Hoopes, M. F., Marchetti, M. P., & Lockwood, J. L., (2013). Progress toward understanding the ecological impacts of nonnative species. *Ecol. Monogr.*, *83*, 263–282.

Savenkova, L., Gercberga, Z., Muter, O., Nikolaeva, V., Dzene, A., & Tupureina, V., (2002). PHB-based films as matrices for pesticides. *Process Biochem.*, *37*, 719–722.

Schmale, D. G., & Bergstrom, G. C., (2003). Fusarium head blight. *Plant Health Instructor.* doi: 10.1094/PHI-I-2003-0612-01.

Serra, A. A., Nuttens, A., Larvor, V., Renault, D., Couée, I., Sulmon, C., & Gouesbet, G., (2013). Low-environmentally relevant levels of bioactive xenobiotics and associated degradation products cause cryptic perturbations of metabolism and molecular stress responses in *Arabidopsis thaliana. J. Experim. Bot.*, *64*, 2753–2766.

Shpaar, D., (2008). *Cereals: Growing, Harvesting, Finalization and Use* (p. 656). Publishing House Agrodelo: Moscow, (in Russian).

Simberloff, D., & Stiling, P., (1996). How risky is biological control? *Ecology*, *77*, 1965–1974.

Sutton, D., Fothergill, A., & Rinaldi, M., (2001). *Key Pathogenic and Opportunistic* Fungi (468 p.). Mir: Moscow (in Russian). ISBN 5-03-003308-4.

Tropin, V. P., (2007). Additional compositions of pesticides and their methods of application. *Protection and Quarantine of Plants, 7*, 32–33 (in Russian).

Tyuterev, S. L., (2005). Etching of cereal crops seeds. *Protection and Quarantine of Plants, 3*, p. 48 (in Russian).

Volova, T. G., Boyandin, A. N., Vasiliev, A. D., Karpov, V. A., Prudnikova, S. V., Mishukova, O. V., et al., (2010). Biodegradation of polyhydroxyalkanoates (PHAs) in tropical coastal waters and identification of PHA-degrading bacteria. *Polym. Degrad. Stab., 95*, 2350–2359.

Volova, T. G., Prudnikova, S. V., & Zhila, N. O., (2018). Fungicidal activity of slow-release P(3HB)/TEB formulations in wheat plant communities infected by *Fusarium moniliforme. Env. Sci. Pollut. Res., 25*, 552–561.

Volova, T. G., Prudnikova, S. V., Zhila, N. O., Vinogradova, O. N., Shumilova, A. A., Nikolaeva, E. D., Kiselev, E. G., & Shishatskaya, I. I., (2017). Efficacy of tebuconazole embedded in biodegradable poly–3-hydroxybutyrate to inhibit the development of *Fusarium moniliforme* in soil microecosystems. *Pest Manag. Sci., 73*, 925–935.

Volova, T., Gladyshev, M., Trusova, M., & Zhila, N. O., (2007). Degradation of polyhydroxyalkanoates in a eutrophic reservoir. *Polym. Degrad. Stab., 29*, 580–586.

Volova, T., Zhila, N., Vinogradova, O., Shumilova, A., Prudnikova, S., & Shishatskaya, E., (2016). Characterization of biodegradable poly–3-hydroxybutyrate films and pellets loaded with the fungicide tebuconazole. *Environ. Sci. Pollut. Res., 23*, 5243–5254.

Watanabe, T., (2002). *Pictorial Atlas of Soil Fungi: Morphologies of Fungi and Key Species* (2nd edn.), CRC Press LLC, Boca Raton.

Yang, D., Wang, N., Yan, X., Shi, J., & Zhang, M., (2014). Microencapsulation of seed-coating tebuconazole and its effect on physiology and biochemistry of maize seedlings. *Colloids Surf. B. Biointerfaces., 114*, 241–246.

Zakharenko, V. A., (2003). Development of plant protection and scientific support. *Agricultural Biology, 1*, 93–104 (in Russian).

Zakharova, N. G., Siraeva, Z. Y., Demidova, I. P., & Egorov, S. Y., (2006). Creation of biological products that are promising for agriculture. *Natural Sciences, 148*, 102–111 (in Russian).

Zhang, B., Zhang, T., Wang, Q., & Ren, T., (2015). Microorganism-based monodisperse microcapsules: Encapsulation of the fungicide tebuconazole and its controlled release properties. *RSC Advances, 32*, 25164–25170.

CHAPTER 6

SLOW-RELEASE FORMULATIONS OF NITROGEN FERTILIZERS AND EVALUATION OF THEIR EFFICACY

The use of mineral and organic fertilizers is a necessary part of modern intensive farming practices, which include growing high-yield crops on the same land for many years. Nutrition is a vital component of the life of each organism, including plants, which supplies structural elements for growth and development (Artyushin, Derzhavin, 1984). Soil is the main source of nutrients for farm crops. However, nutrients taken up by plants from soil organic matter and sparingly soluble mineral compounds may be not enough to produce adequate yields every year. Then, mineral fertilizers need to be applied on a regular basis.

The major elements supplied by mineral fertilizers are nitrogen, phosphorus, and potassium. The organic and inorganic nitrogen-containing fertilizers may be amide, ammonia, and nitrate compounds. Nitrogen compounds have high mobility in soil, which causes nitrogen losses through runoff. Moreover, nitrogen available for plant use is present in the soil in very small amounts, and this limits crop development. Therefore, application of nitrogen fertilizers favorably affects crops.

Most nitrogen fertilizers are readily soluble in water, and they are either not absorbed or poorly absorbed by the soil. The most commonly used agricultural fertilizers are ammonium nitrate and urea, which constitute about 60% of all nitrogen fertilizers. Nitrogen fertilizers may contain three main forms of nitrogen.

1) ammonia bound to a mineral acid; these are ammonia fertilizers (including the widely used ammonium sulfate (Vildflush et al., 2001));
2) nitrate-nitrogen, i.e., salts of nitric acid; these are nitrate fertilizers (such as sodium nitrate or potassium nitrate); and
3) amide nitrogen; these are amide fertilizers.

Ammonium nitrate is a high analysis fertilizer, which is readily soluble in water, containing equal amounts of ammonia and nitrate nitrogen forms. Some fertilizers contain both nitrogen and other vital nutrients (diammonium phosphate, potassium nitrate, etc.).

Application of nitrogen fertilizers by traditional methods has some disadvantages: too high concentration of the fertilizer in soil immediately upon application, which declines over a certain time.

A possible solution may be to bind nitrogen fertilizers with substances that slow down nitrogen release to soil and, thus, maintain relatively stable concentrations of the fertilizer for a time period needed to grow crops. The fertilizers can be bound with biodegradable polymers, which are slowly degraded by soil microflora. As the polymeric matrix is degraded, nitrogen is gradually released to soil. Biodegradable materials can be used to construct slow-release controlled-delivery systems for agriculture (Sopeña, 2005). The matrix for embedding fertilizers must be completely biodegraded in soil, products of its degradation must be nontoxic to biota and humans, and delivery of the fertilizer to plants must be performed in a controlled manner (Shtefan, 1981).

6.1 NEW-GENERATION FORMULATIONS OF NITROGEN FERTILIZERS

The simplest way is to mix fertilizers with organic fillers mechanically. For example, Neata et al. (2012) reported the use of systems containing starch, wood flour, and an NPK fertilizer. The author of another study (Wu, 2008) investigated pressed plates of starch blended with polybutylene succinate and acrylic acid copolymer grafted starch loaded with bacterial fertilizer Bacillus sp. PG01. The study showed that bacterial cells maintained viability in the polymer matrix and were released during polymer degradation. Most studies describe investigations of polymer-coated granular fertilizers. The most popular coating material is polyurethane, which is sometimes blended with materials that speed up its biodegradation – tannin, palm oil, soybean oil, and castor oil (Yang et al., 2012). Other coating materials are sulfur, wax, aldehyde condensates, and resins (Chen et al., 2008a). Plant-derived materials (lignin, cellulose, and starch) are cheaper coatings; moreover, the active ingredient is released quicker (over less than 30 days) (Yang et al., 2012).

Fertilizers are coated by various methods. Jarosiewicz and Tomaszevska (2003) reported a study in which the NPK fertilizer was coated using polysulfone from the dimethylformamide solution, polyacrylonitrile from the dimethylformamide solution, or cellulose acetate from acetone by the phase inversion technique. The slowest release rates were observed for potassium through the coatings made of polysulfone and polyacrylonitrile and for nitrogen – through the cellulose acetate coatings. Zvomuya et al. (2003) developed a polyolefin-coated urea formulation to apply to potato. During the potato growing season, nitrogen leaching was lower from the experimental formulation than from uncoated urea, thus reducing the hazard of groundwater pollution. Lan et al. (2011) described an apparatus for spray coating of polyacrylic acid latex onto urea granules to form film coating. Vashishtha et al. (2010) coated urea by phosphogypsum mixed with neem oil and alkyl benzene. The advantage of that formulation was that sulfur and phosphorus were present in the fertilizer in the forms that were more readily assimilated by plants than sulfur and phosphorus in the uncoated area. Jia et al. (2013) coated potassium-copper phosphate granules with the polydopamine film by oxidative polymerization of dopamine. The granules were incubated in soil, and over 18 days, 1.6 % K_2O, 1 % copper, and 0.4 % P_2O_5 were released from the granules. The coating could be a nitrogen source too. The authors of another study (Anghel et al., 2012) coated a complex fertilizer with a film of polyethylene terephthalate, which was modified to enhance the biodegradability of the film.

Complex coatings retain fertilizers in the granules for longer periods. In a study by Yang et al. (2012), the authors coated urea granules or compressed tablets with polystyrene foam: the fertilizer was sprayed with the ethyl acetate solution of polystyrene or polystyrene foam and additionally coated with wax or castor oil with diphenylmethane diisocyanate. The authors showed that such coatings could be used to prepare slow-release fertilizers, with nitrogen release time reaching 100 days. Liang and Liu (2006) reported preparation of a double-coated fertilizer. Polystyrene was used to make the inner coating, and urea with cross-linked polyacrylamide created the outer coating. Urea granules were placed into the polystyrene solution in tetrahydrofuran and then into the water bath, where the polymer was gelatinized and precipitated. Then, the granules were taken out of the water and immediately, without drying, were shaken together with the polyacrylamide/urea powder. After applying the coating, the granules were sprayed with a 1% solution of epoxy chloropropane in methanol and dried at 70°C. Over 30 days, about 60% nitrogen was released from the double-coated granules to soil.

Another approach to preparing nitrogen formulations is to encapsulate the fertilizer in a polymeric or mineral matrix. Chen et al. (2008b) reported a study in which urea was encapsulated in films of graft-copolymerized starch and poly(L-lactide). The films were produced by evaporation of the dimethyl sulfoxide solution of the copolymer and urea. Urea was completely released from those films in no more than 24 h. Gomez-Martinez et al. (2013) prepared compressed specimens composed of gluten and potassium chloride, with glycerol or PEG used as plasticizers. Jamnongkan and Kaewpirom (2010) prepared complex NH_4NO_3, $NH_4H_2PO_4$, and KNO_3 formulations as polymer hydrogels based on poly(vinyl alcohol) (PVA), chitosan, and PVA/chitosan. The release of the active ingredient was studied in water and soil for 30 days. PVA exhibited the highest percent release into the water while chitosan hydrogel showed the highest percent release into the soil. Saraydin et al. (2000) reported a study in which polymerization of the poly(acrylamide/itaconic acid) system with the herbicide sodium 2,2 dichloropropionate, NH_4NO_3, KNO_3, and $(NH_4)_2SO_4$ was performed by using radiation. Nitrogen encapsulation in a mineral matrix was shown in studies describing the preparation of urea-montmorillonite composites (Pereira et al., 2012; Wanyika, 2014); the fertilizer was released from these formulations very quickly – within 10 days. Hamid et al. (2013) used a urea/chitosan/bentonite system to prepare fertilizer formulations by wet mixing of components followed by drying at 60°C.

Some authors described the preparation of fertilizer formulations that combined embedding of nitrogen into various matrices and coating of the formulation. Ni et al. (2011) developed three types of nitrogen fertilizer formulations, which were based on urea, ammonium sulfate, and ammonium chloride. The fertilizer was mixed with clay, and that mixture was then blended with urea granules and some water. The next step was to spray a mixture of ethyl cellulose and stearic acid (5:2) in ethanol on polymer granules; the formulation was additionally coated with carboxymethyl cellulose/hydroxyethyl cellulose hydrogel. The peak of nitrogen release to soil was observed for five days, followed by slow release over 20 days. Chong (2013) described a study in which urea was embedded in starch. Starch solutions mixed with urea were gelatinized by heating. The time of the release of urea from the resultant formulations was very short – no more than 10 days.

Fertilizers can be embedded in nanoparticles. In a study by Corradini et al. (2010), urea, calcium phosphate, and potassium chloride were incorporated into chitosan nanoparticles crosslinked with methacrylic acid. Hasaneen et al. (2014) loaded $Ca(H_2PO_4)_2 \times H_2O$, urea, and KCl into 17–25-nm-diameter microparticles prepared by polymerizing methacrylic acid in chitosan solution.

Most researchers did not construct fertilizer formulations but rather used commercial preparations. The use of polymer-coated urea increased yields of corn compared with the yields achieved by applying uncoated urea and anhydrous ammonia (Nash et al., 2013). Grant et al. (2012) compared the efficiencies of using ESN – polymer-coated urea (Agrium) containing 44% nitrogen – and ammonium sulfate, which is used to grow wheat, canola, barley, and corn. The authors of that study did not find any differences between the efficiencies of the use of the coated and uncoated urea in the semi-arid region but reported higher efficiency of the polymer-coated urea in moister regions. The authors attributed this to delays in the release of urea from coated granules in moist soil. ESN used to grow corn was found to be more effective than broadcast urea (Nelson et al., 2013). Soon et al. (2011) noted that the application of ESN increased nitrogen availability and decreased nitrous oxide (N2O) losses in the growth period compared to the use of conventional urea. ESN was used to grow cotton plants in the basin of the Yangtze River (Wang et al., 2013), and in most plots, the efficiency of ESN was higher than that of conventional urea, but in some of the plots, no difference was noted between the two types of urea formulations. Polyolefin-coated commercial fertilizers Meister 70 and Meister 270 – urea with surfactants coated with thermoplastic resin – were used to grow cotton (Chen et al., 2008a). Nitrogen release from Meister 270 was too slow. Meister 70 maintained a higher nitrogen concentration in the soil than uncoated urea and, in some cases, increased the yields.

Release of eight commercial fertilizers (sulfur-coated urea, resin-coated ammonium nitrate-phosphate, polymer-sulfur-coated urea, reactive layer-coated urea, a polyolefin-coated complex fertilizer consisting of ammonium nitrate – ammonium phosphate – potassium nitrate, isobutylene urea, and Ureaform – a complex fertilizer consisting of urea (12%), a slowly soluble nitrogen source (18%), and an insoluble nitrogen source (70%) as well as activated biosoil) into model soil was evaluated by using a soil incubation column leaching procedure for 180 days (Medina et al., 2014). Over 180 days, the highest percentage of nitrogen was released from resin-coated ammonium nitrate-phosphate while Ureaform exhibited the slowest release rate.

Huett and Gogel (2000) measured release of active ingredients of commercial controlled-release NPK formulations – Nutricote, Apex Gold, Osmocote, and Macrocote – in sand at different temperatures and found that 90% release of the nutrients occurred over much shorter time spans than the nominated release periods and that a 10°C temperature rise (e.g., from

30 to 40°C) could speed up the nutrient release by 20%. A similar study was conducted by Adams et al. (2013) with complex fertilizers containing micronutrients Polyon, Nutricote, and Osmocote. No significant difference was found between nutrient release rates in water and in moist solid substrate. The effectiveness of the fertilizer Osmocote was low because of the too quick release of active ingredients. The optimal release rates of the components from Nutricote were achieved at temperatures between 20 and 30°C, accelerating too much at higher temperatures. At decreased temperatures, release rates were sufficiently low, suggesting good winter storage of the fertilizers. In a study by Du et al. (2006), commercial NPK fertilizers encapsulated in polyurethane-like coating 0.0065-cm and 0.0096-cm thick were incubated in moistened sand and water to evaluate the release of the active ingredients. At 20°C, 60% of nitrates and 15–25% of phosphates were released over 70 days. Medina et al. (2008) described laboratory and field experiments with complex NPK fertilizers CitriBlen (films), Agrocote Type A (granules), and Agrocote Type C (granules), and the nitrogen fertilizer Agrocote Poly-S (granules) used to fertilize citrus trees. The study showed the effectiveness of using these fertilizers, CitriBlen in particular. Nutrient release patterns matched well with the recommended citrus fertilization strategy. The effectiveness of the commercial nitrogen formulation Meister programmed release N fertilizer T15 for cotton production was shown in field experiments in Arkansas and Tennessee (Oosterhius and Howard, 2008). However, the use of Meister 7, urea with dicyandiamide, and polyolefin-coated urea to fertilize barley, potatoes, and corn did not show any positive effect of urea encapsulation compared with the use of conventional urea (Shoji et al., 2001).

A number of studies were devoted to modeling release kinetics of fertilizers from slow-release formulations. Shaviv et al. (2003) proposed a comprehensive model describing the release of nutrients from a single granule of polymer-coated fertilizer, consisting of three stages. Stage I is a lag period, during which insignificant amounts of the fertilizer are released or there is no release at all. During this stage, water vapor penetrates the coating of the granule, hydrating the fertilizer. The authors noted that the length of this stage might be determined by the time needed for internal cavities of the granule to be hydrated or for the equilibrium between water inflow and solution outflow to be established. The hydrostatic pressure within the coating increases due to water absorption and an increase in the mass. Stage II is a period of linear release during which the gradient of nutrient release remains constant due to the undissolved fertilizer, which is slowly dissolved, maintaining solution concentration at a steady level. The authors noted that the movement of nutrients through the coating might be enhanced by the pressure gradient. Finally, during Stage III, a period of decaying release, the release of the nutrient decreases as the gradient goes down because of the decreasing fertilizer concentration in the granule. A similar study was conducted with polymer-coated urea by Trihn et al. (2013).

The main obstacle to the use of slow-release fertilizers is their high cost. Polymer-coated urea now costs 4–8 times more than ordinary urea (Lammel, 2005), mainly due to the cost of the coating (Yang et al., 2012).

Thus, various approaches are being developed to producing fertilizers embedded in different materials. Since nitrogen has high mobility in soil, it is very important to develop effective controlled-release formulations of nitrogen fertilizers.

6.2 POLYHYDROXYALKANOATES AS A DEGRADABLE BASIS FOR CONSTRUCTING SLOW-RELEASE FORMULATIONS OF NITROGEN FERTILIZERS

Polyhydroxyalkanoates, which are biodegradable in natural media and have physicochemical properties making them suitable for processing by various techniques, are promising materials for constructing carriers for slow-release fertilizer systems. However, very few studies have investigated the possibility of using them as a matrix or coating for nitrogen fertilizers. In the study by Costa et al. (2013), a solution of poly–3-hydroxybutyrate in chloroform was used to coat urea. Urea granules were either immersed into the solution and then dried on the glass surface or sprayed with the solution and dried. However, urea was released to distilled water in about five minutes, suggesting low effectiveness of such coatings. Obviously, more complex approaches should be developed for effective use of P3HB as a carrier of nitrogen fertilizers, which will enable slower release of nitrogen.

6.2.1 CONSTRUCTING SLOW-RELEASE UREA FORMULATIONS

Three types of urea formulations were described by Volova et al. (2016): films and compressed pellets with embedded urea and polymer-coated granular urea. Films loaded with urea were prepared as follows. A P3HB/urea dichloromethane solution was prepared by adding urea to the P3HB solution to reach different urea concentrations (10, 20 and 30% of the polymer weight). After that, the P3HB/urea solution was cast in Teflon-coated metal molds, which were then kept in a dust-free drying cabinet for 3 days, until solvent evaporation took place. The films were cut into 5×5 mm squares, which were then used in experiments. Pellets were prepared from the mixture of P3HB and urea powders by cold pressing, using a Carver Auto Pellet 3887/4387 press (Carver, U.S.) under pressing force of 100 kgf/cm2. The resulting pellets were 3 mm in diameter and 1.5 mm high. Urea granules were soaked in a dichloromethane solution of poly–3-hydroxybutyrate. After that, to affix the polymer coating to the granules, they were placed into a hexane bath and, then, dried at ambient temperature. Photos of the formulations are shown in Figure 6.1.

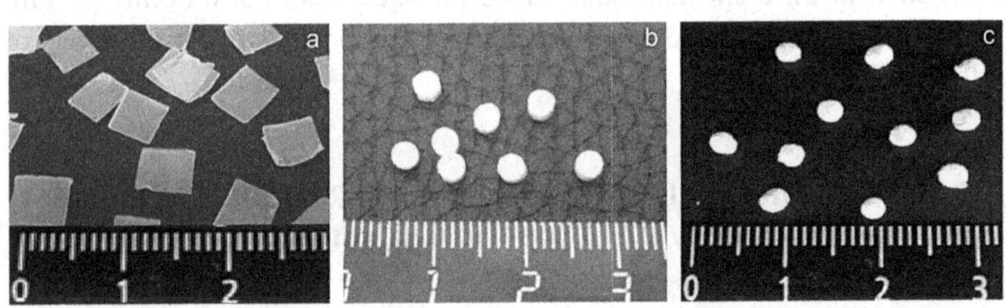

FIGURE 6.1 Experimental formulations of urea embedded in the P3HB matrix: a – films, b – compressed pellets, and c – polymer-coated granules. (Reprinted with permission from Volova, T. G., Prudnikova, S. V., & Boyandin, A. N., (2016). Biodegradable poly–3-hydroxybutyrate as a fertiliser carrier. J. Sci. Food. Agr., 96, 4183–4193. © 2016 John Wiley and Sons.)

To compare release kinetics of nitrogen in soil and degradation of the polymer matrix, the experimental P(3HB)/urea formulations were placed in laboratory soil micro-ecosystems (Figure 6.2). Soil (200 g) was placed into 250-mm^2 plastic containers; P3HB/urea specimens were weighed, placed in close-meshed gauze bags (3 specimens of the same formulation per bag), and buried in the soil at a depth of 2 cm. The containers with the soil and specimens were incubated in a temperature-controlled cabinet at a constant temperature of 21 ± 0.1°C and soil moisture content of 50%. Films loaded with different amounts of urea (10, 20, and 30% of the polymer weight) were used to study nitrogen release behavior (Figure 6.2b).

Urea release from the films occurred at a uniform rate, and by the end of this experiment (30 days), 28, 33, and 45% of the encapsulated urea had been released from the films initially loaded with different amounts of the fertilizer. By contrast, the release from the granules in control reached 78% after 5 days, and after 15 days, urea granules in control were completely dissolved. Nitrogen concentration in the soil increased as P3HB polymer matrix was gradually degraded (Figure 6.2a).

At the beginning of the exposure (for up to 5 days), the polymer was not degraded: as we wrote previously (Prudnikova, Volova, 2012), it takes some time for PHA degrading microorganisms to get attached to the surface of the polymer samples and adapt their enzyme systems to the polymer as a substrate. Polymer degradation during 30 days occurred at a uniform rate, which, however, depended on the degree to which the films were loaded with urea. Films with 30% urea loading were degraded with the fastest rate: by the end of the experiment, residual P3HB in them had decreased to 56% of the initial polymer. Undegraded P3HB in the formulations containing 10 and 20% urea reached 73 and 66%, respectively. Release kinetics of nitrogen embedded in pellets and their degradation behavior were similar to those observed in the experiment with films, but the rates of both processes were slower. After 30 days, release of nitrogen from the pellets with 10, 20, and 30% urea loading reached 20, 28, and 35%, and that was 10–15% lower than the amount released from the films (Figure 6.2d). Degradation of P3HB pellets in the soil was also somewhat slower. Thus, after 30 days of incubation in soil, the undegraded polymer in the pellets loaded with 10, 20, and 30% urea reached 67, 78, and 85% of its initial amounts, respectively (Figure 6.2c).

Release kinetics of nitrogen from the granular fertilizer coated with a layer of P3HB and polymer biodegradation are shown in Figures 6.2e and 6.2f. Urea release from the polymer-coated granules occurred at a higher rate than urea release from the pellets and was comparable to the release from the films (Figure 6.2). After 30 days, about 50% of the nitrogen contained in polymer-coated granules was released into the soil. The uncoated granular urea, which was used as the control, was completely dissolved after 2 weeks. At the beginning of the soil exposure, degradation of the polymer coating occurred in a similar way to degradation of the formulations described above, i.e., it was rather slow on the first 3 to 5 days. Then, polymer biodegradation rate increased, and after 30 days, the mass of undegraded polymer was no more than 20% of its initial mass (Figure 6.2f).

Electron microscopy of the P3HB/urea specimens incubated in soil revealed differences in the surface microstructure of the initial specimens and changes that occurred during their degradation (Figure 6.3). The surface of the initial specimens had an indistinct pattern, without defects or fractures, with few 2- to 4-μm micropores (Figure 6.3, 1a). As the films were degraded and their mass decreased, the surface became rough and uneven, and the films became thinner. We observed formation of small pores, which had not been detected on the initial specimens. These pores grew larger as the experiment proceeded. After 30 days of degradation, the film surface became more

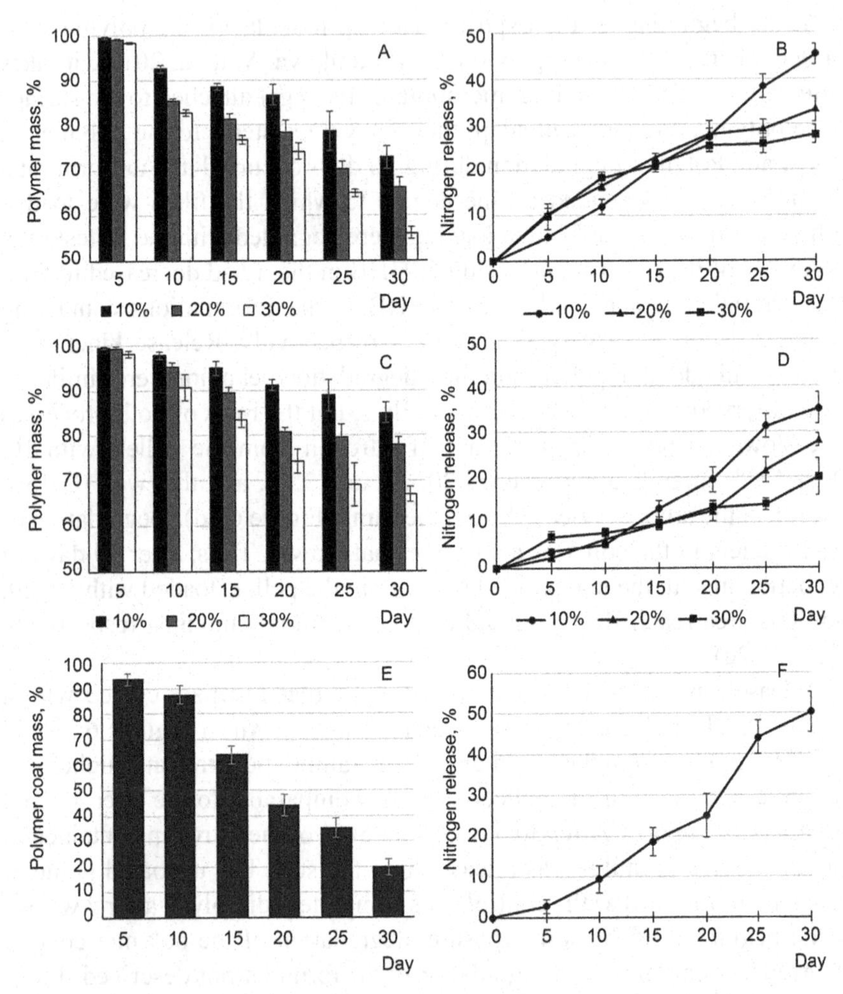

FIGURE 6.2 P3HB degradation behavior and kinetics of nitrogen release to soil expressed as % of its initial content in the formulation; a, b – urea embedded in films; c, d – urea embedded in pellets; e, f – P3HB-coated granules. (Reprinted with permission from Volova, T. G., Prudnikova, S. V., & Boyandin, A. N., (2016). Biodegradable poly–3-hydroxybutyrate as a fertiliser carrier. J. Sci. Food. Agr., 96, 4183–4193. © 2016 John Wiley and Sons.)

nonuniform, with chaotically positioned sharply defined structures varying in shape and pores larger than 2 μm in size, probably as a result of leaching of the polymer amorphous phase (Figure 6.3, 1b). The surface of the pellets, which degrade with slower rates and retain their mass and shape for longer periods of time, showed formation of pores and shallow pits, as some regions on the surface were degraded, but, generally, the specimens remained better preserved (Figure 6.3, 2a and 2b). These results are in good agreement with the data previously reported by other authors (Molitoris et al., 1996; Tsuji and Suzuyoshi, 2002), suggesting that degradation rates of PHA specimens are related to their surface morphology. Molitoris et al. (1996) showed that hydrolysis started at the surface and at lesions and proceeded to the inner part of the polymer. Thus, pellets, which originally have denser structure and smoother surface than films, are less "attractive" to soil microflora that colonizes the surface of the specimens, and hence, are degraded at slower rates.

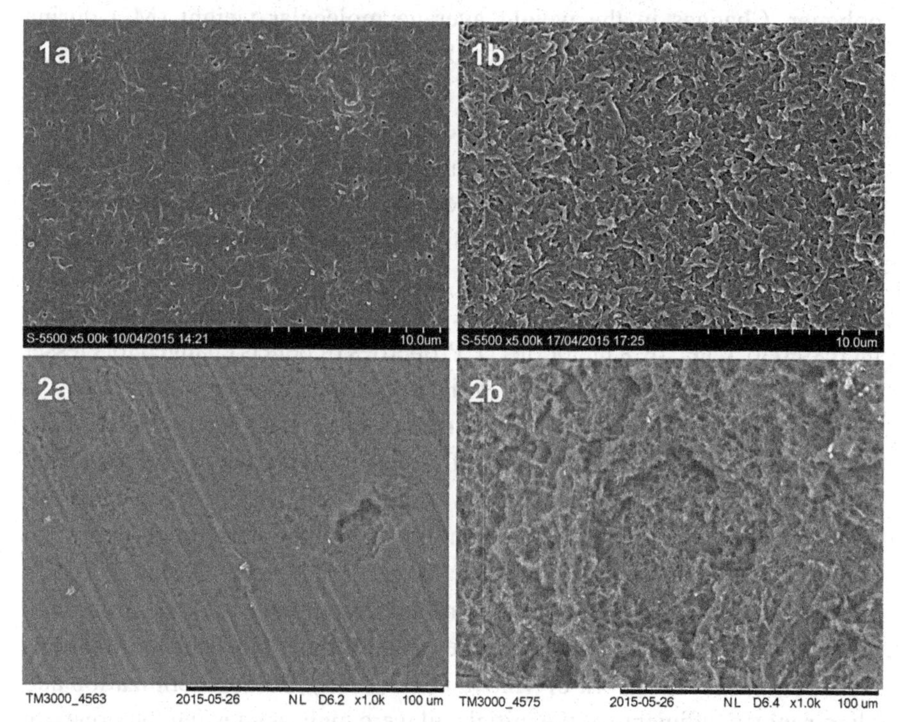

FIGURE 6.3 SEM micrographs of P3HB/urea films (1) and pellets (2) before (a) and after (b) incubation in soil. (Reprinted with permission from Volova, T. G., Prudnikova, S. V., & Boyandin, A. N., (2016). Biodegradable poly-3-hydroxybutyrate as a fertiliser carrier. J. Sci. Food. Agr., 96, 4183–4193. © 2016 John Wiley and Sons.)

One of the factors influencing PHA biodegradation is the degree of crystallinity of a PHA: the lower the degree of crystallinity, i.e., the larger the amorphous region of the polymer, the higher the rate of polymer degradation. Therefore, in the specimens undergoing degradation, the ratio of amorphous to crystalline regions changes, and the degree of crystallinity must change too (Kunioka et al., 1989). Our results are consistent with this reasoning.

The degree of crystallinity (C_x) of the initial P3HB specimens was 70%. The P3HB/urea specimens that had been incubated in soil and partially degraded showed an increased C_x, and the more the specimen was degraded, the more the degree of crystallinity was increased.

The degree of crystallinity of the most degraded P3HB/urea films increased to 82%, and in the pellets, which were less degraded, to 76%. In addition to changes in the degree of crystallinity during degradation, P3HB/urea specimens also showed different molecular weight properties of the polymer. Changes in the weight average molecular weight (M_w) during degradation were directly related to the degree to which the specimen was degraded. After 7 days of incubation, for the films, which were degraded at a higher rate, the values of the molecular weight were significantly lower and polydispersity (Đ) significantly higher than in the initial polymer. The increase in polydispersity also proved the occurrence of changes in the polymer structure and formation of fragments with different carbon chain length. After 30 days of incubation in soil, the weight average weight of the polymer dropped from 700 to 320 Da, i.e., by more than half; polydispersity increased to 3, i.e., by a factor of 1.5. Properties of pellets changed to a lesser extent: M_w decreased to 200 kDa, while polydispersity increased to 2.6.

PHAs are biodegraded by microorganisms that have extracellular PHA-depolymerases. This process influences the structure of soil microbial community, and it is related to many factors (polymer chemical composition, soil structure, environmental conditions, and the structure of microbial community) (Jendrossek, Handrick, 2002; Prudnikova, Volova, 2012). Therefore, both the initial soil microbial community and the microbial community after incubation of the prepared nitrogen fertilizer formulations were analyzed in the work of Volova et al. (2016). Both mineralization and oligotrophy coefficients of the initial soil were high, suggesting the presence of mature microbial community and completion of mineralization (Table 6.1). Plants grown on this soil consumed mineral nutrients, and that caused a decrease of about 50% in the total counts of bacteria in the control, where no nitrogen fertilizer was applied, compared to the initial soil microbial counts. Under nitrogen deficiency, the number of nitrogen-fixing bacteria in the soil increased by a factor of 2.

Application of urea as a commercial fertilizer formulation (positive control) increased the amount of available nitrogen in the soil and facilitated the growth of all groups of oligotrophic bacteria, while the number of nitrogen-fixing bacteria decreased.

TABLE 6.1 Ecological-Trophic Groups of Microorganisms in Samples of Soil Receiving Different Nitrogen Formulations.

Soil samples	Number of microorganisms, million CFU in 1 g				Mineralization coefficient	Oligotrophy coefficient
	Copiotrophs	Prototrophs	Oligotrophs	Nitrogen-fixing bacteria		
Initial soil	Soil before the experiment					
	16.3 ± 5.1	24.7 ± 7.1	190.9 ± 70.7	26.1 ± 4.7	1.52	11.71
Soil samples after 30 days of experiment						
Negative control*	9.5 ± 3.2	13.4 ± 4.5	81.1 ± 5.9	55.3 ± 17.8	1.41	8.54
Positive control**	18.3 ± 2.7	30.6 ± 11.2	110.6 ± 18.4	38.1 ± 12.2	1.67	6.04
Embedded in films (30%)	182.5 ± 24.3	65.5 ± 10.4	75.5 ± 16.3	53.2 ± 13.6	0.36	0.41
Embedded in pellets (30%)	129.2 ± 13.4	48.3 ± 12.4	84.3 ± 15.7	62.4 ± 14.7	0.37	0.65
Polymer-coated granules	115.4 ± 21.7	27.9 ± 8.1	92.8 ± 23.6	49.5 ± 11.5	0.24	0.80

*Fertilizer-free soil

**With a commercial urea formulation

(Reprinted with permission from Volova, T. G., Prudnikova, S. V., & Boyandin, A. N., (2016). Biodegradable poly–3-hydroxybutyrate as a fertiliser carrier. J. Sci. Food. Agr., 96, 4183–4193. © 2016 John Wiley and Sons.)

P3HB/urea formulations and polymer-coated granules delivered not only nitrogen but also carboniferous compounds, which, after P3HB hydrolysis by PHA degrading microorganisms, became an attractive substrate for the growth of soil microbes. Application of P3HB/urea formulations increased the number of copiotrophic bacteria by a factor of 7 to 11 and the number of nitrogen-fixing bacteria by a factor of 1.9 to 2.4 as compared with their number in the initial soil and decreased the number of oligotrophic bacteria by a factor of 2 to 2.5. Mineralization coefficient was decreased to 0.24–0.37 and oligotrophy coefficient – to 0.4–0.8. These changes were indicative of the high rate of mineralization and the presence of the readily available organic matter resulting from P3HB hydrolysis in the soil. The greatest increase in the number of bacteria was observed in the soil with P3HB/urea films. As films had a larger surface area than pellets, the substrate was more readily available to microbial cells, which facilitated their growth.

In the initial soil, the taxonomic composition of the microbial community was dominated by Actinobacteria (up to 80% strains), including Nocardia, Arthrobacter, Corynebacterium, Terrabacter, and other genera (Figure 6.4). These microorganisms are representatives of the autochthonic microflora, and they can exist at a low concentration of available organic matter. This microflora usually dominates mature soil microbial communities.

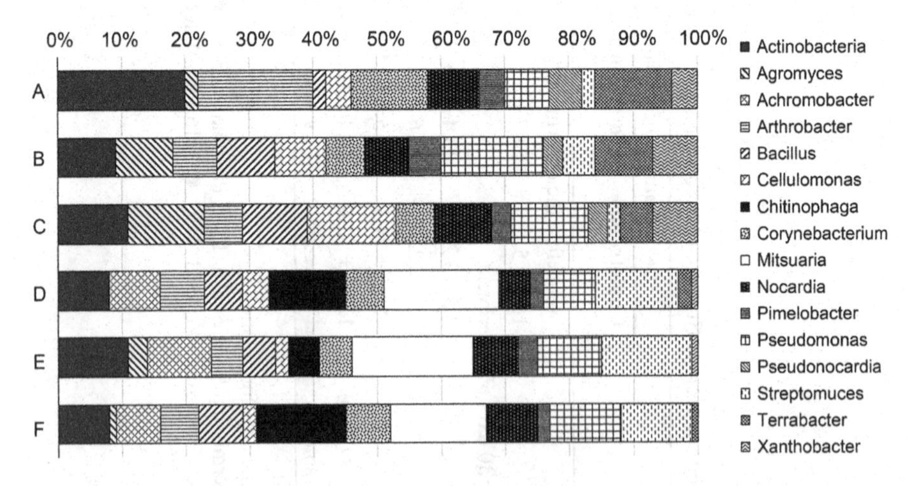

FIGURE 6.4 Taxonomic composition of soil microorganisms with free and encapsulated urea; A – initial soil, B – control without fertilizer, C – commercial urea formulation, D – urea embedded in PHB films, E – urea embedded in pellets, F – PHB-coated urea granules. (Reprinted with permission from Volova, T. G., Prudnikova, S. V., & Boyandin, A. N., (2016). Biodegradable poly–3-hydroxybutyrate as a fertiliser carrier. J. Sci. Food. Agr., 96, 4183–4193. © 2016 John Wiley and Sons.)

Our study showed a considerable increase in the number of bacteria possessing PHB-depolymerases in the soil with urea embedded in P3HB matrices (Figure 6.5). Identification of the major PHB-degrading microorganisms based on their morpho-physiological and molecular-genetic traits showed the presence of such bacteria as Mitsuaria sp., Chitinophaga sp., Achromobacter sp., and Streptomyces albolongus.

Root exudates of growing plants attract soil microorganisms, and they are quickly assimilated by organotrophic bacteria. In the control, where Agrostis stolonifera was grown in the fertilizer-free soil (negative control), proportions of microorganisms changed very little: Actinobacteria remained the major group, but the number of spore-forming bacteria of the genus Bacillus and Gram-negative bacilli of the genus Pseudomonas increased. Free urea was readily dissolved in soil, and in 30 days after its application, the structure of the soil microbial community was similar to that in the control, as nitrogen was quickly assimilated by plants and microorganisms. The application of the P3HB/urea films and pellets changed the proportions of taxonomic groups in the microbial community. The percentage of Gram-negative bacilli increased to 45–50%, and that was 1.5–2 times more than in the controls. The percentage of bacteria of the genus Streptomyces was also 2–3 times higher than in the control.

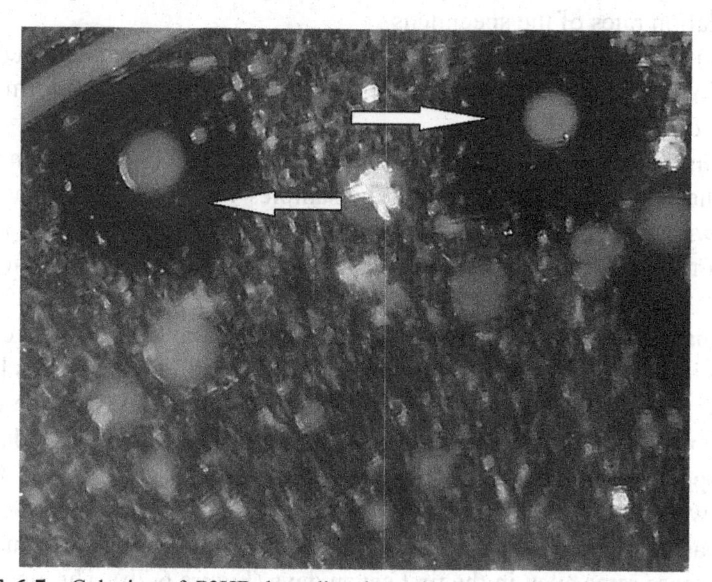

FIGURE 6.5 Colonies of P3HB-degrading bacteria on the medium containing P3HB. Arrows point to the regions of polymer hydrolysis. (Reprinted with permission from Volova, T. G., Prudnikova, S. V., & Boyandin, A. N., (2016). Biodegradable poly–3-hydroxybutyrate as a fertiliser carrier. J. Sci. Food. Agr., 96, 4183–4193. © 2016 John Wiley and Sons.)

6.2.2 EMBEDDING AMMONIUM NITRATE IN THE P(3HB) AND COMPOSITE MATRICES

Ammonium nitrate, in comparison with urea, contains both ammonium and nitrate nitrogen forms. Along with pure P3HB, P3HB composites with wood (birch) flour, polyethylene glycol (PEG) or polycaprolactone (PCL) were also tested.

Previously, we evaluated soil degradation of P3HB composites that did not contain a fertilizer, based on the weight loss of the specimens (Boyandin et al., 2016). Compressed specimens were placed in containers with soil with constant (50%) moisture content. The experiment was conducted for 35 days at a constant temperature of 20°C. Degradation behavior of four types of the specimens (pure polymer and composites) was studied (Figure 4.13). P3HB/PEG specimens had the highest degradation rate, and their residual weight at Day 35 of the experiment constituted 37.7% of their initial weight. Some of the weight loss could be caused by PEG being dissolved in water and washed away. At the end of the experiment, the weight of P3HB/PCL specimens constituted 50.6% of their initial weight, and the weight of P3HB and P3HB/wood flour specimens decreased to 69.7% and 72.9% of their initial weight, respectively. Thus, the blending of P3HB with fillers affected degradation rates of the specimens.

Then, the composite matrix was used to construct pressed 3D forms loaded with ammonium nitrate at 25% of the total mass of the formulation (Boyandin et al., 2017).

To measure the weight loss of the samples and kinetics of nitrogen release from the polymer matrix into soil, the samples were placed into containers filled with soil. The soil moisture content was maintained at 50% and the temperature at 20°C. The process of sample degradation is shown in Figure 6.6.

Dynamics of changes in the weight of formulations and nitrogen concentration in the soil are shown in Figure 6.7. P(3HB)/PEG specimens had the highest degradation rate, and their residual weight at Day 7 of the experiment constituted 56.2% of their initial weight. The rates of degradation of other specimens were slower, and their residual weight was 70.3–74% of the initial weight.

Measurements of ammonium nitrogen concentrations in soil indicated high release rates even in the first seven days (Figure 6.7b). Throughout the experiment, ammonium nitrogen concentration in soil increased gradually, with no dramatic changes and with relatively similar patterns observed for formulations with different compositions.

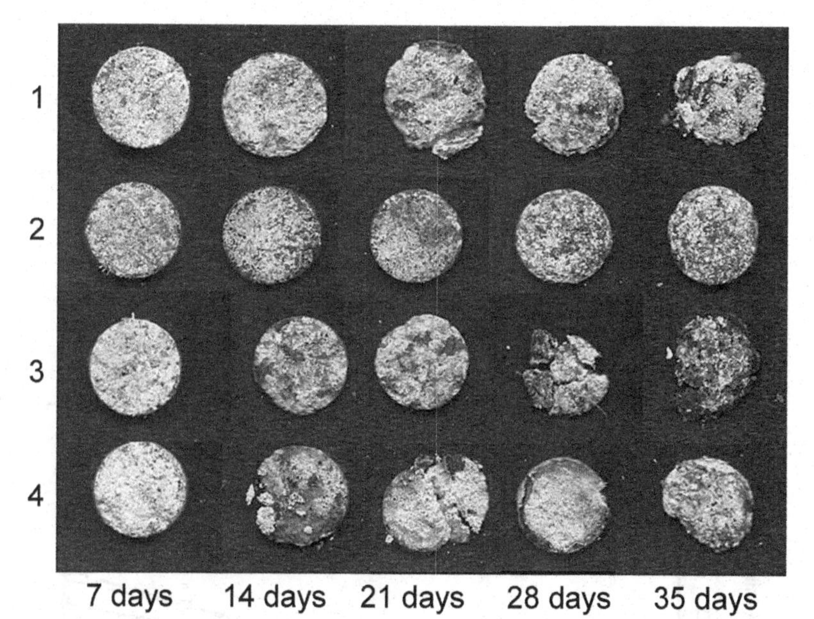

FIGURE 6.6 Ammonium nitrate embedded in the polymer matrices of different compositions during degradation: 1 – P3HB; 2 – P3HB/wood flour; 3 – P3HB/PCL; 4 – P3HB/PEG. (Reprinted with permission from Boyandin, A. N., Kazantseva, E. A., Varygina, D. E., & Volova, T. G., Constructing slow-release formulations of ammonium nitrate fertilizer based on degradable poly(3-hydroxybutyrate). J. Agr. Food Chem., 65, 6745–6752. © 2016 American Chemical Society.)

Degradation of polymer/fertilizer formulations investigated in this study occurred with the highest rate, as suggested by different changes in the weights of specimens with and without the fertilizer (Figure 4.13 , 6.7). This study proved that both pure poly–3-hydroxybutyrate and blends thereof could be used to prepare slow-release fertilizer formulations. Although P3HB and composite matrices were comparable in their degradation and fertilizer release rates, the use of less expensive fillers (such as wood flour) can reduce the cost of slow-release fertilizer formulations.

The total counts of copiotrophic bacteria in the soil with the P(3HB)/nitrogen formulation were higher than in the control soil (without fertilizers) by a factor of ten. In the soil with P(3HB)PCL, the total counts of bacteria were 3 times higher and in the soil with P(3HB)/PEG 2 times lower than in the control soil (Figure 6.8a). The abundance of prototrophic bacteria in all treatments was lower than in the control soil (Figure 6.8b). That could be associated with the input of organic matter of polymeric formulations and the increase in the population of copiotrophic bacteria. The abundance of

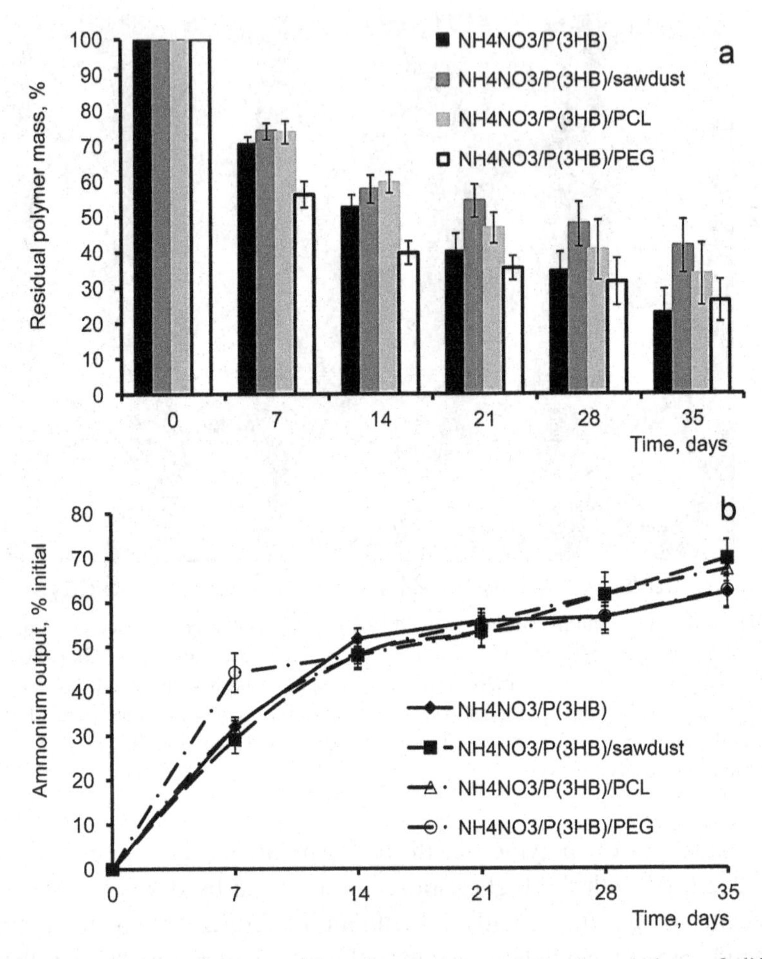

FIGURE 6.7 Changes in the weight of ammonium nitrate formulations of different composition (a) and ammonium nitrogen release to soil (b). (Reprinted with permission from Boyandin, A. N., Kazantseva, E. A., Varygina, D. E., & Volova, T. G., Constructing slow-release formulations of ammonium nitrate fertilizer based on degradable poly(3-hydroxybutyrate). J. Agr. Food Chem., 65, 6745–6752. © 2016 American Chemical Society; Boyandin's data).

oligotrophic bacteria was the highest in soil samples collected from the surface of the P(3HB)/wood flour and P(3HB)/PCL formulations. It was 60 and 100 times higher, respectively, than the abundance of oligotrophic bacteria in the control (Figure 6.8c). In the P(3HB) and P(3HB)/PEG treatments, the counts of oligotrophic bacteria did not differ significantly but were higher than in the control. In all treatments with fertilizer formulations embedded in the polymer or composite matrices, the counts of nitrogen-fixing bacteria were more than 10 times higher than in the control (Figure 6.8d).

To slow down fertilizer release from the experimental formulations, we coated them with a layer of P3HB. In the treatment with uncoated specimens, 30–35% of the embedded nitrogen was released to soil in two weeks (Figure 6.7b), while the amount of the fertilizer released from the coated formulations was no greater than 15–20% after six weeks of incubation in soil (Figure 6.9).

Thus, the study shows that slow-release nitrogen formulations can be constructed using degradable poly–3-hydroxybutyrate as a matrix and that the rate of nitrogen release into soil can be regulated by varying the type of formulation and production technique. The experimental formulations prepared and tested in this study had significantly greater longevity than those previously described by other researchers. The time necessary for complete release of the fertilizer varied from 24 h (Chen et al., 2008) to 10 days (Wanyika, 2014). Only one study (Hamid et al., 2013) reported an about 15% weight loss over 20 days for the urea/chitosan/bentonite system.

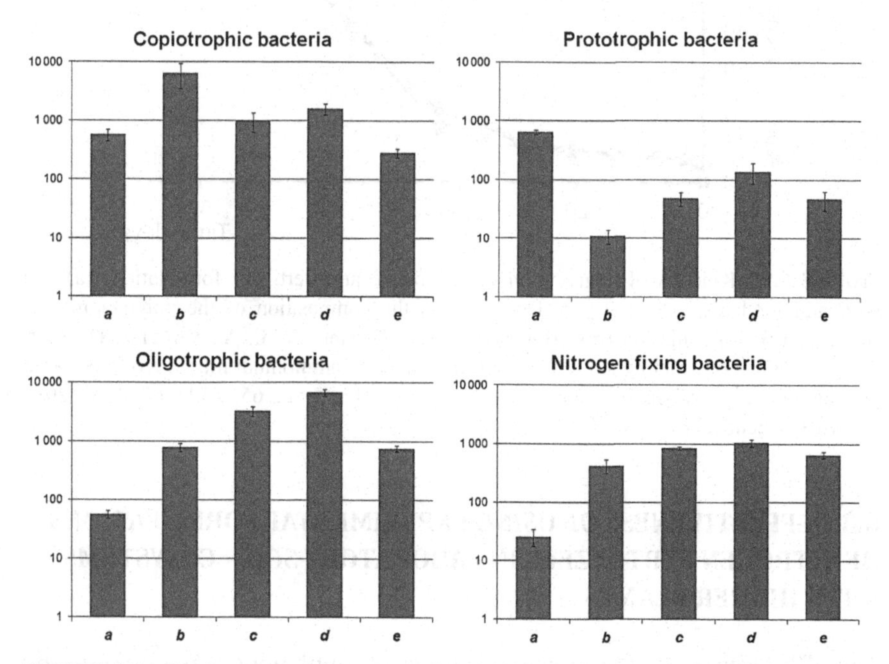

FIGURE 6.8 Total counts of bacteria (10^6 CFU in 1 g of air-dry soil): a – control (with no fertilizer); b – P(3HB); c – P(3HB) + wood flour; d – P(3HB)+PCL; e – P(3HB)+PEG (Zhila's data).

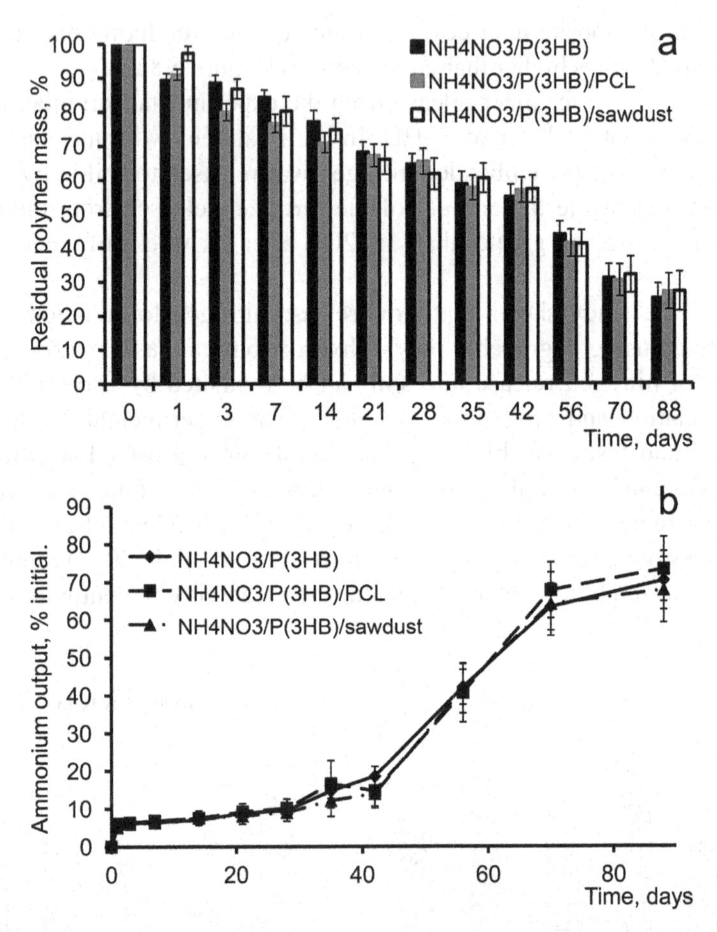

FIGURE 6.9 Kinetics of degradation of polymer-coated fertilizer formulations (a) and ammonium release to soil (b) as dependent on the composition of the polymer matrix. (Reprinted with permission from Boyandin, A. N., Kazantseva, E. A., Varygina, D. E., & Volova, T. G., Constructing slow-release formulations of ammonium nitrate fertilizer based on degradable poly(3-hydroxybutyrate). J. Agr. Food Chem., 65, 6745–6752. © 2016 American Chemical Society.)

6.3 EFFECTIVENESS OF USING EXPERIMENTAL FORMULATIONS OF NITROGEN FERTILIZERS IN LABORATORY SOIL ECOSYSTEMS WITH HIGHER PLANTS

The effectiveness of the prepared nitrogen formulations was evaluated by Volova et al., 2016, in experiments on growing plants under laboratory conditions. The effectiveness of the formulations was tested on the creeping bentgrass (*Agrostis stolonifera* L.) (a perennial grass) and lettuce (Latuca sativa) cv. Moskovskii. Field soil was placed into 150- to 500-cm^3 plastic containers, and specimens of the urea formulations and seeds of the plants were simultaneously buried in the soil. In the positive control, we used uncoated urea; the negative control was the soil with no nitrogen added to it. The number of specimens buried in the soil was determined in such a way that concentrations of the active ingredient were comparable in the treatment and the control. The loading of the polymer matrix with urea was determined in accordance with the recommended norms for nitrogen applied to soil: 0.15 mg/g soil for the creeping bentgrass and 3.7 mg/g soil for lettuce (Volova et al., 2016).

The creeping bentgrass (*Agrostis stolonifera* L.) was grown in summer, under natural light; lettuce (*Latuca sativa*) was grown under artificial light (in a growing chamber) under continuous light, 100 W/m^2), using the technology developed at the Laboratory of Controlled Photosynthesis of Phototrophic Organisms at the Institute of Biophysics SB RAS. The effectiveness of the formulations was determined from the amounts of the harvested vegetative parts of the plants, on an oven-dry weight basis.

The results are illustrated in Table 6.2 and Figures 6.10–6.11.

TABLE 6.2 The Effect of the Way of Nitrogen Delivery on Plant Biomass Growth

Agrostis stolonifera L (dry biomass, mg)					
Plant age, days	**Initial soil (negative control)**	**Commercial urea (positive control)**	**P3HB/urea films (nitrogen loading (% of polymer mass)**		
			10	**20**	**30**
7	3.3 ± 0.1	19.1 ± 1.4	17.6 ± 2.3	18.8 ± 1.5	15.5 ± 1.7
15	20.6 ± 2.6	41.0 ± 3.4	26.4 ± 1.54	39.8 ± 3.5	38 ± 4.2
22	72.6 ± 6.8	128.4 ± 14.7	133.8 ± 10.5	146.8 ± 12.4	135 ± 10.3

Latuca sativa (dry biomass, mg)

Plant age, days	**Initial soil (negative control)**	**Commercial urea (positive control)**	**P3HB/urea (nitrogen loading (20% of polymer mass)**	
			films	**pellets**
14	27.8 ± 0.3	48.1 ± 2.3	49.9 ± 1.7	57.7 ± 3.1
21	181 ± 6	262 ± 7	297 ± 9	243 ± 4
28	319 ± 13	420 ± 22	578 ± 10	602 ± 11

(Reprinted with permission Volova, T. G., Prudnikova, S. V., & Boyandin, A. N., (2016). Biodegradable poly–3-hydroxybutyrate as a fertiliser carrier. J. Sci. Food. Agr., 96, 4183–4193.

FIGURE 6.10 Visual analysis of the growth of creeping bentgrass (Agrostis stolonifera L.): 1 – negative control, 2 – positive control, free fertilizer, 3 – films with 10% urea, 4 – 20% urea 5 – 30% urea. (Reprinted with permission Volova, T. G., Prudnikova, S. V., & Boyandin, A. N., (2016). Biodegradable poly–3-hydroxybutyrate as a fertiliser carrier. J. Sci. Food. Agr., 96, 4183–4193. © 2016 John Wiley and Sons.)

FIGURE 6.11 Photographs of the lettuce Latuca sativa grown on the soil with urea delivered in different forms: 1 – initial unfertilized soil, 2 – granular urea, 3 – urea embedded in films, and 4 – urea embedded in pellets (20% loading). (Reprinted with permission Volova, T. G., Prudnikova, S. V., & Boyandin, A. N., (2016). Biodegradable poly-3-hydroxybutyrate as a fertiliser carrier. J. Sci. Food. Agr., 96, 4183–4193. © 2016 John Wiley and Sons.)

Results of our study showed the adequacy of P3HB/urea films and pellets and P3HB-coated urea granules as fertilizer formulations. Then, we conducted experiments with higher plants to estimate the effectiveness of these formulations; results are given in Table 6.2. A favorable effect of the nitrogen fertilizer embedded in P3HB films was observed in experiments with growing Agrostis stolonifera L. plants as compared to the positive control – application of commercial granular urea (Figure 6.10). This effect was quite evident at day 7 after the application of the fertilizer. Then, differences were even more pronounced. After 22 days of experiment, the biomass of the plants grown on the soil fertilized with formulations containing different amounts of urea reached between 135 and 146 mg, and that was higher than in the negative control (72.6 mg). The use of the films with 10% urea loading yielded the biomass comparable with that produced in the experiment with unmodified granular urea (128.4 mg). The amounts of the plant biomass produced in experiments with the films with different loadings of urea did not significantly differ from each other. The effectiveness of nitrogen embedded in P3HB films and pellets was also studied in experiments with the cultivated plant lettuce (Latuca sativa), in growing chambers (Table 6.2, Figure 6.11). The amounts of the harvested lettuce determined on an oven-dry biomass basis were similar in experiments with the films and pellets (580–600 mg), and they were only slightly higher than in the positive control (420 mg) but considerably, by a factor of 1.8, higher than in the negative control (unfertilized field soil).

We compared nitrogen concentrations in soil and drain water to determine the loss of nitrogen added to the soil as free urea and as formulations of urea embedded in the polymer. Nitrogen concentrations in soil differed considerably between experiments with urea formulations. In the positive control, nitrogen concentration in the soil increased dramatically in the first 5–7 days, reached its maximum at day 8 (15 mg/g soil), and, then, as the plants were growing, decreased, but stayed within a range of 6 to 8 mg/g until between day 15 and day 28. Nitrogen concentration in the unfertilized soil (negative control) was somewhat lower (6–8 mg/g) in the first half of the experiment, dropping to 3–4 mg/g in its second half. By contrast, in experiments with P3HB/urea formulations, nitrogen concentration in the soil increased gradually, reaching 8 and 10 mg/g at day 10 for pellets and films, respectively. Then, as lettuce biomass increased considerably, nitrogen release continued, and its concentration in the soil was stable and comparable with that of the positive control.

Measurements of nitrogen concentrations in drain water showed that in experiments with the free fertilizer, a considerable amount of nitrogen was removed from the system with drain water rather quickly, within the first 7 days, and, thus, plants could not use it. Losses of nitrogen embedded in polymer films and, especially, pellets were lower. The differences in nitrogen supply correspond to the differences in the productivity of plants (Table 6.2).

Thus, the use of the biodegradable poly–3-hydroxybutyrate both as a carrier of nitrogen and as a coating for urea granules is suitable for slowing down the release of the nitrogen fertilizer into soil. To prepare polymer-coated granules is simpler than to produce films and pellets with urea embedded in them, and the use of the coating method requires smaller amounts of the polymer. Our study showed that by varying the geometry of the polymer carrier and its loading with the fertilizer, one can achieve gradual (lasting several dozen days) leaching of nitrogen into soil with different release rates. The use of the prepared urea formulations favored the growth of the model plants and considerably reduced nitrogen loss with drain water.

Slow-release formulations with polymer-embedded ammonium nitrate, NH_4NO_3, were tested in a similar experiment in wheat stands. Experimental fertilizers were prepared as 3D formulations. Poly–3-hydroxybutyrate powder and a composite of P(3HB) and birch wood flour (1:1) were used as polymer matrices. The powdered components were mixed at a 25% fertilizer to 75% polymer ratio, the total weight of the mixture being 120 mg, and subjected to cold pressing. To further decrease the fertilizer release rate, some of the specimens were coated by a layer of P(3HB) by dipping them in a 5% polymer solution and drying. The specimens were placed into containers with the field soil. Plants were grown in a Conviron A1000-AR environmental chamber, under conditions that simulated diurnal variations in temperature and lighting (Table 6.3). Plants were watered when the soil was dry, usually twice a week, with 50 ml of tap water. Every 7 days, we used 150 ml of water to irrigate the plants and collected the irrigation solution. Kinetics of ammonium nitrogen release to soil was determined based on its concentration in the irrigation solution, and the total ammonium released to soil was calculated for the total volume of the solution. Every 7–14 days, we removed plants and (if present) tablet formulations from some of the containers of each group (positive control, negative control, coated, and uncoated fertilizer formulations based on P3HB and P3HB/wood flour). Plants were dried at 105°C and tablet formulations at 40°C for 24 h, until they reached constant weight. The aboveground and belowground parts of the plants and the remaining formulations were weighed. Soil was analyzed for dissolved ammonium nitrogen. Polymer biodegradation was evaluated from the weight loss of the specimens.

TABLE 6.3 Conditions in the Conviron Environmental Chamber.

	Time	t, °C, Weeks 1–7	t, °C, Weeks 8–12	Minimal humidity, %	Irradiance, µmol/m²/s
1	0:00	10°C	14°C	50	0
2	6:00	12°C	16°C	50	100
3	9:00	14°C	18°C	40	200
4	12:00	18°C	22°C	20	300
5	16:00	16°C	20°C	30	200
6	20:00	13°C	17°C	30	100
7	23:00	10°C	14°C	40	0

(Reprinted with permission from Boyandin, A. N., Kazantseva, E. A., Varygina, D. E., & Volova, T. G., Constructing slow-release formulations of ammonium nitrate fertilizer based on degradable poly(3-hydroxybutyrate). J. Agr. Food Chem., 65, 6745–6752. © 2016 American Chemical Society.)

Biodegradation rates of formulations based on pure P3HB and P3HB/wood flour tested in this study were similar to each other (Figure 6.12a) while the rates of degradation of coated and uncoated formulations differed considerably. Degradation rates of uncoated formulations were substantially slower than in the previous experiment, without plants. A possible reason for that may different temperature conditions of the experiments: in the first experiment, specimens were incubated at a constant temperature of 20°C while in the second experiment, the temperature was 2–10°C lower than that for the greater part of the 24-h period.

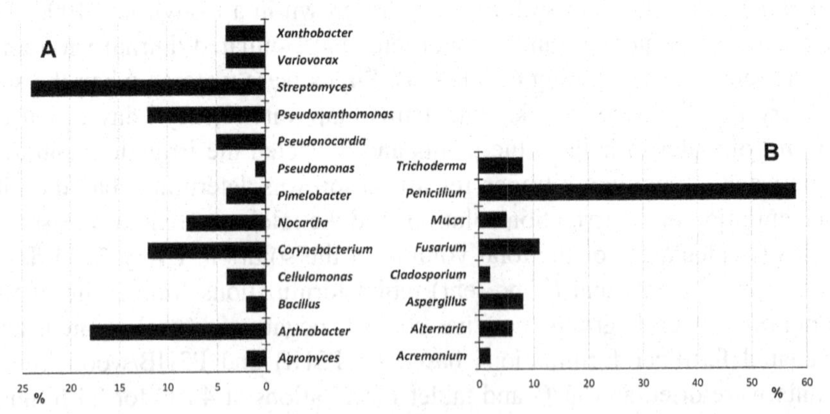

FIGURE 6.12 Changes in the weight of fertilizer tablets (a) and in the amounts of ammonium in soil (b) and kinetics of ammonium release to the irrigation solution (c) during the experiment with wheat. (Reprinted with permission from Boyandin, A. N., Kazantseva, E. A., Varygina, D. E., & Volova, T. G., Constructing slow-release formulations of ammonium nitrate fertilizer based on degradable poly(3-hydroxybutyrate). J. Agr. Food Chem., 65, 6745–6752. © 2016 American Chemical Society.)

Measurements of soil nitrogen concentrations, performed after Week 1, showed that it gradually decreased in both the controls and soil with the uncoated experimental formulations (Figure 6.12b). In the treatments with coated formulations, ammonium release was noticeably delayed. The ammonium concentration peak was observed at Weeks 4–6 of the experiment. Although ammonium concentration in soil gradually decreased in all laboratory ecosystems, as the fertilizer was washed out and consumed by plants, the level of ammonium nitrogen remained an order of magnitude higher in the treatments than in the positive control throughout the rest of the experiment.

Nitrogen concentration in soil correlated with the amount of nitrogen released to the irrigation solution (Figure 6.12c). During Week 1 of the experiment, the highest concentration of ammonium nitrogen was observed in the positive control and much lower concentration – in the treatment with uncoated formulations. In the treatment with coated formulations, initial concentration of ammonium in the irrigation solution was comparable with that in the negative control. In the treatment with uncoated formulations, ammonium noticeably dropped at Weeks 3–4 of the experiment. In the treatments with the experimental formulations, initial nitrogen concentration in the irrigation water was considerably lower, but at Week 7, it began to increase, reaching levels one or two orders of magnitude higher than in the positive control, which suggested that the fertilizer was still being released to soil.

The 12-weeks experiment with spring wheat proved that the experimental formulations were more effective than the commercial formulation (Table 6.4, Figure 6.13). In the treatments with uncoated formulations, the biomass

TABLE 6.4 Increase in the Total Dry Biomass of Soft Wheat (Triticum Aestivum L.) cv. Altaiskaya 70 in the Experiment With Different Formulations of Ammonium Nitrate (mg per plant). (Reprinted with permission from Boyandin, A. N., Kazantseva, E. A., Varygina, D. E., & Volova, T. G., Constructing slow-release formulations of ammonium nitrate fertilizer based on degradable poly(3-hydroxybutyrate). J. Agr. Food Chem., 65, 6745–6752. © 2016 American Chemical Society.)

Week	Negative control	Positive control	P(3HB), uncoated	P(3HB), coated	P(3HB)/wood flour, uncoated	P(3HB)/wood flour, coated
2	27.4 ± 0.3	34.8 ± 0.5	32.3 ± 1.6	36.8 ± 0.6	30.8 ± 1.3	33.1 ± 0.9
3	42.3 ± 2.9	45.0 ± 2.3	52.0 ± 2.2	47.3 ± 3.3	55.8 ± 1.9	44.9 ± 2.6
4	54.7 ± 3.1	69.3 ± 4.3	77.9 ± 4.1	51.7 ± 3.9	81.8 ± 3.8	61.0 ± 2.1
5	54.8 ± 3.7	127.1 ± 6.9	142.1 ± 5.3	101.4 ± 8.1	145.2 ± 10.6	110.0 ± 6.4
6	86.0 ± 6.2	138.0 ± 8.7	169.8 ± 6.2	123.4 ± 10.0	179.6 ± 12.1	150.4 ± 8.8
8	106.0 ± 9.8	195.0 ± 11.1	239.0 ± 10.9	257.2 ± 16.9	193.2 ± 10.9	275.0 ± 13.9
10	179.2 ± 11.2	295.8 ± 14.2	356.1 ± 10.7	454.4 ± 26.8	333.1 ± 12.3	412.7 ± 19.5
12	180.8 ± 16.7	362.6 ± 12.6	393.3 ± 15.8	463.8 ± 23.3	392.8 ± 16.3	460.0 ± 29.0

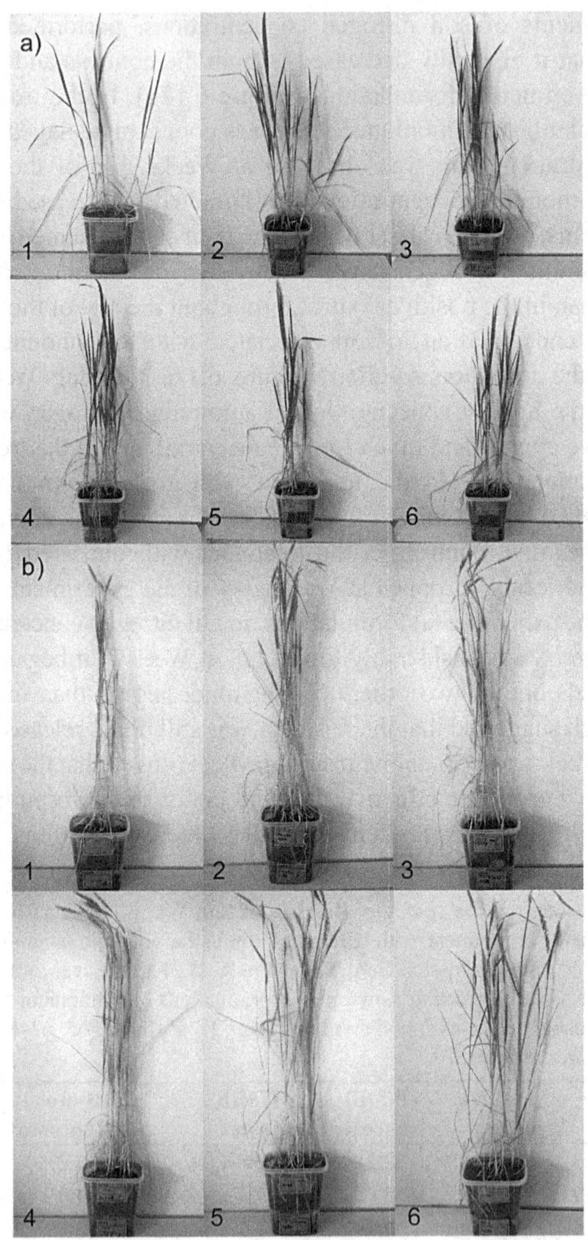

FIGURE 6.13 A photograph of soft spring wheat (cv. Altaiskaya 70) stands with different nitrogen formulations: 1 – soil without a fertilizer; 2 – ammonium nitrate granules; 3 – uncoated 3D P(3HB)/NH_4NO_3; 4 – polymer-coated P(3HB)/NH_4NO_3; 5 – uncoated 3D P(3HB)/wood flour/NH_4NO_3; and 6 – polymer-coated 3D P(3HB)/wood flour/NH_4NO_3; a) – 8 weeks; b) – 12 weeks (Boyandin's photographs).

increase was 12–20% higher than in the positive control at Week 4, and that was the greatest difference observed; by the end of the experiment, it had decreased to 8%. The effect of the coated formulations became evident later, by Week 8, but it was more pronounced, and in these treatments, at the end of the experiment, the biomass increase was more than 25% higher than in the positive control. Thus, the use of slow-release formulations alleviated nitrogen deficiency in soil in the second half of the experiment, when the soil that had not been supplemented with nitrogen (negative control) was actually nitrogen-depleted.

Thus, controlled slow-release fertilizer formulations can be prepared by embedding nitrogen fertilizers into matrices of poly-3-hydroxybutyrate and composites based on it. These formulations are capable of maintaining higher concentrations of nitrogen in soil compared to fertilizers applied by usual methods over long time periods (up to several months). These experimental formulations can be used as prototypes for developing controlled-release agricultural formulations.

KEYWORDS

- **amide nitrogen**
- **ammonia fertilizers**
- **nitrate fertilizers**
- **soil microflora**

REFERENCES

Adams, C., Frantz, J., & Bugbee, B., (2013). Macro- and micronutrient-release characteristics of three polymer-coated fertilizers: Theory and measurements. *J. Plant Nutr. Soil Sci.*, *176*, 76–88.

Anghel, A., Lăcătuşu, A. R., Lăcătuşu, R., Iancu, S., Lungu, M., Lazăr, R., et al., (2012). Testing kinetic of nutrients release from complex mineral fertilizers coated with co-polyester films from pet waste recycling and effect on soil chemical properties. *Scientific Papers. Series A. Agronomy*, *55*, 13–18.

Artyushin, A. M., & Derzhavin, L. M., (1984). *Kratkiy Slovar po Udobreniyam (Concise Dictionary of Fertilizers)* (2nd edn., p. 208). Kolos: Moscow (in Russian).

Boyandin, A. N., Kazantseva, E. A., Varygina, D. E., & Volova, T. G., (2017). Constructing slow-release formulations of ammonium nitrate fertilizer based on degradable poly(3-hydroxybutyrate). *J. Agr. Food Chem.*, *65*, 6745–6752.

Boyandin, A. N., Zhila, N. O., Kiselev, E. G., & Volova, T. G., (2016). Constructing slow-release formulations of metribuzin based on degradable poly(3-hydroxybutyrate). *J. Agr. Food Chem.*, *64*, 5625–5632.

Chen, D., Freney, J. R., Rochester, I., Constable, G. A., Mosier, A. R., & Chalk, P. M., (2008a). Evaluation of a polyolefin coated urea (Meister) as a fertilizer for irrigated cotton. *Nutr. Cycl. Agroecosyst.*, *81*, 245–254.

Chen, L., Xie, Z., Zhuang, X., Chen, X., & Jing, X., (2008b). Controlled release of urea encapsulated by starch-g-poly(L-lactide). *Carbohydr. Polym.*, *72*, 342–348.

Chong, W. S., (2013). *Investigation of Degradation of Starch Complex for CRF Application* (p. 51). Bachelor dissertation in Chemical Engineering. Universiti Teknologi Petronas: Tronoh, Malaysia, May.

Corradini, E., De Moura, M. R., & Mattoso, L. H. C., (2010). A preliminary study of the incorporation of NPK fertilizer into chitosan nanoparticles. *Express Polym. Lett.*, *4*, 509–515.

Costa, M. M., Cabral-Albuquerque, E. C., Alves, T. L., Pinto, J. C., & Fialho, R. L., (2013). Use of polyhydroxybutyrate and ethyl cellulose for coating of urea granules. *J. Agr. Food Chem.*, *61*, 9984–9991.

Du, C. W., Zhou, J. M., & Shaviv, A., (2006). Release characteristics of nutrients from polymer-coated compound controlled release fertilizers. *Polym. Environ.*, *14*, 223–230.

Gómez-Martínez, D., Partal, I., Martínez, P., & Gallegos, C., (2013). Gluten-based bioplastics with modified controlled-release and hydrophilic properties. *Ind. Crops Prod.*, *43*, 704–710.

Grant, C. A., Wu, R., Selles, F., Harker, K. N., Clayton, G. W., Bittman, S., Zebarth, B. J., & Lupwayi, N. Z., (2012). Crop yield and nitrogen concentration with controlled release urea and split applications of nitrogen as compared to non-coated urea applied at seeding. *Field Crops Res.*, *127*, 170–180.

Hamid, N. N. A., Mohamad, N., Hing, L. Y., Dimin, M. F., Azam, M. A., Hassan, M. H. C., Ahmad, M. K. S. M., & Shaaban, A., (2013). The effect of chitosan content to physical and degradation properties of biodegradable urea fertilizer. *Int. J. Sci. Innov. Res.*, *2*, 893–902.

Hasaneen, M. N. A., Abdel-Aziz, H. M. M., El-Bialy, D. M. A., & Omer, A. M., (2014). Preparation of chitosan nanoparticles for loading with NPK fertilizer. *Afr. J. Biotechnol.*, *13*, 3158–3164.

Huett, D. O., & Gogel, B. J., (2000). Longevities and nitrogen, phosphorus, and potassium release patterns of polymer-coated controlled-release fertilizers at 30°C and 40°C. *Commun. Soil Sci. Plant Anal.*, *31*, 959–973.

Jamnongkan, T., & Kaewpirom, S., (2010). Potassium release kinetics and water retention of controlled-release fertilizers based on chitosan hydrogels. *J. Polym. Environ.*, *18*, 413–421.

Jarosiewicz, A., & Tomaszewska, M., (2003). Controlled-release NPK fertilizer encapsulated by polymeric membranes. *J. Agric. Food Chem.*, *51*, 413–417.

Jendrossek, D., & Handrick, R., (2002). Microbial degradation of polyhydroxyalkanoates. *Ann. Rev. Microbiol.*, *56*, 403–432.

Jia, X., Ma, Z., Zhang, G., Hu, J., Liu, Z., Wang, H., & Zhou, F., (2013). Polydopamine film coated controlled-release multielement compound fertilizer based on mussel-inspired chemistry. *J. Agr. Food Chem.*, *61*, 2919–2924.

Kunioka, M., Kawaguchi, Y., & Doi, Y., (1989). Production of biodegradable copolyesters of 3-hydroxybutyrate and 4-hydroxybutyrate by *Alcaligenes eutrophus*. *Appl. Microbiol. Biotechnol.*, *30*, 569–573.

Lammel, J., (2005). *Cost of the Different Options Available to the Farmers: Current Situation and Prospects* (pp. 28-30). In: Proceedings of the IFA International Workshop on

Enhanced-Efficiency Fertilizers: Frankfurt, Germany, 28–30 June 2005. International Fertilizer Industry Association: Paris.

Lan, R., Liu, Y., Wang, G., Wang, T., Kan, C., & Jin, Y., (2011). Experimental modeling of polymer latex spray coating for producing controlled-release urea. *Particuology, 9*, 510–516.

Liang, R., & Liu, M., (2006). Preparation and properties of a double-coated slow-release and water-retention urea fertilizer. *J. Agr. Food Chem., 54*, 1392–1398.

Medina, L. C., Obreza, T. A., Sartain, J. B., & Rouse, R. E., (2008). Nitrogen release patterns of a mixed controlled-release fertilizer and its components. *Hort. Technology, 18*, 475–480.

Medina, L. C., Sartain, J. B., Obreza, T. A., Hall, W. L., & Thiex, N. J., (2014). Evaluation of a soil incubation method to characterize nitrogen release patterns of slow- and controlled-release fertilizers. *J. AOAC Int., 97*, 643–660.

Molitoris, H. P., Moss, S. T., De Koning, G. J. M., & Jendrossek, D., (1996). Scanning electron microscopy of polyhydroxyalkanoate degradation by bacteria. *Appl. Microbiol. Biotechnol., 46*, 570–579.

Nash, P. R., Nelson, K. A., & Motavalli, P. P., (2013). Corn yield response to polymer and non-coated urea placement and timings. *Int. J. Plant Prod., 7*, 373–392.

Neata, G., Popa, M., Mitelut, A., Guidea, S., & Pukanski, B., (2012). Bio-based composite use in fertilization of petunia and carnation culture. *Scientific Bulletin. Series F. Biotechnologies, 16*, 30–35.

Nelson, K. A., Nash, P. R., & Dudenhoeffer, C. J., (2013). Effect of nitrogen source and weed management systems on no-till corn yields. *J. Agr. Sci., 5*, 87–96.

Ni, B., Liu, M., Lü, S., Xie, L., & Wang, Y., (2011). Environmentally friendly slow-release nitrogen fertilizer. *J. Agr. Food Chem., 59*, 10169–10175.

Oosterhuis, D. M., & Howard, D. D., (2008). Evaluation of slow-release nitrogen and potassium fertilizers for cotton production. *Afr. J. Agric. Res., 3*, 68–73.

Pereira, E. I., Minussi, F. B., Da Cruz, C. C. T., Bernardi, A. C. C., & Ribeiro, C., (2012). Urea–montmorillonite-extruded nanocomposites: A novel slow-release material. *J. Agric. Food Chem., 60*, 5267–5272.

Prudnikova, S. V., & Volova, T. G., (2012). *Ecological Role of Polyhydroxyalkanoates – An Analog of Synthetic Plastics: Biodegradation Behavior in Natural Environments and Interaction With Microorganisms (Ekologicheskaya rol Poligidroksialkanoatov – Analoga Sinteticheskikh Plastmass: Zakonomernosti Biorazrusheniya v Prirodnoi Srede i Vzaimodeistviya s Mikroorganizmami)* (p. 183). Krasnoyarskii Pisatel: Krasnoyarsk, (in Russian).

Saraydin, D., Karadag, E., & Güven, O., (2000). Relationship between the swelling process and the releases of water-soluble agrochemicals from radiation cross-linked acrylamide/itaconic acid copolymers. *Polym. Bull., 45*, 287–294.

Shaviv, A., Raban, S., & Zaidel, E., (2003). Modeling controlled nutrient release from polymer-coated fertilizers: Diffusion release from single granules. *Environ. Sci. Technol., 37*, 2251–2256.

Shoji, S., Delgado, J., Mosier, A., & Miura, Y., (2001). Use of controlled release fertilizers and nitrification inhibitors to increase nitrogen use efficiency and to conserve air and water quality. *Commun. Soil Sci. Plant Anal., 32*, 1051–1070.

Shtefan, V. K., (1981). *Plant Life and Fertilizers (Zhizn Rasteniy i Udobreniya)* (p. 240). Moskovskiy Rabochiy: Moscow (in Russian).

Soon, Y. K., Malhi, S. S., Lemke, R. L., Lupwayi, N. Z., & Grant, C. A., (2011). Effect of polymer-coated urea and tillage on the dynamics of available N and nitrous oxide emission from Gray. *Luvisols Nutr. Cycl. Agroecosyst.*, *90*, 267–279.

Sopeña, F., Cabrera, A., Maqueda, C., & Morillo, E., (2005). Controlled release of the herbicide norflurazon into the water from ethylcellulose formulations. *J. Agr. Food Chem.*, *53*, 3540–3547.

Trinh, T. H., Shaari, K. Z. K., Shuib, A. S. B., & Ismail, L. B., (2013). *Modeling of urea release from coated urea for prediction of coating material diffusivity.* In: Proceedings of the 6th International Conference on Process Systems Engineering (PSE ASIA), 25–27 June 2013, Kuala Lumpur, Malaysia. URI: http://epublication.cheme.utm.my/id/eprint/128.

Tsuji, H., & Suzuyoshi, K., (2002). Environmental degradation of biodegradable polyesters. 1. Poly(e-caprolactone), poly[(R)–3-hydroxybutyrate], and poly(L-lactide) films in controlled static seawater. *Polym. Degrad. Stab.*, *75*, 347–355.

Vashishtha, M., Dongara, P., & Singh, D., (2010). Improvement in properties of urea by phosphogypsum coating. *Int. J. Chem. Tech Res.*, *2*, 36–44.

Vildflush, I. R., Kukresh, S. P., Ionas, V. A., Kamasin, S. M., Kalikinsky, A. A., Bogdevich, I. M., & Lapa, V. V., (2001). *Agrokhimiya (Agrochemistry)* (p. 488). Uradzhay: Minsk, (in Russian).

Volova, T. G., Prudnikova, S. V., & Boyandin, A. N., (2016). Biodegradable poly–3-hydroxybutyrate as a fertilizer carrier. *J. Sci. Food. Agr.*, *96*, 4183–4193.

Wang, S., Li, X., Lu, J., Hong, J., Chen, G., Xue, X., Li, J., Wei, Y., Zou, J., & Liu, G., (2013). Effects of controlled-release urea application on the growth, yield and nitrogen recovery efficiency of cotton. *Agricultural Sciences*, *4*, 33–38.

Wanyika, H., (2014). Controlled release of agrochemicals intercalated into montmorillonite interlayer space. *Sci. World J.*, *2014*, 656287.

Wu, C. S., (2008). Controlled release evaluation of bacterial fertilizer using polymer composites as matrix. *J. Control. Release*, *132*, 42–48.

Yang, Y. C., Zhang, M., Li, Y., Fan, X. H., & Geng, Y. Q., (2012). Improving the quality of polymer-coated urea with recycled plastic, proper additives, and large tablets. *J. Agr. Food Chem.*, *60*, 11229–11237.

Zvomuya, F., Rosen, C. J., Russelle, M. P., & Gupta, S. C., (2003). Nitrate leaching and nitrogen recovery following application of polyolefin-coated urea to potato. *J. Environ. Qual.*, *32*, 480–489.

CONCLUSION

Traditional use of products of chemical synthesis derived from nonrenewable natural resources leads to excessive and increasing accumulation of non-recyclable wastes, coming into conflict with environmental protection activities and posing a global environmental problem. One way out of this conundrum is to expand the use of tools and methods of biotechnology, which, on the one hand, protect beneficial biota and enhance productivity in agriculture and, on the other, reduce toxic impacts on individual ecosystems and the entire biosphere.

This book addresses a very important scientific subject: reducing the risk of uncontrolled distribution of chemicals released in the technosphere and their accumulation in the biosphere by providing a scientific basis for constructing new-generation pesticides and fertilizers and using them in agriculture.

That was the underpinning of the project prepared by a team of researchers at the Institute of Biophysics SB RAS and submitted for a grant of the Russian Science Foundation intended for support of existing laboratories in 2014. Implementation of the project "Scientific basis for constructing and using new-generation agrochemicals" (No. 14–26–00039) resulted in development of the scientific foundation for the new line in ecological and agricultural biotechnology, including development of technologies for constructing new-generation slow-release formulations, construction, and investigation of a series of fungicidal and herbicidal formulations and nitrogen fertilizers embedded in degradable polymer matrix.

This book was written to summarize results of research implemented for the RSF grant: establishment of the scientific basis for constructing and using slow-release environmentally friendly formulations of fertilizers and pesticides to protect crops against pathogens and weeds and to supply nitrogen fertilizers to them. The book describes the authors' studies on the properties of the new herbicide and fungicide formulations and their efficacy in laboratory ecosystems with higher plants infected by fusarium wilt and weeds.

The following results were obtained:

- the data on interactions between chemicals and polyhydroxyalkanoates (PHAs) in different phase states and, based on this, development of methods of loading of the chemicals intended for suppressing agents of plant diseases and killing weeds and fertilizers into variously shaped polymer matrices;
- a series of experimental slow-release formulations produced using different methods; results of studying their structure and physical/mechanical properties;
- data on the release kinetics of the active ingredients from the polymer matrix obtained in laboratory soil microecosystems as dependent on the geometry of formulations, concentrations of the chemicals in them, chemical composition of the soil, and the type of soil microbial community;
- positive evaluation of the effectiveness of using the slow-release chemical formulations in laboratory conditions: in soil ecosystems with known properties containing higher plants infected by plant pathogens and weeds.

During Phase 1, based on the review of the modern literature and preliminary studies, we selected the agrochemicals that were sufficiently effective, could be used on a large scale, were suitable for soil applications, were compatible with the polymeric matrix in different phase states, were stable in nonpolar solvent solutions, and could be analyzed by spectrophotometric methods (chromatography-mass spectrometry, HPLC, and IR spectroscopy). These were herbicides Magnum Super and Sencor Ultra, fungicide Vial Trust (tebuconazole), and nitrogen fertilizers – granular urea and ammonium nitrate. Then, we prepared polymer/active ingredient two-phase systems in the form of solutions, emulsions, and powders, which were investigated by using HPLC, DSC, X-Ray, and IR spectroscopy.

These polymer/active ingredient two-phase systems in the form of solutions, emulsions, and powders were used to construct slow-release formulations of the selected chemicals, which were shaped as microparticles, microgranules, films, and pellets and were loaded with different amounts of agrochemicals. These were:

- herbicide (metribuzin, 2,4-chloroacetic acid, tribenuron) formulations loaded with the active ingredient to different extents in the form of films prepared from the solution of the polymer of 3-hydroxybutyric acid [P(3HB)], granules prepared from the polymer solution, 3D constructs prepared from a mixture of powdered polymer and MET, and MET-loaded microparticles;

- formulations of fungicide tebuconazole in the form of films, microgranules, and 3D pressed constructs, loaded with the active ingredient to different extents;
- four slow-release formulations of nitrogen fertilizer based on urea and ammonium nitrate in the form of films, 3D constructs, core/shell polymer-coated urea granules, and pressed 3D constructs coated with several layers of the polymer.

Investigation of the initial substances (polymer and chemicals) and the experimental formulations by methods of IR spectroscopy, DSC, and X-Ray showed that loading of the polymer matrix with the chemicals did not produce any noticeable effect on the physicochemical properties of the polymer and, hence, its performance and that no chemical binding of the components occurred.

To evaluate the experimental slow-release formulations, we constructed and characterized laboratory soil microecosystems with two types of agro-transformed soil (field and garden soils). We examined the chemical composition of the soils and the structure of soil microbial communities and revealed the dominant microbial species, including primary degraders of PHAs of different chemical compositions. We studied the relationships between the release profiles of the fungicide, herbicide, and nitrogen fertilizer and the geometry of the forms, the type of the embedded chemical, and the degree of loading of the matrix, taking into account degradation rate of the polymer matrix. The embedding of the chemicals into the degradable PHAs enabled slow (over 60 days for TEB and MET and over 30 days for nitrogen) and burst-free release of the active ingredients into soil.

To influence PHA degradation and to increase the availability of these polymers, we prepared P(3HB) composites with polyethylene glycol, polycaprolactone, and birch wood chips. The composition of the matrix had a substantial effect on the release of the agrochemicals. By varying the composition of the matrix and by coating nitrogen fertilizer with a polymer, one can regulate release rates of the active ingredient within a wide range. The outcomes of the studies conducted during implementation of the project were the development of the methods and techniques for construction of the novel formulations of agrochemicals based on degradable PHAs. The studies addressed the effects of various factors (geometry of the form, the degree of loading, the chemical composition of the polymer matrix, the type of soil and its microbial community) on degradation of the formulations in the soil and release profiles of the active ingredients and determined the major factors that could be used to regulate those processes. The results proved that these formulations were suitable for slow release of herbicides, fungicides, and nitrogen fertilizers.

In the final stage of the project implementation, we studied the effectiveness of the formulations developed during the research in laboratory ecosystems with plants, which contained the following model weeds and crops:

- weeds: a perennial grass species *Agrostis stolonifera* – creeping bentgrass, *Setaria macrocheata* – foxtail millet, *Chenopodium album* – lamb's quarters, *Melilotus albus* – white sweet clover, *Amaranthus retroflexus* – red-root amaranth;
- crops: lettuce *Latuca sativa*, radish *Raphanus sativus*, wheat *Triticum aestivum*;
- soil microecosystems infected by plant pathogens *F. moniliforme* and *F. solani* –root rot agents;
- crop *Triticum aestivum* infected by *Fusarium moniliforme* and weeds, used to evaluate the efficacy of fungicides.

The studies showed that the slow-release herbicide and fungicide formulations developed during implementation of this project were effective against weeds and plant pathogens and that their effects were comparable with or stronger than those of free forms of agrochemicals. Gradual release of nitrogen from the experimental nitrogen fertilizer formulations significantly reduced nitrogen losses.

Positive results of the research suggest that the use of natural polyesters, PHAs, as degradable polymer matrix for constructing slow-release formulations of agrochemicals is an effective approach and that the next step should be field tests of the experimental formulations.

Research performed to implement the project has produced new data on degradable PHAs as material for constructing slow-release formulations of agrochemicals. The scientific basis has been provided for constructing environmentally friendly and targeted controlled-release formulations of fertilizers and pesticides to protect crops against pests and pathogens. The development of this line of research will help mitigate the risk of accumulation and uncontrolled spread of xenobiotics in the environment and replace dead-end synthetic plastics by degradable materials capable of joining the biospheric cycles.

INDEX

B

C

D

H

I

K

L

T

Printed in the United States
by Baker & Taylor Publisher Services